Supramolecular Catalysis

Edited by
Piet W. N. M. van Leeuwen

Related Titles

Laguna, A. (ed.)

Modern Supramolecular Gold Chemistry

Gold-Metal Interactions and Applications

2009

ISBN: 978-3-527-32029-5

Vögtle, F., Richardt, G., Werner, N., Rackstraw, A.J.

Dendritic Molecules

2009

ISBN: 978-3-527-32066-0

Diederich, F., Stang, P. J., Tykwinski, R. R. (eds.)

Modern Supramolecular Chemistry

Strategies for Macrocycle Synthesis

2008

ISBN: 978-3-527-31826-1

Cragg, P.

A Practical Guide to Supramolecular Chemistry

2005

ISBN: 978-0-470-86654-2

Supramolecular Catalysis

Edited by
Piet W. N. M. van Leeuwen

WILEY-VCH Verlag GmbH & Co. KGaA

The Editor

Professor Piet van Leeuwen
Institut of Chemical Research
of Catalonia (ICIQ)
Av. Paisos Catalans 16
43007 Tarragona
Spain

All books published by Wiley-VCH are carefully produced. Nevertheless, authors, editors, and publisher do not warrant the information contained in these books, including this book, to be free of errors. Readers are advised to keep in mind that statements, data, illustrations, procedural details or other items may inadvertently be inaccurate.

Library of Congress Card No.: applied for

British Library Cataloguing-in-Publication Data
A catalogue record for this book is available from the British Library.

Bibliographic information published by the Deutsche Nationalbibliothek
Die Deutsche Nationalbibliothek lists this publication in the Deutsche Nationalbibliografie; detailed bibliographic data are available on the Internet at http://dnb.d-nb.de

© 2008 WILEY-VCH Verlag GmbH & Co. KGaA, Weinheim

All rights reserved (including those of translation into other languages). No part of this book may be reproduced in any form – by photoprinting, microfilm, or any other means – nor transmitted or translated into a machine language without written permission from the publishers. Registered names, trademarks, etc. used in this book, even when not specifically marked as such, are not to be considered unprotected by law.

Cover Adam Design, Weinheim
The molecule on the front cover is a 3-D adaptation of Fujita's complex kindly supplied by Dr. Pablo Ballester of the Institució Catalana de Recerca i Estudis Avançats
Typesetting Thomson Digital, Noida, India
Printing Strauss GmbH, Mörlenbach
Binding Litges & Dopf Buchbinderei GmbH, Heppenheim

Printed in the Federal Republic of Germany
Printed on acid-free paper

ISBN: 978-3-527-32191-9

Contents

Preface *XI*
List of Authors *XIII*

1 **Introduction to Supramolecular Catalysis** *1*
Pablo Ballester and Anton Vidal-Ferran
1.1 Introduction *1*
1.2 Design Approaches to Supramolecular Catalysis *3*
1.2.1 Molecular Receptors that Place a Binding Site Close to a Catalytic Center *3*
1.2.2 Molecular Receptors that Promote the Reaction of two Simultaneously Complexed Reactants *7*
1.2.3 Preparation of the Catalyst Backbone via Supramolecular Interactions *16*
1.3 Artificial Biomacromolecules for Asymmetric Catalysis *22*
1.4 Summary and Outlook *24*
References *24*

2 **Supramolecular Construction of Chelating Bidentate Ligand Libraries through Hydrogen Bonding: Concept and Applications in Homogeneous Metal Complex Catalysis** *29*
Bernhard Breit
2.1 Introduction *29*
2.2 Emulation of Chelation through Self-Assembly of Monodentate Ligands *30*
2.3 Tautomeric Self-Complementary Interligand Hydrogen Bonding *33*
2.3.1 Hydroformylation *33*
2.3.2 Room Temperature/Ambient Pressure Hydroformylation *37*
2.3.3 Asymmetric Hydrogenation *37*
2.4 A-T Base Pair Analogous Complementary Hydrogen Bonding for the Construction of Heterodimeric Self-Assembling Ligands *38*

Supramolecular Catalysis. Edited by Piet W. N. M. van Leeuwen
Copyright © 2008 WILEY-VCH Verlag GmbH & Co. KGaA, Weinheim
ISBN: 978-3-527-32191-9

2.4.1	Aminopyridine/Isoquinolone Platform 38
2.4.1.1	Hydroformylation 38
2.4.1.2	Hydration of Alkynes 39
2.4.1.3	Hydration of Nitriles 45
2.4.1.4	Asymmetric Hydrogenation 46
2.4.2	Platform Variation 49
2.4.2.1	Hydroformylation 49
2.5	Conclusion and Outlook 52
	References 52
3	**Bis-Azolylazine Derivatives as Supramolecular Synthons for Copper and Silver [2 × 2] Grids and Coordination Polymers** 57
	Félix A. Jalón, Blanca R. Manzano, M. Laura Soriano, and Isabel M. Ortiz
3.1	Introduction 57
3.2	"Planar" and "Non-Planar" Azolyl Azines 58
3.2.1	Synthesis 59
3.2.2	Crystallographic Evidence for the Planarity 61
3.3	Preparation of [2 × 2] Grids with Cu(I) or Ag(I) 63
3.3.1	Synthesis 64
3.3.2	X-Ray and other Techniques for Structural Characterization in the Solid State 65
3.3.3	Structural Characterization in Solution by NMR 69
3.3.4	Anion Exchange in the Solid State 71
3.4	Preparation of Coordination Polymers with 2,3-Pyrazolylquinoxalines or 2,3-Pyrazolylpyrazines and Cu(I) or Ag(I) 72
3.4.1	Preparation and Characterization of Dinuclear Building Blocks and Coordination Polymers 72
3.4.2	X-Ray and other Techniques for Structural Characterization 74
3.5	Preparation of Supramolecular Structures with 2,4-Diamino-6-R-1,3,5-triazines and Ag(I) 79
3.5.1	Synthesis 80
3.5.2	X-Ray Structure Determination 80
3.5.3	Structural Characterization in Solution by NMR 84
3.6	Conclusions 85
	References 86
4	**Chiral Metallocycles for Asymmetric Catalysis** 93
	Wenbin Lin
4.1	Introduction 93
4.2	Thermodynamically-Controlled Metallocycles 94
4.3	Kinetically-Controlled Metallocycles 95
4.4	General Synthetic Strategies for Chiral Metallocycles 96
4.5	Self- and Directed-Assembly of Chiral Pt-Alkynyl Metallocycles 101

4.6	Chiral Pt-Alkynyl Metallocycles for Asymmetric Catalysis	107
4.7	Concluding Remarks	109
	References	110

5 Catalysis of Acyl Transfer Processes by Crown-Ether Supported Alkaline-Earth Metal Ions 113
Roberta Cacciapaglia, Stefano Di Stefano, and Luigi Mandolini

5.1	Introduction	113
5.2	Basic Facts and Concepts	113
5.2.1	Reactivity of Alkaline-Earth Metal Alkoxides	114
5.2.2	The Influence of Crown Ethers	115
5.2.3	Preorganized Systems	116
5.2.3.1	Selected Examples	116
5.3	Nucleophilic Catalysts with Transacylase Activity	118
5.3.1	Calixcrowns	119
5.3.1.1	Catalytic Efficiency vs. Ester Reactivity	121
5.3.1.2	Trifunctional Catalysis	123
5.3.1.3	*p-tert*-Butylcalix[5]arene Derivatives	123
5.3.2	Thiol-Pendant Crown Ethers	124
5.4	Bimetallic Catalysts	128
5.4.1	Azacrown Ligating Units	129
5.4.1.1	Azacrown Decorated Calixarenes	133
5.4.2	Stilbenobis(18-Crown-6) Ligands	133
5.4.3	A Phototunable Dinuclear Catalyst	135
5.4.4	Effective Molarity and Catalytic Efficiency	136
5.5	Concluding Remarks	139
	References	140

6 Bio-Inspired Supramolecular Catalysis 143
Johannes A.A.W. Elemans, Jeroen J.L.M. Cornelissen, Martinus C. Feiters, Alan E. Rowan, and Roeland J.M. Nolte

6.1	Introduction	143
6.2	Host–Guest Catalysis	144
6.2.1	Rhodium-based Receptors	145
6.2.2	Copper-based Receptors	146
6.2.3	Porphyrin-based Receptors	149
6.3	Cytochrome P450 Mimics	153
6.3.1	Membrane-based Catalysts	153
6.3.2	Single Molecule Studies on Epoxidation Catalysts	155
6.4	Biohybrid Catalytic Systems	157
6.4.1	Bioamphiphiles	157
6.4.2	Single Enzyme Catalysis	159
6.5	Outlook	161
	References	162

7		**Selective Stoichiometric and Catalytic Reactivity in the Confines of a Chiral Supramolecular Assembly** *165*
		Michael D. Pluth, Robert G. Bergman, and Kenneth N. Raymond
	7.1	Introduction *165*
	7.2	Chemistry of Organometallic Guests *167*
	7.3	The Assembly as a Catalyst *175*
	7.3.1	Electrocyclic Rearrangements *175*
	7.3.2	Acid-Catalyzed Reactions *183*
	7.4	Conclusions and Outlook *191*
		References *191*
8		**New Supramolecular Approaches in Transition Metal Catalysis; Template-Ligand Assisted Catalyst Encapsulation, Self-Assembled Ligands and Supramolecular Catalyst Immobilization** *199*
		Joost N.H. Reek
	8.1	Introduction *199*
	8.2	Template-Ligand Assisted Catalyst Encapsulation *200*
	8.3	Self-Assembled Ligands in Transition Metal Catalysis *210*
	8.3.1	Template Approach *212*
	8.3.2	Direct Approach *217*
	8.4	Supramolecular Anchoring of Catalysts to Support *225*
	8.5	Conclusion *228*
		References *229*
9		**Chirality-Directed Self-Assembly: An Enabling Strategy for Ligand Scaffold Optimization** *235*
		James M. Takacs, Shin A. Moteki, and D. Sahadeva Reddy
	9.1	Introduction *235*
	9.2	The Need for New Catalyst Systems *235*
	9.3	A Typical Modular Approach to Chiral Bidentate Ligand Design *236*
	9.4	A Further Rationale for Developing Combinatorial Approaches to Scaffold Optimization *237*
	9.5	Approaches to Scaffold Optimization *238*
	9.6	A Convergent Approach to the Formation of Heterobimetallic Catalyst Systems *239*
	9.7	Chirality-Directed Self-Assembly: Selective Formation of Neutral, Heteroleptic Zinc(II) Complexes *240*
	9.8	*In situ* SAL Preparation *244*
	9.9	Ligand Scaffold Optimization in Palladium-Catalyzed Asymmetric Allylic Amination *244*
	9.10	What has been Learned? *246*
	9.11	Why such Wide Variation in Enantiomeric Excess given the Relatively Small Changes in Scaffold Structure? *248*
	9.12	Ligand Scaffold Optimization in Rhodium-Catalyzed Asymmetric Hydrogenation *248*

9.13	Concluding Remarks *250*	
	References *251*	
10	**Supramolecular Catalysis: Refocusing Catalysis** *255*	
	Piet W. N. M. Van Leeuwen and Zoraida Freixa	
10.1	Introduction: A Brief Personal History *255*	
10.2	Secondary Phosphines or Phosphites as Supramolecular Ligands *258*	
10.3	Host–Guest Catalysis *263*	
10.4	Ionic Interactions as a Means to Form Heterobidentate Assembly Ligands *269*	
10.5	Ditopic Ligands for the Construction of Bidentate Phosphine Ligands *276*	
10.6	Conclusions and Outlook *289*	
	References *291*	

Index *301*

Preface

In recent years supramolecular, homogeneous catalysis has undergone a renaissance and the activities in this area are growing rapidly. In the seventies, supramolecular catalysis was largely equivalent to mimicking enzymes via host-guest catalysis and the reactions studied were mainly of the types also occurring in enzymes, such as hydrolysis or oxidation reactions. Occasionally enormous accelerations were noted, or changes in selectivity, but applications in synthetic chemistry remained elusive. Of the non-enzymatic reactions, the Diels-Alder reaction was also studied successfully. Progress into other directions, amongst them organometallic catalysis, was slow mainly due to the tedious synthesis of host molecules equipped with catalytic entities. Organometallic catalysts have played a key role in the syntheses of chemical commodities as well as fine chemicals since the late 1960s and in the last decade its contribution to fine chemical syntheses has rapidly grown. In view of the required better use of feedstocks and the change in feedstocks, the role of selective catalysis will become even more important. Enzymes remain a source of inspiration, but more convenient routes to catalyst systems based on organometallic catalysts and containing supramolecular features, such as host-guest interaction, are needed.

In the last decade several new approaches have been introduced which avoid the use of elaborate syntheses. Both cavities and ligands are prepared via assembly processes which speed up the process enormously and assembly also leads to a large number of catalyst systems where only a limited number of building blocks have to be synthesized. In just a few years time this has led to an outburst of "supramolecular" catalysts, of astounding beauty, with unprecedented selectivities, or with high practicality. Of the latter group a few hold even promise for industrial application.

To highlight the recent advances in supramolecular catalysis, the Catalan Institution for Research and Advanced Studies (ICREA foundation) and the Institute of Chemical Research of Catalonia (ICIQ) organized the Conference on Supramolecular Approaches to Catalysis (SUPRAcat, March 2008, Barcelona). The conference brought together some of the leading and internationally recognized researchers in the field to discuss the development of these novel supramolecular catalysts and to

Supramolecular Catalysis. Edited by Piet W. N. M. van Leeuwen
Copyright © 2008 WILEY-VCH Verlag GmbH & Co. KGaA, Weinheim
ISBN: 978-3-527-32191-9

identify future directions for this exciting area of research. This book was inspired by the conference and a selection of the presenting speakers has contributed to this work. The organizers of the conference, Pablo Ballester and Anton Vidal, wrote the introductory chapter, thus highlighting basic concepts, different approaches, and a few of the many successes. The remaining nine chapters of the book give a cross section of the field and many aspects of modern supramolecular catalysis are dealt with.

I am very grateful to all contributors, their fast responses, and their willingness to participate in this project on such a short notice.

Thanks should also go to the publisher, Wiley-VCH, for the support provided in the compilation of this book. I would especially like to thank the team of Dr. Manfred Kohl, Lesley Belfit and Axel Eberhard for all their hard work.

January 2008

Piet W.N.M. van Leeuwen
Tarragona and Amsterdam

List of Authors

Pablo Ballester
Institució Catalana de Recerca i Estudis
Avançats and Institut Català
d'Investigació Química
Avgda. Països Catalans 16
43007 Tarragona
Spain

Robert G. Bergman
University of California
Department of Chemistry
Berkeley
CA 94720-1460
USA

Bernhard Breit
Albert-Ludwigs-Universität Freiberg
Institut für Organische Chemie und
Biochemie
Albertstraße 21
79104 Freiburg
Germany

Roberta Cacciapaglia
IMC-CNR
Sezione Meccanismi di Reazione
c/o Dipartimento di Chimica
Universita' La Sapienza
Piazzale Aldo Moro 5
00185 Roma
Italy

Jeroen J.L.M. Cornelissen
Radboud University Nijmegen
Institute for Molecules and Materials
Toernooiveld 1
6525 ED Nijmegen
The Netherlands

Stefano Di Stefano
Universita' La Sapienza
Dipartimento di Chimica
Piazzale Aldo Moro 5
00185 Roma
Italy

Johannes A.A.W. Elemans
Radboud University Nijmegen
Institute for Molecules and Materials
Toernooiveld 1
6525 ED Nijmegen
The Netherlands

Martinus C. Feiters
Radboud University Nijmegen
Institute for Molecules and Materials
Toernooiveld 1
6525 ED Nijmegen
The Netherlands

Supramolecular Catalysis. Edited by Piet W. N. M. van Leeuwen
Copyright © 2008 WILEY-VCH Verlag GmbH & Co. KGaA, Weinheim
ISBN: 978-3-527-32191-9

List of Authors

Zoraida Freixa
ICIQ – Institute of Chemical Research
of Catalonia
Av. Països Catalans 16
43007 Tarragona
Spain

Félix A. Jalón
Universidad de Castilla-La Mancha
Facultad de Químicas – IRICA
Av. Camilo José Cela 10
13071 – Cuidad Real
Spain

Wenbin Lin
University of North Carolina
Department of Chemistry
CB#3290
Chapel Hill
NC 27516
USA

Luigi Mandolini
Universita' La Sapienza
Dipartimento di Chimica
Piazzale Aldo Moro 5
00185 Roma
Italy

Blanca R. Manzano
Universidad de Castilla-La Mancha
Facultad de Químicas-IRICA
Av. Camilo José Cela 10
3071 – Cuidad Real
Spain

Shin A. Moteki
University of Nebraska-Lincoln
Department of Chemistry
Brace Laboratory
Lincoln
NE 68588
USA

Roeland J.M. Nolte
Radboud University Nijmegen
Institute for Molecules and Materials
Toernooiveld 1
6525 ED Nijmegen
The Netherlands

Isabel M. Ortiz
Universidad de Castilla-La Mancha
Facultad de Químicas-IRICA
Av. Camilo José Cela 10
13071 – Cuidad Real
Spain

Michael D. Pluth
University of California
Department of Chemistry
Berkeley
CA 94720-1460
USA

Kenneth. N. Raymond
University of California
Department of Chemistry
Berkeley
CA 94720-1460
USA

D. Sahadeva Reddy
University of Nebraska-Lincoln
Department of Chemistry
Brace Laboratory
Lincoln
NE 68588
USA

Joost N. H. Reek
University of Amsterdam
van't Hoff Institute for Molecular
Sciences
Nieuwe Achtergracht 166
1018 WV Amsterdam
The Netherlands

Alan E. Rowan
Radboud University Nijmegen
Institute for Molecules and Materials
Toernooiveld 1
6525 ED Nijmegen
The Netherlands

M. Laura Soriano
Universidad de Castilla-La Mancha
Facultad de Químicas-IRICA
Av. Camilo José Cela 10
13071 – Cuidad Real
Spain

James M. Takacs
University of Nebraska-Lincoln
Department of Chemistry
Brace Laboratory
Lincoln
NE 68588
USA

Piet W. N. M. Van Leeuwen
ICIQ – Institute of Chemical Research
of Catalonia
Av. Països Catalans 16
43007 Tarragona
Spain

Anton Vidal-Ferran
Institució Catalana de Recerca i Estudis
Avançats (ICREA)
Pg. Lluís Companys 2
08010 Barcelona
Spain

1
Introduction to Supramolecular Catalysis
Pablo Ballester and Anton Vidal-Ferran

1.1
Introduction

Much of the inspiration for the design of supramolecular catalytic system arises from the observation and understanding of enzyme catalysis [1–3]. However, synthetic models usually contain only one or few of the features that are present in the biologically enzymatic systems. In contrast, supramolecular enzyme model systems are smaller and structurally simpler than enzymes. The fact that the synthetic systems are simpler than the biological ones does not limit the detail of questions that may be investigated; instead, using these simpler systems it is possible to estimate the relative importance of different factors contributing to catalysis. A further advantage of the use of supramolecular models for studying catalysis is that the compounds can be synthetically manipulated to study a specific property. In the biological realm of catalysis it is a tremendous task to discern a particular factor responsible for the catalytic efficiency of the enzyme.

Supramolecular systems can be considered as new tools of modern physical organic chemistry. The study of catalytic processes using supramolecular model systems aims to explain the observed rate enhancement in terms of structure and mechanism. In some cases, the model systems may even provide a simplified simulation of the action of an enzyme and lead to further understanding of the different mechanism by which enzymes are able to achieve impressive reaction rate accelerations and turnover numbers.

Catalysis is a longstanding proposed application of supramolecular chemistry and the production of supramolecular systems capable of mimicking the catalytic ability of natural enzymes is one of the ultimate goals of self-assembly research. Supramolecular chemists have approached these challenging endeavors from different perspectives. On the one hand, many model systems have been designed that make use of the binding energy to achieve catalysis. Within this approach two types of systems can be differentiated: (a) molecular receptors in which a catalytic site is placed close to a binding site that has been designed to bind selectively the reactant,

Supramolecular Catalysis. Edited by Piet W. N. M. van Leeuwen
Copyright © 2008 WILEY-VCH Verlag GmbH & Co. KGaA, Weinheim
ISBN: 978-3-527-32191-9

and (b) molecular receptors that promote the reaction of two simultaneously complexed reactants, forming a multimolecular (ternary or higher order) complex, which is held together by weak and reversible interactions.

When two reacting functionalities are brought in close proximity, i.e., by binding to a template/receptor or by inclusion into a molecular vessel, the observed rate acceleration may be a simple effect due to an increase in the effective local concentration [4]. The entropic advantage of an "intramolecular" over an intermolecular reaction can be quantified by measuring the effective molarity EM = $k_{intramol.}/k_{intermol.}$ [3]. The EM value can tell us how catalytic efficiency relates to structure in a system designed to bring functional groups together in close proximity. Although, Page and Jenks have estimated that the entropy changes in solution may have an effective concentration of about 6×10^8 M, and high EMs have been measured for simple cyclization reactions, with rare exceptions the simple approximation of reactants caused by synthetic supramolecular catalytic systems achieve rate accelerations that are tiny by comparison with enzymes (EM < 10 M). The low efficiency of this mechanism of catalysis for two-substrate reactions is probably because the molecules of the bimolecular or ternary complexes are not tightly bound – there is residual entropy in the complex due to vibrations. As soon as the molecules become linked in the TS there is still a large loss of entropy not overcome by the binding energy.*

A key feature of enzymes is their ability to bind, and thus stabilize, selectively the transition state and intermediates for a particular reaction. So the problem of catalysis can be defined in terms of the molecular recognition of transition states. We will see below that many supramolecular models often fail to reproduce the turnover ability of enzymes due to inhibition by strongly-bound products. More sophisticated synthetic hosts have been designed to achieve catalysis not only by placing converging binding sites in such a way that reactant molecules are brought together in close proximity (entropy of activation is reduced or partially compensated by the favorable binding energy) but also by stabilization of the intermediate or transition state.† However, even from these more elaborated structures very few efficient supramolecular catalysts have emerged [5]. Sanders has pointed out that it is probably the fear of entropy that has taken supramolecular chemist too far in the direction of rigidity and preorganization. Rigid structures with a slightly mismatch to the TS will not be effective catalysts. Furthermore, the use of large and rigid molecular components in the construction of catalytic supramolecular systems makes it difficult to achieve the sub-ångström adjustments required for perfect TS stabilization. However, to the best of our knowl-

---

* In enzyme catalysis entropy is probably one of the most important factors. Enzyme reactions take place with substrates that are nanoconfined in the active sites and form a very tight enzyme–substrate complex. The catalytic groups are part of the same molecule as the substrate so there is no loss of transition or rotational entropy in the TS.
† Simply bringing together the reactant groups of the molecules makes productive encounters more probable but the most effective system will be one in which the flexibility of different binding geometries is eliminated and only the transition state like geometry is prescribed. The binding forces for the formation of a bond between two reactants are the intrinsic reactivity of the functional groups and the way the groups are brought together – this second factor is responsible of the efficiency of the catalysis.

edge, this point has not been tested so far, mainly due to the difficult design and synthesis effort required. The construction of self-assembled catalyst in which all recognition motifs, both between the catalyst subunits and between substrate and catalyst, are kinetically labile could provide a feasible approximation to approach the issue of rigidity of the catalytic supramolecular system [6]. The better fit to the transition state accomplished by a flexible supramolecular system is achieved at the expense of a larger loss of vibrational and rotational entropy on going from a not rigid and conformationally unrestricted complex to the fixed geometry of the TS.

Desolvation of the reacting polar groups is also a mechanism by which enzymes achieve rate accelerations. The desolvation of functional groups takes places during the inclusion of the reactants within the catalytic apparatus of the active site. This is another way of looking at the selective stabilization of the TS, in terms of a "specific" solvation of the reactants by the enzyme residues of the active site, which replaces the random solvation by the solvent molecules and the inherent enthalpic and entropic cost associated with their reorganization in the TS.

1.2
Design Approaches to Supramolecular Catalysis

1.2.1
Molecular Receptors that Place a Binding Site Close to a Catalytic Center

Early examples of two-substrate supramolecular catalysis emerge from the work of Bender et al., who studied the hydrolysis of m-tert-butylphenyl acetate in the presence of 2-benzimidazoleacetic acid with α-cyclodextrin [7]. Breslow, Knowles and others further extended the use of cyclodextrins as enzyme models. In this regard, Breslow's group synthesized a β-cyclodextrin (cycloheptaamylose) carrying two imidazole groups to model ribonuclease A [8]. More recently, Kim et al. have used a series of functionalized β-cyclodextrins with different polyazamacrocycles [9]. The Zn-complexes of these molecules are carboxypeptidase mimics, with the hydrophobic cavity of the cyclodextrin acting as a binding site for the aromatic residue of p-nitrophenyl acetate and the Zn(II) metal center complexed by the azamacrocycle being the catalytic site. The doughnut-shaped structure of the cyclodextrins is rather inflexible and it has twelve hydroxyl groups on the top side and six primary hydroxyl groups at the bottom. The torus is slightly more open on the side of the secondary OH groups, but β-cyclodextrins with a cavity diameter of about 7 Å display an efficient binding of the substrate from the side of the primary OH groups. The binding geometry of the complex places the reactive acetate group of the organic substrate close to a nucleophilic water molecule bound to the chelated Zn(II). The reported EMs are in the range 0.2–0.3 M and are based on measurements of $k_{cat}=k_{intra}$ for the decomposition of the productive ternary complex shown in Figure 1.1 and k_{inter} for the reaction of p-nitrophenyl acetate and the corresponding Zn(II) complex of the polyazamacrocycle. Two basic assumptions are implicit in the reliability of the calculated EMs: (a) the pK_a and

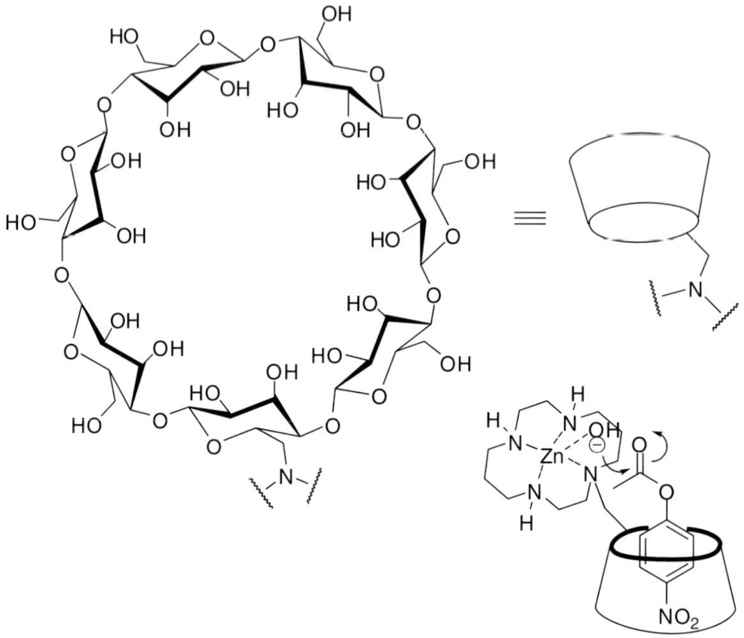

Figure 1.1 Molecular structure of functionalized β-cyclodextrin and the corresponding binary complex with p-nitrophenyl acetate.

nucleophilicity of the bound water is unchanged upon linking the macrocycle to the cyclodextrin and (b) the reactivity of the p-nitrophenyl acetate does not change on complexation.

Recent examples of this kind of methodology can be found, for example, in the work of Rebek et al. [10] The catalyst used is a cavitand armed with a Zn salen-type complex (Figure 1.2). The cavitand adopts a vase-like conformation that is stabilized by a seam of hydrogen bonds provided by the six secondary amides. The structure of the catalyst permits a slow dynamic exchange between free and bound guest (reactant) on the ^{1}H NMR time-scale that is controlled by the folding and unfolding of the cavitand.

When the guest used is p-nitrophenylcholine carbonate (PNPCC) the Lewis acid zinc(II) activates the well-positioned carbonyl group in the PNPCC@Zn-cavitand towards reactions with external nucleophiles. The energy minimized structure of the PNPCC@Zn-cavitand complex shows that cation–π interactions and C=O···Zn coordination bond occurs simultaneously.

Kinetic studies revealed that the hydrolysis of PNPCC by water present in commercial CH_2Cl_2 buffered with $CF_3CO_2H/EtN(i-Pr)_2$ was catalyzed in the presence of the Zn-cavitand. The hydrolysis of the carbonate is slow under these reaction conditions and only ca. 30% pf PNPCC is decomposed after 5 h. The acceleration of the reaction rate is more than 50-fold when 1 equiv of cavitand is present.

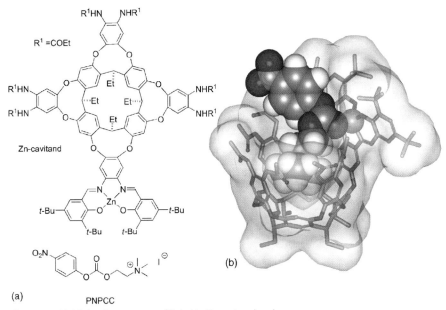

Figure 1.2 (a) Molecular structure of Rebek's Zn-cavitand and p-nitrophenylcholine carbonate (PNPCC). (b) CAChe minimized structure of the PNPCC@Zn-cavitand complex.

Another example of a synthetic catalyst capable of orienting the substrate towards the reaction center has been described recently by Crabtree and coworkers (Figure 1.3) [11]. The authors combined molecular recognition through hydrogen bonds and C−H activation to obtain high turnover catalytic regioselective functionalization of sp^3 C−H bonds remote from the −COOH recognition group. The catalyst contains a di-μ-oxo dimanganese catalytic core and two ligands that are based on the covalent connection of a Kemp's triacid unit with a terpy group through a phenylene linker. The Kemp's triacid unit provides a well-known U-turn motif having a −COOH group suitably oriented for the molecular recognition of another −COOH function.

Molecular modeling studies allowed to predict from the proposed geometry of the H-bonded complex with ibuprofen – 2-(4-isobutylphenyl)propionic acid – which C−H (indicated by to arrows) in the substrate would be expected to come closest to the active site and consequently became oxidized. If the oxidation operates via the catalyst–substrate complex predicted by the model complex, then the regioselective product should be the major component of the reaction mixture. When ibuprofen was treated with the catalyst, the selectivity for the regioselective product (97.5 : 2.5) was raised more than 10-fold when compared to the value obtained with a catalyst lacking the −COOH group (77 : 23). Oxidation of an alkyl carboxylic acid using the same catalysts led not only to regioselective oxygenation but also to diastereoselection of a single isomer. With a 0.1% catalyst-to-substrate ratio, a total turnover number of 580 was attained without loss of regioselectivity.

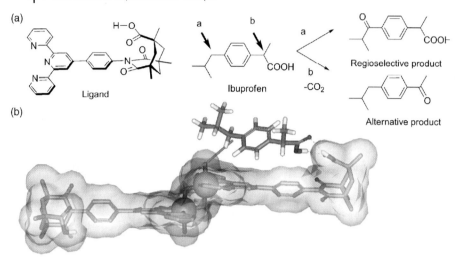

Figure 1.3 (a) Molecular structure of Crabtree's ligand and oxidation products of ibuprofen with the synthetic di-μ-oxo dimanganese catalyst. (b) Molecular model of the supramolecular catalyst (CAChe minimized) docked with ibuprofen.

Wärnmark et al. [12] have reported the formation of a dynamic supramolecular catalytic system involving a hydrogen bonding complex between a Mn(III) salen and a Zn(II) porphyrin (Figure 1.4). The salen sub-unit acts as the catalytic center for the catalytic epoxidation of olefins while the Zn-porphyrin component performs as the binding site. The system exhibits low selectivity for pyridine-appended styrene derivatives over phenyl-appended derivatives in a catalytic epoxidation reaction. The

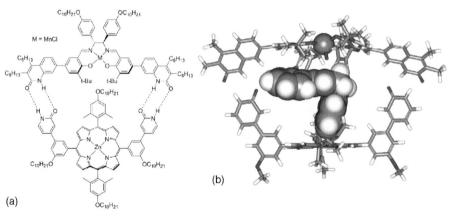

Figure 1.4 (a) Molecular structure of the supramolecular macrocyclic heterodimer catalyst used by Wärnmark et al. (b) CAChe minimized 3D representation of a substrate bound inside the hydrogen bonded macrocycle. Long alkyl chains are reduced to a methyl group for clarity. The substrate and the Mn atom (purple) are shown in CPK representation.

authors provide evidence that substrates bound to the supramolecular receptor and the non-bonded substrates are epoxidized by two different catalytic species. They also established the main reasons for the low observed selectivity.

In a somewhat related work, Nolte, Rowan et al. [13] described in 2003 a rotaxane complex that mimics the ability of processive enzymes to catalyze multiple rounds of reaction while the polymer substrate stays bound. The catalyst, which consists of a substrate-binding cavity incorporating a manganese(III) porphyrin complex acting as the catalytic center, can oxidize alkenes complexed within the toroid cavity, provided a ligand has been attached to the outer face of the toroid to both activate the porphyrin complex and prevent it from being able to oxidize alkenes outside the cavity.

1.2.2
Molecular Receptors that Promote the Reaction of two Simultaneously Complexed Reactants

The design of supramolecular systems capable of catalyzing bimolecular reactions is challenging: the supramolecular host first needs to recognize the reagents (which requires sufficiently strong binding) and, second, needs to correctly orient the two reagents and bring them together. Kirby [2] has coined the term "matchmakers" to describe synthetic hosts that perform these functions.

If the host has a higher affinity for the product arising from the bimolecular reaction than for the reagents, an additional problem comes into play: inhibition of turnover of the catalyst by the product. Although a host can not be regarded as truly catalytic if this occurs (stoichiometric amounts of host are required to achieve full conversion), the host can still accelerate the rate of the reaction and, interestingly, may even influence the outcome of the reaction.

Sanders and coworkers have designed and prepared a series of cyclic Zn(II) porphyrin trimers that are noteworthy not only because they accelerate the Diels–Alder reaction of a pyridine-substituted diene and a dienophile (Figure 1.5) [14] but because they also influence the stereochemical outcome of the cycloaddition reaction. The cyclic porphyrin oligomers can accommodate the diene and dienophile inside the cavity, thus lowering the activation energy by holding both reagents in close proximity. The larger 2,2,2-porphyrin trimer catalyzes the formation of the thermodynamically more favored *exo*-adduct up to 1000× faster than the formation of the *endo*-isomer. This rate acceleration corresponds to an EM of ca. 20 M, which is quite high for an artificial system. The key effect of the binding process inside the cavity should be recalled at this point, which is capable of reversing the stereochemistry of the reaction and mediating the formation of the unexpected *exo*-product (the *endo*-compound is produced in the absence of catalyst by kinetic control).

The computer generated model in Figure 1.6 shows the perfect fit of the *exo*-adduct inside the 2,2,2-trimer cavity. The smaller 1,1,2-trimer showed a stereoselective bias towards the endo-product at 30 °C; however, this stereoselectivity is lost at higher temperatures (a mixture of both the *endo*- and *exo*-products was observed at 60 °C). The reversal of the stereochemical outcome of the Diels–Alder reaction between the two cyclic dimers (30 °C) appears to lie firstly in a large (500-fold) *endo* acceleration

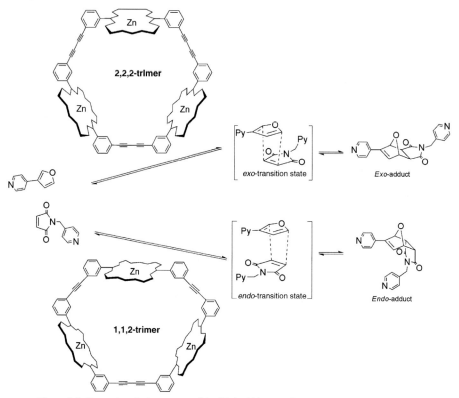

Figure 1.5 Stereochemical outcome of the Diels–Alder reactions under the influence of the 2,2,2- and 1,1,2-trimers.

induced by the smaller 1,1,2-dimer and, secondly, in a lack of complementarity at 30°C of the less flexible 1,1,2-trimer for the *exo*-product. Given its greater flexibility, the 2,2,2-trimer is better suited to accommodate at room temperature the transition state that leads to the *exo*-product.

Sanders and coworkers have extended the use of these cyclic porphyrin systems in the catalysis of acyl-transfer reactions [15] and hetero-Diels–Alder reactions [16].

Kelly and coworkers devised a two binding-site host that accelerates an S_N2 reaction between a primary aliphatic amine and an alkyl bromide [17]. Kelly's host (Figure 1.7) acted as template for the two reactants that were able to form three hydrogen bonds to each aminopyridone from the host. The reactants were the aminomethyl- and bromomethylnaphthyridines indicated in Figure 1.7, which bound strongly ($K > 10^4\,M^{-1}$) to the aminopyridone moieties from the host. A sixfold acceleration of the S_N2 reaction was observed, but turnover could not be demonstrated (the product precipitated from the $CHCl_3$ solution as the HBr salt).

An unavoidable limitation of Kelly's host was that the two binding sites were identical, which allowed non-productive binding of two identical substrates. Kelly therefore designed a new receptor with two different binding motifs that selectively

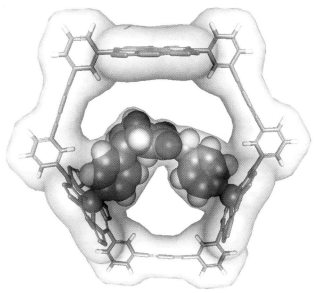

Figure 1.6 CAChe minimized 3D structure of the bound exo-adduct inside the cavity of 2,2,2-trimer (porphyrin substituents omitted for clarity).

recognize each of the two reactants. Furthermore, an additional phenyl group was introduced in the molecule to ensure that the bromide was productively directed towards the amino group. The reaction rate obtained with this variant was twice as high as that obtained with the symmetrical host. These results clearly illustrate the potential of supramolecular catalysis to provide substantial rate accelerations even in reactions with strict stereoelectronic requirements for a linear transition state.

Self-replication offers a direct path to supramolecular catalysis. Rebek *et al.* reported an impressive self-replicating system, which involves the assembly of suitably designed components by hydrogen bonding and π–π-stacking [18]. In the first step, the naphthalene ester and a heterocyclic amine self-assemble under the influence of hydrogen bonding and π–π-stacking interactions. The assembly is preorganized to facilitate the aminolysis of the neighboring ester group (Figure 1.8). The resulting *cis*-amide undergoes isomerization towards the less strained trans-isomer. This compound can then bind the two original reagents to form a ternary complex, which is again preorganized for nucleophilic attack of the ester group to give a dimer. Self-replication is thus elegantly achieved; the only drawback being the high stability of the dimer, which inhibits its role as a template for further self-replication.

Catalysis has also been induced by encapsulation or inclusion of molecules into cavities of molecular or supramolecular hosts. In both cases, the cavity should be large enough to accommodate concurrently at least two reactants in such a way that the reaction among them is favored (productive geometry). Probably, in these examples, not only the binding energy is used to induce catalysis by reducing or compensating the unfavorable entropy of activation but other specific mechanisms

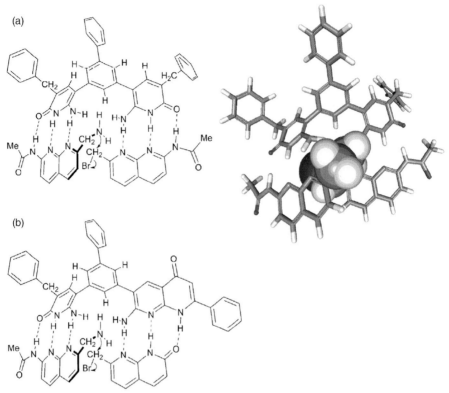

Figure 1.7 Nucleophilic reaction catalyzed by Kelly's hosts; S_N2 reaction in the host with two symmetrical (a) or two asymmetrical (b) hydrogen bonding recognition motifs.

may also come to play. The inner surface of the molecular cage can be complementary to the transition state (TS) and not also to the ground states of the reactants. It is also worth mentioning that, in some cases, when the reactants are encapsulated they become completely isolated from the solvent. This "specific" solvation eliminates the enthalpic and entropic cost of reorganizing the solvent molecules in the TS. Furthermore, the encapsulation or inclusion of the reactants into molecular vessels may produce a two-fold positive contribution to catalysis: (1) by exerting some "strain" to the molecular receptor, i.e., the minimum energy geometry of the vessels is slightly distorted due to the inclusion of the reactants, and (2) by inducing certain "stress" to the included reactants. The "strain" and "stress" energies are expected to be eliminated or reduced upon reaction. The amount of "stress" than can be forced on the included reactants of course depends on the strength of the interactions used to hold together the molecular components that constitute the host. For self-assembled supramolecular nanovessels the interactions are of "noncovalent" nature (usually hydrogen bonds and coordination bonds).

An early study of catalysis induced by inclusion of reactants in molecular vessels derives from the work of Mock et al. [19–21]. The authors reported that the

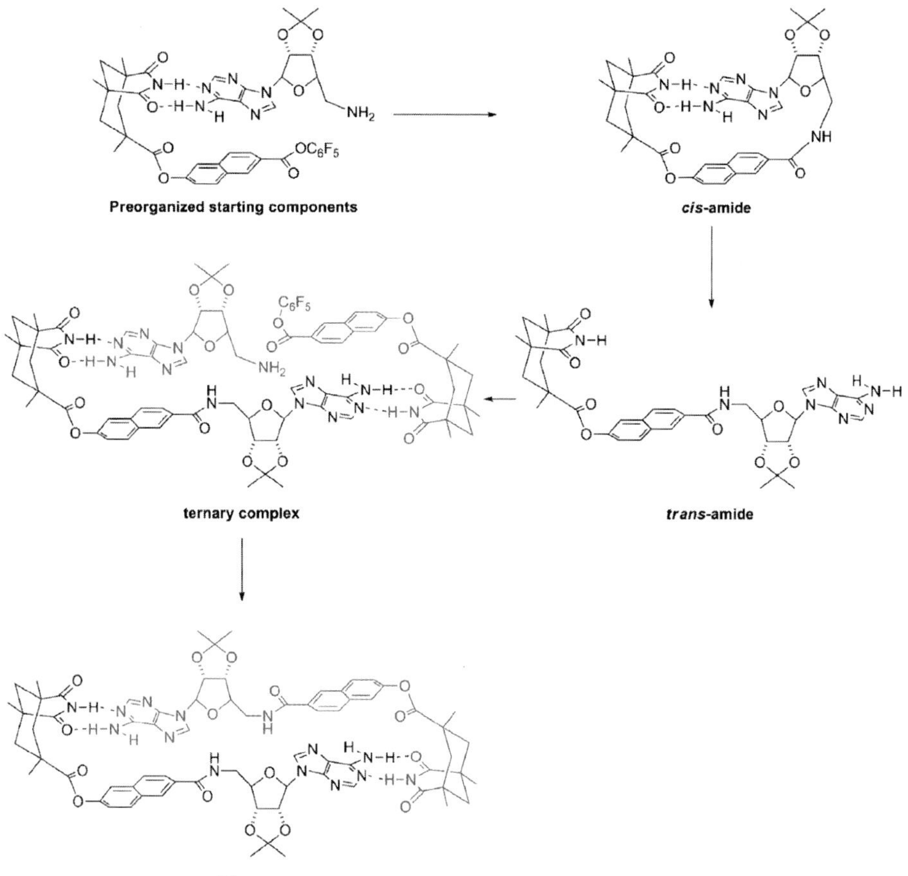

Figure 1.8 Rebek's self-replicating system.

intermolecular 1,3 dipolar cycloaddition between an azide and an alkyne, both of them substituted with an ammonium group, was substantially accelerated (ca. 10^5-fold or EM $= 1.6 \times 10^4$ M based on Ref. [3]) and became highly regioselective in the presence of cucurbit[6]uril (Figure 1.9). The calculated cavity volume of cucurbit[6]uril (164 Å^3) translates into an encapsulated reactant concentration of 10 M. The reaction kinetics also show catalytic saturation behavior, substrate inhibition and slow product release. The presence of a tertiary complex azide-alkyne@cucurbit[6]uril is documented by kinetic analysis. The simultaneous binding of both alkyne and azide (with the NH$_3^+$ group bound to each set of the carbonyls and with the substituents extended inside the cavity) aligns the reactive groups within the cavity of the host in a productive geometry that catalyzes the exclusive formation of the 1,4-substituted triazole. The authors state that cucurbit[6]uril accomplished catalysis through more than one mechanism. Curcurbit[6]uril not only eliminates or reduces the entropic constrains of the reaction by bringing both reactants together but also its cavity is more adequate for the geometry of the TS cycloaddition than for the bound reactants. Recent

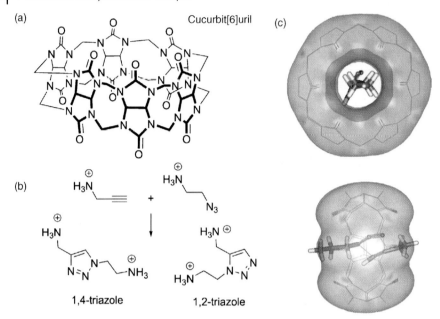

Figure 1.9 (a) Structure of cucurbit[6]uril; (b) 1,3-dipolar cycloaddition; (c) the reaction ternary complex for the formation of the 1,4-triazole.

computational investigations performed on the same system by Maseras and Carlqvist [22] suggest that the main catalytic effect is the elimination of the entropic cost of bringing the reactants together in the ternary complex and turning the addition reaction into a unimolecular one. Maseras and Carlqvist did not find computational evidences for transition state stabilization by the cucurbit[6]uril host.

Rebek has reported a new synthetic molecular vessel that can accelerate a 1,3-dipolar cycloaddition between an alkyne and an azide (Figure 1.10) [23]. The catalytic molecular capsule is formed by dimerization of two resorcinarene derivative subunits. The capsule has a roughly cylindrical cavity capable of accommodating two different aromatic guests. The orientation of the encapsulated guests is constrained to edge to edge approaches, and only the peripheral substituents make contacts. This arrangement seems to be appropriate for catalyzing the reaction between peripheral substituents of substrates anchored in the capsule by their respective aromatic groups. In particular, the cycloaddition under study involves phenylacetylene and phenyl azide. These compounds react very slowly to give equal amounts of the two regioisomers in the absence of the capsule ($k_{out} = 4.3 \times 10^{-9} \, M^{-1} \, s^{-1}$, $v = 1.3 \times 10^{-15} \, M \, s^{-1}$). In the presence of the capsule only the 1,4-isomer is formed. Control experiments have shown that only the 1,4-isomer is encapsulated. The local concentration of each reactant inside the capsule is 3.7 M (capsule volume $\approx 450 \, Å^3$). Consequently, the rate inside the capsule due to just an increase of the effective local concentration, and assuming that the productive geometry can be achieved, would be $v \sim 6 \times 10^{-8} \, M \, s^{-1} = k_{out} [alkyne]_{in}[azide]_{in}$, a value larger than the initial rate observed ($v = 1.3 \times 10^{-9}$

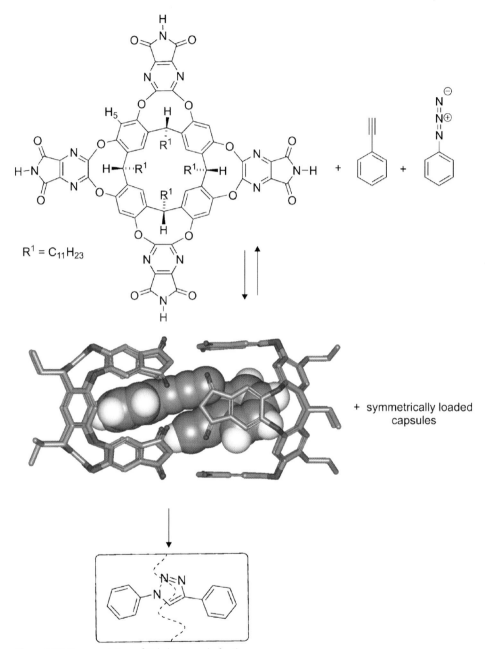

Figure 1.10 Representation of Rebek's capsule for the acceleration of a 1,3-dipolar cycloaddition between an alkyne and an azide.

M s^{-1}). The reaction rate outside the capsule at the concentrations used for the experiment should be v = k_{out} [alkyne] [azide] = 4.3 × 10^{-9} M^{-1} s^{-1} × 0.05 M × 0.025 M = 5.4 × 10^{-12} M s^{-1}. Thus, a rate acceleration of some 240 times can be estimated for the reaction taken place inside the capsule (EM = 120 M value taken from Ref. [3]). Probably, the reactants in the capsule are not positioned in the ideal geometry to achieve the TS and there are no reasons to believe that the TS is bound better than reactants. The most striking result of this work is the direct observation of the Michaelis complex, which simplifies the kinetic analysis. Unfortunately, the product is the best guest in the system and gradually the capsule is filled with it and the reaction is slowed by product inhibition.

In the latter two examples the EM values calculated from rate data are higher than the actual concentration of encapsulated reactants. This result indicates a favorable directional correlation of functionalities in the complex, but it is apparent that Rebek's capsule is less efficient than cucurbit[6]uril in promoting the productive geometry or stabilizing the TS. The molecular capsules not only concentrate reactants but also increase the time that the reacting function reside within critical distance (productive geometry) – both factors are determinant in achieving effective catalysis [24].

In striking contrast with these two previous examples of 1,3-dipolar cycloaddition catalyzed by encapsulation, the EMs calculated for particular examples of Diels–Alder reactions catalyzed by Rebek's softball [25] or by Sanders' [26] cyclophane, which was selected from a dynamic combinatorial library (DCL), are lower than the actual reactant concentration calculated from the volumes of the molecular cavities. Probably, the Diels–Alder reactions have more stringent orientational requirements than the 1,3-dipolar cycloaddition. The reactants of the Diels–Alder reactions, when encapsulated or included, spend a significant amount of time in ternary complexes displaying a non-productive mutual orientation.

Fujita et al. [27]. have used a self-assembled octahedral [M$_6$L$_4$] (M = {Pd^{II}(TME-DA)}, L = tris-(3pyridyl)triazine) cage to catalyze and impose geometry constrains onto the Diels–Alder reaction of anthracene and maleimide to give an unexpected regioselectivity (Figure 1.11). As commented above, in the two previous examples of 1,3-dipolar additions, the geometric requirements of the interior of the molecular capsules control the most favorable arrangement of the reacting groups, producing a change of the reaction selectivity observed in solution. In Fujita's example, the two reactants, N-cyclohexylmaleimide and several anthracene derivatives, adopt a fixed orientation within the capsule in a way that the cycloaddition can only take place with the 1,4-positions of the anthracene molecule. The Diels–Alder reaction of anthracene in the absence of the capsule yields and adduct bridging the center ring (9,10-positions) of the anthracene framework as a consequence of the high localization of π-electron density at that site. Also, when sterically less demanding N-propylmaleimide was used, only the 9,10-adduct was formed. In the presence of the M$_6$L$_4$ capsule the yield of the reaction after 5 hours at 80°C was 98% and only the *syn*-1,4-adduct was detected. In contrast, in the absence of the capsule the reaction gave only the conventional 9,10-adduct in 44% yield. However, the reaction using this capsule is not catalytic as the product is bound better.

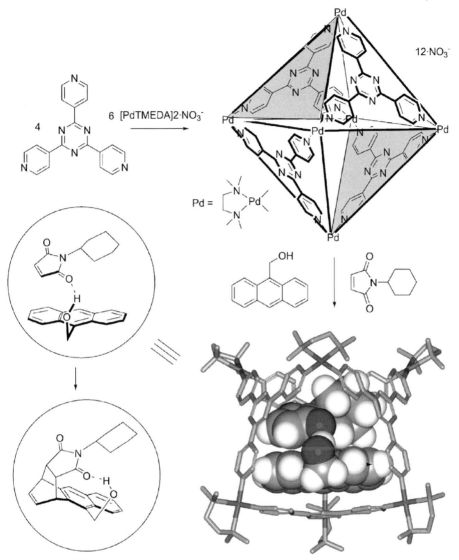

Figure 1.11 Fujita's self-assembled coordination cage, which is prepared by simple mixing of an exo-tridentate organic ligand and end-capped Pd(II) ion in a 4 : 6 ratio. Cache optimized structure of the ternary complex anthracene-maleimide@metallocage. Molecular structure of the syn-isomer of the 1,4-Diels–Alder adduct.

Raymond et al. [28] have employed a cavity-containing tetrahedral metal ligand assembly (M_4L_6) as a catalytic host for several reactions, e.g., the 3-aza-Cope rearrangement of allyl enammonium cations (Figure 1.12). Unstable organic species can also be stabilized within the same capsule [29]. The self-assembled cage preferentially hosts cationic guest over neutral ones as the cage is negatively charged.

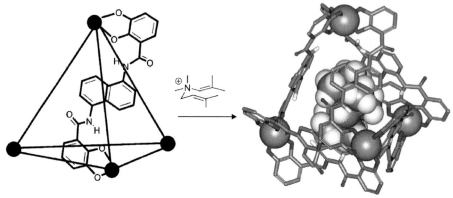

Figure 1.12 Schematic representation of the M_4L_6 cage and molecular structure of an enammonium substrate for the 3-aza Cope reaction. CAChe model of the M_4L_6 supramolecular assembly, encapsulating the enammonium substrate.

Upon binding the rates of rearrangement are accelerated for all substrates studied. The measured activation parameters of the reactions reveal that the supramolecular host can reduce both the entropic and the enthalpic barriers for the rearrangement. The capsule acts as a true catalyst, since release and hydrolysis facilitates turnover.

1.2.3
Preparation of the Catalyst Backbone via Supramolecular Interactions

In the earliest examples of supramolecular catalysis, a full arsenal of noncovalent interactions (electrostatic, hydrogen bonding, π–π stacking, and van der Waals interactions, hydrophobic and solvatophobic effects) and other reversible interactions, such as metal–ligand interactions, was employed to generate chemical assemblies between the reagent(s) and the catalytic system. By virtue of this supramolecular assembly, the reagents are more readily transformed into the final products (rate acceleration) or can even undergo reactions that are otherwise unfavored. All of the representative examples shown so far in this introduction demonstrate how supramolecular interactions have been involved in the directing and/or facilitating the catalytic event.

A completely different approach to achieve catalysis using supramolecular strategies has been developed in recent years. Noncovalent interactions normally lead to the rapid formation of bonds in an effective, reversible and strikingly apparently simple way. Chemists have exploited these features and have elegantly used reversible interactions (noncovalent effects and metal–ligand chemistry) to generate the backbones of a myriad supramolecular catalysts by bringing together the starting building blocks. These molecules contain the functional groups required for desired catalysis, and the motifs required for the assembly process of the supramolecular catalyst, via noncovalent and metal–ligand interactions. Furthermore, this novel methodology has enabled synthesis of libraries of structurally diverse supramolecular ligands with unprecedented ease compared to the use of standard covalent chemistry. The design and synthesis of chiral supramolecular catalysts appear to be very straightforward with this strategy, as the chiral components of the starting building blocks are simply incorporated into the supramolecular catalysts within the assembly process. Early examples in this field emerge from the pioneering work of Breit and coworkers [30] (Chapter 2), who reported the *in situ* generation of a library of bidentate ligands from the tautomeric 2-pyridone/2-hydroxypyridine pair. The dimerization of the tautomeric pair is governed by hydrogen bonding and is strongly favored in aprotic solvents. In addition, when the monomers contain a phosphorus donor group (PPh$_2$ in Figure 1.13), which can coordinate to a metal centre, the formation of the bidentate ligand is favored through chelation. This approach was used to prepare suitable catalysts for the rhodium-catalyzed

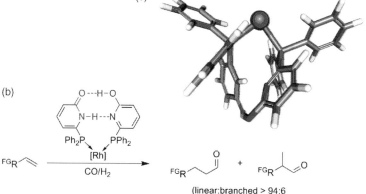

Figure 1.13 Generation of rhodium-based supramolecular catalysts by assembly of pyridine/hydroxypyridine pairs: (a) Self-assembly modes of pyridine-based phosphines. (b) Alkene hydroformylation with supramolecular rhodium-diphosphine catalysts; (c) CAChe minimized 3D structure of the rhodium-diphosphine complex (other ligands from the metal omitted for clarity).

hydroformylation of various structurally diverse terminal olefins, including arenes, alcohols, aldehydes, acetals, amides, esters, ethers and carbamates. The catalyst itself was generated by self-assembly of the pyridone and hydroxypyridine tautomers of 6-(diphenyl-phosphino)pyridine-2(1H)-one (6-DPPon), and a rhodium precursor. Notably, the reactions were generally carried out at room temperature and under ambient pressure to afford the linear aldehydes in good yields and with high regioselectivities [31,32]. Bear in mind that the catalytic properties of the metal center have been tuned by the use of ditopic ligands with a molecular skeleton based on supramolecular interactions. Studies on the complexing of 6-DPPon with different metals and ligands revealed that the catalytically less active polynuclear complexes is also formed, depending on the co-ligands and the complexing conditions [33].

The development of new, highly active chiral catalysts is an ever challenging area of research. Breit and coworkers have expanded their aforementioned supramolecular strategy to the preparation of asymmetric chiral bis-phosphorus substituted derivatives (Figure 1.14) [34]. They have also reported the generation of a library of heterodimeric chelating ligands based on the adenine-thymine (A-T) base pair, which serves in biological systems for a high precision self-assembly of complementary species by hydrogen bonding [35,36]. The preliminary pyridine/hydroxypyridine tautomers used are not amenable to the preparation of defined heterodimeric systems, since the mixing of two differently substituted pyridones would lead to statistical mixtures of homo- and heterodimeric dimers. Therefore, aminopyridine and isoquinolone moieties, which were substituted with several phosphorus-containing functional groups, were used as adenine and thymine analogues in an attempt to utilize a precise self-assembly strategy. As Breit and coworkers aimed to develop chiral supramolecular bidentate ligands for asymmetric catalysis, the adenine and thymine-like building blocks incorporated chiral phosphino or phosphonite groups. Hence, the assembled heterodimers could be described as chiral chelating bis-phosphorus substituted derivatives (Figure 1.14). Rhodium complexes derived from this supramolecular species were studied in the asymmetric hydrogenation of functionalized olefins. The bisphosphonite based on the BINOL-substituted aminopyridine and isoquinolone (Figure 1.14) yielded remarkably high enantiomeric excesses (up to 99%) in the hydrogenation of substituted acrylic acid derivatives. To further investigate the substrate scope of the newly prepared catalyst, cinnamic and itaconic acid derivatives were also hydrogenated, with enantiomeric excesses of 90% and 94% obtained, respectively.

The advancements in supramolecular catalysis are not limited to transitions-metal catalyzed reactions. Clarke and coworkers recently reported the preparation of a library of organocatalysts and their application in the asymmetric Michael addition of ketones to nitroalkenes [37]. They proposed use of a supramolecular catalyst formed

Figure 1.14 Generation of supramolecular catalysts for asymmetric hydrogenation: (a) Assembly of heterodimeric chelating ligands. (b) Structure of the optimal rhodium-diphosphonite complex for asymmetric hydrogenation (other ligands from the metal center omitted for clarity). (c) Enantioselective hydrogenation of functionalized alkenes.

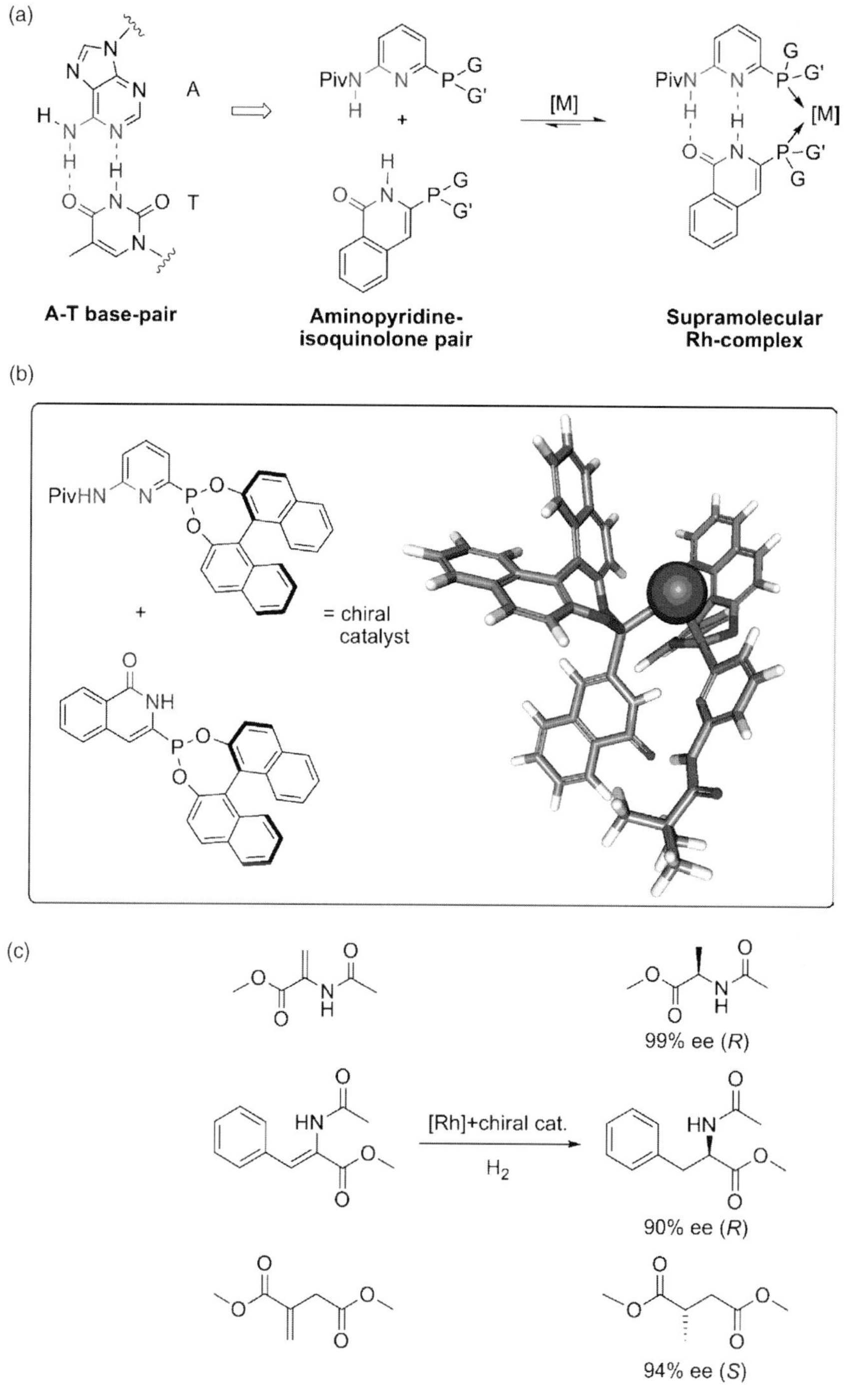

1.2 Design Approaches to Supramolecular Catalysis | 19

(a)

A-T base-pair

Aminopyridine-isoquinolone pair

Supramolecular Rh-complex

(b)

= chiral catalyst

(c)

99% ee (R)

[Rh]+chiral cat.
H$_2$

90% ee (R)

94% ee (S)

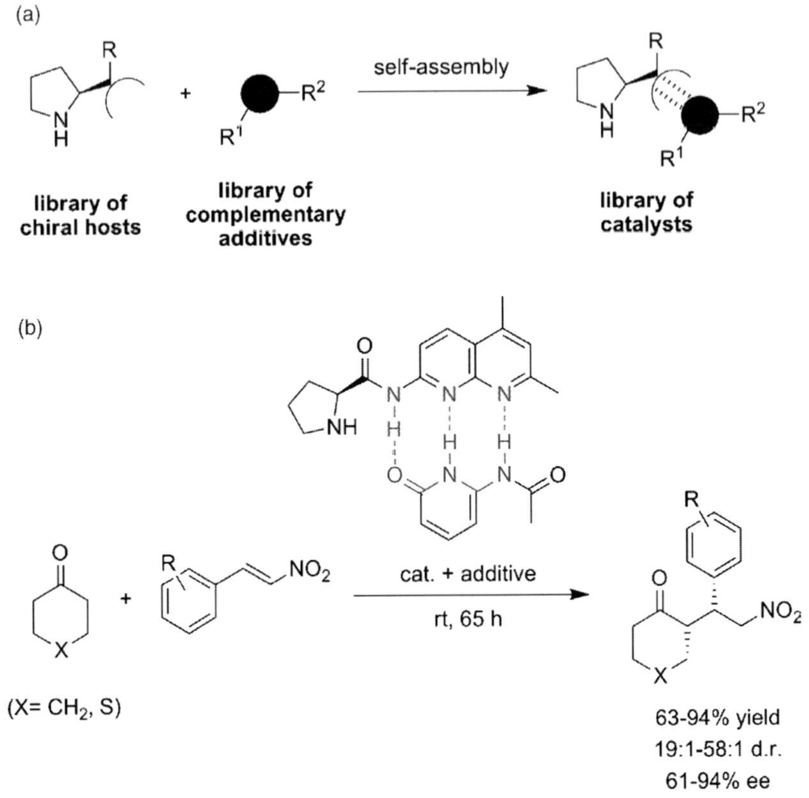

Figure 1.15 Supramolecular catalysts for organocatalysis: (a) Route to catalyst libraries; (b) asymmetric nitro-Michael reaction.

by complementary hydrogen bonding between a first unit, containing a proline fragment responsible for enantioselective catalysis, and a second, hydrogen bonding-complementary, unit bearing an achiral additive to influence the steric environment at the catalytic site (Figure 1.15).

Since amidonaphthyridines and pyridinones were known to be highly complementary recognition motifs, and were also easily accessible, they were chosen as the complementary units to achieve precise self-assembly by hydrogen bonding. It was observed that any catalyst arising from the combination of a chiral and an achiral fragment was more effective than the simple proline catalyst – a general enhancement in the diastereoselectivity and enantioselectivity of the reaction was observed when an additive was used. The catalyst/additive combination indicated in Figure 1.15 turned out to be best suited for the Michael reaction between six-membered cyclic ketones and several nitroalkenes.

Hydrogen bonding is not the only supramolecular strategy for efficiently generating the catalyst backbone by assembly of suitable constituent blocks: metal-directed self-assembly has also been employed. Reek, van Leeuwen and coworkers (Chapter 8) have exploited the use of a nitrogen donor group that binds to Zn(II)-porphyrins to construct a library of chiral supramolecular phosphorus-containing bidentate ligands for asymmetric catalysis [38]. Takacs and coworkers (Chapter 9) have prepared libraries of chiral bidentate P,P-ligands via metal-directed self-assembly, and used them in asymmetric allylation and hydrogenation reactions [39]. Their route to the supramolecular catalysts was based upon the self-assembly of two bisoxazoline-containing units around a zinc(II) center to form a tetrahedral (box)$_2$Zn complex. The final complex contained a second set of ligating groups equipped to bind a second metal to form a catalytic site. All of these examples of supramolecular catalysts formed via metal–ligand interactions will be discussed in more detail in other chapters of this book. Rather than designing two complementary constituent blocks that are capable of assembling with each other, Reek, van Leeuwen and coworkers reported in their pioneering work from 2003 [40] a very elegant strategy to generate supramolecular catalysts in which the constituent units of the final entity are assembled around a template molecule. The supramolecular bisphosphite indicated in Figure 1.16 was generated by the binding of two pyridine-phosphite units onto a bis-Zn(II) porphyrin template. The preferential binding affinity of Zn(II)-porphyrins for nitrogen donor atoms rather than phosphorus groups was crucial and allowed generation of a supramolecular catalyst with two free phosphite groups that could be further utilized in rhodium-catalyzed transformations. The supramolecular rhodium-complexes indicated in Figure 1.16 were studied in the hydroformylation of 1-octene and styrene. With 1-octene, the supramolecular complex showed slightly

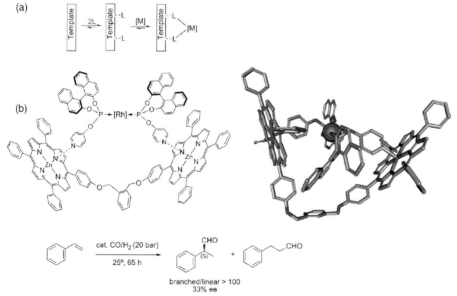

Figure 1.16 Transition metal Rh-catalysts formed by self-assembly: (a) Schematic representation of the self-assembly process. (b) Hydroformylation of styrene with supramolecular Rh-catalysts.

lower activity than that of the complex based on the analogue monodentate ligand, but it exhibited much higher selectivity for the linear product than for the branched product. These results indicated that the supramolecular bisphosphite ligands developed by Reek, van Leeuwen and coworkers had catalytic properties characteristic of bidentate ligands. In terms of enantioselectivity, mediocre results were obtained in the hydroformylation of styrene (33% ee), but these results are better than those obtained with the monomeric building blocks, thus reaffirming the bidentate character of the supramolecular phosphoros-containing catalyst.

Reek, van Leeuwen et al. have also reported the template-induced generation of bidentate ligands onto a rigid bis-zinc(II) salphen platform. Homobidentate [41] or heterobidentate [42] P,P-ligands have been prepared and tested in hydroformylation and hydrogenation reactions.

1.3
Artificial Biomacromolecules for Asymmetric Catalysis

In the fledgling period of the field of asymmetric catalysis, Whitesides devised a supramolecular catalytic system that relied on the generation of an artificial enzyme by incorporating a metal-containing fragment into a host protein [43]. In this pioneering "chemical mutation" of a protein, Whitesides took advantage of a very strong noncovalent interaction between a protein (avidin) and a small molecule (biotin). Functionalization of biotin with a rhodium diphosphine afforded a metal fragment with high affinity for the host protein. The strength of the avidin–biotin interaction ($K =$ ca. $10^{15}\,\text{M}^{-1}$) ensured quantitative binding of the rhodium diphosphine moiety into the chiral protein environment. Thus, an artificial metalloenzyme for the asymmetric hydrogenation of alkenes had been created. Whitesides and coworkers showed that the hydrogenation of N-acetamidoacrylate with catalytic metalloenzyme amounts rendered (S)-N-acetamidoalanine in 41% ee and with full conversion (Figure 1.17). Whitesides' approach remained untouched for 20 years until, in 1999, Chan and coworkers took it up again by linking a chiral diphosphine to biotin [44] and studied the catalytic properties of the resulting complex in the hydrogenation of itaconic acid. Methylsuccinic acids could be prepared with this strategy with moderate enantioselectivity. Ward and coworkers achieved a spectacular breakthrough in the stereoselectivity of this metalloenzyme-catalyzed hydrogenation of acetamidoacrylate derivatives [45] by combining the biotinylated diphosphine (Figure 1.17) with mutated streptavidin (WT Sav), rather than the original host protein (avidin). Quantitative yield and an excellent enantioselectivity (94%) were achieved using this new catalytic system. Using the same approach, Ward and coworkers have identified efficient artificial transfer hydrogenases [46]. Feringa and coworkers have developed a related supramolecular catalytic system that is based on DNA as the biomacromolecule, which provides the chiral environment, and 9-aminoacridine-modified Cu(II) complexes as the metal fragment with high DNA affinity. A hybrid DNAzyme was thus generated, and its activity in an asymmetric Diels–Alder reaction of cyclopentadiene with a

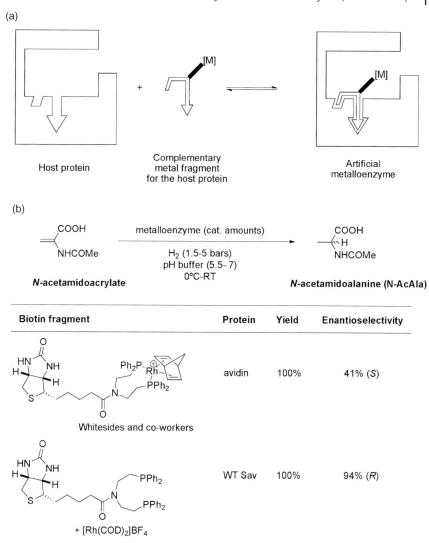

Figure 1.17 Artificial metalloenzymes: (a) Strategy to incorporate a catalytically active metal fragment within a host protein. (b) Alkene hydrogenation based on the biotin–(strept)avidin technology.

dienophile that binds the Cu(II) ion through a pyridyl group was studied. High endo:exo diastereoselectivities (91 : 9) and enantioselectivities (90%) were achieved in the optimal case [47]. Lastly, Kamer, van Leeuwen and coworkers have developed phosphine-containing oligonucleotides and have studied their application in asymmetric allylic aminations [48].

1.4
Summary and Outlook

The reader is reminded that this overview of supramolecular catalysis was not meant to be exhaustive, rather it is an introductory chapter. Definitive examples that illustrate the underlying principles of the design of these supramolecular systems have been described. The synthetic chemist has often looked to nature for inspiration in the design of efficient supramolecular catalysts. However, enzymes have evolved to be reaction specific, and supramolecular catalysis continues to be aimed at surpassing this limitation by generating not only effective catalysts for a particular reaction but also for entire classes of related reactions. To this end, the current understanding in the way supramolecular interactions (electrostatic, hydrogen bonding, π–π stacking, and van der Waals interactions, hydrophobic and solvatophobic effects, and metal–ligand interactions) operate, the arsenal of efficient synthetic methodologies and the wide variety of three-dimensional architectures available should foster the design and preparation of novel supramolecular catalysts with ever greater efficiency (i.e., turnover, product inhibition, chemo- and stereoselectivities, etc.).

Acknowledgments

The authors gratefully acknowledge the ICIQ and ICREA foundations.

References

1 Fersht, A. (1977) *Enzyme Structure and Mechanism*, W.H. Freeman and Company, Reading and San Francisco, pp. 371.
2 Kirby, A.J. (1996) Enzyme mechanisms, models, and mimics. *Angew. Chem., Int. Ed. Engl.*, **35**, 707–724.
3 Cacciapaglia, R., Di Stefano, S. and Mandolini, L. (2004) Effective molarities in supramolecular catalysis of two-substrate reactions. *Acc. Chem. Res.*, **37**, 113–122.
4 Mulder, A., Auletta, T., Sartori, A., Del Ciotto, S., Casnati, A., Ungaro, R., Huskens, J. and Reinhoudt, D.N. (2004) Divalent binding of a bis(adamantyl)-functionalized calix[4]arene to beta-cyclodextrin-based hosts: An experimental and theoretical study on multivalent binding in solution and at self-assembled monolayers. *J. Am. Chem. Soc.*, **126**, 6627–6636; Huskens, J., Mulder, A., Auletta, T., Nijhuis, C.A., Ludden, M.J.W. and Reinhoudt, D.N. (2004) A model for describing the thermodynamics of multivalent host–guest interactions at interfaces. *J. Am. Chem. Soc.*, **126**, 6784–6797.
5 Sanders, J.K.M. (1998) Supramolecular catalysis in transition. *Chem.–Eur. J.*, **4**, 1378–1383.
6 Jonsson, S., Odille, F.G.J., Norrby, P.-O. and Warnmark, K. (2005) A dynamic supramolecular system exhibiting substrate selectivity in the catalytic epoxidation of olefins. *Chem. Commun.*, 549–551.
7 Komiyama, M., Breaux, E.J. and Bender, M.L. (1977) The use of cycloamylose to probe the "charge-relay" system. *Bioorg. Chem.*, **6**, 127–136.

8 Breslow, R., Doherty, J.B., Guillot, G. and Lipsey, C. (1978) β-Cyclodextrinyl-bisimidazole, a model for ribonuclease. *J. Am. Chem. Soc.*, **100**, 3227–3229.

9 Kim, D.H. and Lee, S.S. (2000) Origin of rate-acceleration in ester hydrolysis with metalloprotease mimics. *Bioorg. Med. Chem.*, **8**, 647–652.

10 Richeter, S. and Rebek, J. Jr. (2004) Catalysis by a synthetic receptor sealed at one end and functionalized at the other. *J. Am. Chem. Soc.*, **126**, 16280–16281; Zelder, F.H. and Rebek, J. Jr. (2006) Cavitand templated catalysis of acetylcholine. *Chem. Commun.*, 753–754; Purse, B.W. and Rebek, J. Jr. (2005) Supramolecular structure and dynamics special feature: Functional cavitands: Chemical reactivity in structured environments. *Proc. Natl. Acad. Sci. U.S.A.*, **102**, 10777–10782.

11 Das, S., Incarvito, C.D., Crabtree, R.H. and Brudvig, G.W. (2006) Molecular recognition in the selective oxygenation of saturated C–H bonds by a dimanganese catalyst. *Science*, **312**, 1941–1943.

12 Jonsson, S., Odille Fabrice, G.J., Norrby, P.-O. and Warnmark, K. (2006) Modulation of the reactivity, stability and substrate- and enantioselectivity of an epoxidation catalyst by noncovalent dynamic attachment of a receptor functionality – aspects on the mechanism of the Jacobsen-Katsuki epoxidation applied to a supramolecular system. *Org. Biomol. Chem.*, **4**, 1927–1948; Jonsson, S., Odille Fabrice, G.J., Norrby, P.-O. and Warnmark, K. (2005) A dynamic supramolecular system exhibiting substrate selectivity in the catalytic epoxidation of olefins. *Chem. Commun.*, 549–551.

13 Thordarson, P., Bijsterveld, E.J.A., Rowan, A.E. and Nolte, R.J.M. (2003) Epoxidation of polybutadiene by a topologically linked catalyst. *Nature*, **424**, 915–918.

14 Walter, C.J., Anderson, H.L. and Sanders, J.K.M. (1993) exo-Selective acceleration of an intermolecular Diels–Alder reaction by a trimeric porphyrin host. *Chem. Commun.*, 458–460; Walter, C.J. and Sanders, J.K.M. (1995) Free-energy profile for a host-accelerated Diels–Alder reaction: The sources of exo selectivity. *Angew. Chem., Int. Ed. Engl.*, **34**, 217–219; Clyde-Watson, Z., Vidal-Ferran, A., Twyman, L.J., Walter, C.J., McCallien, D.W.J., Fanni, S., Bampos, N., Stephen Wylie, R. and Sanders, J.K.M. (1998) Reversing the stereochemistry of a Diels–Alder reaction: Use of metalloporphyrin oligomers to control transition state stability. *New J. Chem.*, **22**, 493–502; Clyde-Watson, Z., Bampos, N. and Sanders, J.K.M. (1998) Mixed cyclic trimers of porphyrins and dioxoporphyrins: geometry vs. electronics in ligand recognition. *New J. Chem.*, **22**, 1135–1138.

15 Mackay, L.G., Wylie, R.S. and Sanders, J.K.M. (1994) Catalytic acyl transfer by a cyclic porphyrin trimer - efficient turnover without product inhibition. *J. Am. Chem. Soc.*, **116**, 3141–3142.

16 Marty, M., Clyde-Watson, Z., Twyman, L.J., Nakash, M. and Sanders, J.K.M. (1998) Acceleration of a hetero-Diels–Alder reaction by cyclic metalloporphyrin trimers. *Chem. Commun.*, 2265–2266.

17 Kelly, T.R., Zhao, C. and Bridger, G.J. (1989) A bisubstrate reaction template. *J. Am. Chem. Soc.*, **111**, 3744–3745; Kelly, T.R., Bridger, G.J. and Zhao, C. (1990) Bisubstrate reaction templates. Examination of the consequences of identical versus different binding sites. *J. Am. Chem. Soc.*, **112**, 8024–8034.

18 Tjivikua, T., Ballester, P. and Rebek, J. Jr. (1990) Self-replicating system. *J. Am. Chem. Soc.*, **112**, 1249–1250; Nowick, J.S., Feng, Q., Tjivikua, T., Ballester, P. and Rebek, J. Jr. (1991) Kinetic studies and modeling of a self-replicating system. *J. Am. Chem. Soc.*, **113**, 8831–8839.

19 Mock, W.L. (1995) Cucurbituril. *Top. Curr. Chem.*, **175**, 1–24.

20 Mock, W.L., Irra, T.A., Wepsiec, J.P. and Adhya, M. (1989) Catalysis by cucurbituril. The significance of bound-substrate

destabilization for induced triazole formation. *J. Org. Chem.*, **54**, 5302–5308.

21 Mock, W.L., Irra, T.A., Wepsiec, J.P. and Manimaran, T.L. (1983) Cycloaddition induced by cucurbituril. A case of Pauling principle catalysis. *J. Org. Chem.*, **48**, 3619–3620.

22 For a recent computational study to elucidate the catalytic function of cucurbit[6]uril in the 1,3-dipolar cycloaddition between an azide and an alkyne see: Carlqvist, P. and Maseras, F. (2007) A theoretical analysis of a classic example of supramolecular catalysis. *Chem. Commun.*, 748–750.

23 Chen, J. and Rebek, J. Jr. (2002) Selectivity in an encapsulated cycloaddition reaction. *Org. Lett.*, **4**, 327–329.

24 Menger, F.M. (1985) On the source of intramolecular and enzymatic reactivity. *Acc. Chem. Res.*, **18**, 128–134.

25 Kang, J., Santamaria, J., Hilmersson, G. and Rebek, J. Jr. (1998) Self-assembled molecular capsule catalyzes a Diels–Alder reaction. *J. Am. Chem. Soc.*, **120**, 7389–7390. Kang, J., Hilmersson, G., Santamaria, J. and Rebek, J. Jr. (1998) Diels–Alder reactions through reversible encapsulation. *J. Am. Chem. Soc.*, **120**, 3650–3656.

26 Brisig, B., Sanders, J.K.M. and Otto, S. (2003) Selection and amplification of a catalyst from a dynamic combinatorial library. *Angew. Chem., Int. Ed.*, **42**, 1270–1273.

27 Yoshizawa, M., Tamura, M. and Fujita, M. (2006) Diels–Alder in aqueous molecular hosts: Unusual regioselectivity and efficient catalysis. *Science*, **312**, 251–254.

28 Fiedler, D., van Halbeek, H., Bergman, R.G. and Raymond, K.N. (2006) Supramolecular catalysis of unimolecular rearrangements: Substrate scope and mechanistic insights. *J. Am. Chem. Soc.*, **128**, 10240–10252.

29 Dong, V.M., Fiedler, D., Carl, B., Bergman, R.G. and Raymond, K.N. (2006) Molecular recognition and stabilization of iminium ions in water. *J. Am. Chem. Soc.*, **128**, 14464–14465.

30 Breit, B. and Seiche, W. (2003) Hydrogen bonding as a construction element for bidentate donor ligands in homogeneous catalysis: Regioselective hydroformylation of terminal alkenes. *J. Am. Chem. Soc.*, **125**, 6608–6609.

31 Seiche, W., Schuschkowski, A. and Breit, B. (2005) Bidentate ligands by self-assembly through hydrogen bonding: A general room temperature/ambient pressure regioselective hydroformylation of terminal alkenes. *Adv. Synth. Catal.*, **347**, 1488–1494.

32 Breit, B. (2005) Supramolecular approaches to generate libraries of chelating bidentate ligands for homogeneous catalysis. *Angew. Chem., Int. Ed.*, **44**, 6816–6825.

33 Birkholz, M.-N., Dubrovina, N.V., Jiao, H., Michalik, D., Holz, J., Paciello, R., Breit, B. and Boerner, A. (2007) Enantioselective hydrogenation with self-assembling rhodium phosphane catalysts: Influence of ligand structure and solvent. *Chem.–Eur. J.*, **13**, 5896–5907.

34 Weis, M., Waloch, C., Seiche, W. and Breit, B. (2006) Self-assembly of bidentate ligands for combinatorial homogeneous catalysis: Asymmetric rhodium-catalyzed hydrogenation. *J. Am. Chem. Soc.*, **128**, 4188–4189.

35 Breit, B. and Seiche, W. (2006) Self-assembly of bidentate ligands for combinatorial homogeneous catalysis based on an A-T base pair model. *Pure Appl. Chem.*, **78**, 249–256.

36 Breit, B. and Seiche, W. (2005) Self-assembly of bidentate ligands for combinatorial homogeneous catalysis based on an A-T base-pair model. *Angew. Chem., Int. Ed.*, **44**, 1640–1643.

37 Clarke, M.L. and Fuentes, J.A. (2007) Self-assembly of organocatalysts: Fine-tuning organocatalytic reactions. *Angew. Chem., Int. Ed.*, **46**, 930–933.

38 Slagt, V.F., Roeder, M., Kamer, P.C.J., Van Leeuwen, P.W.N.M. and Reek, J.N.H. (2004) Supraphos: A supramolecular strategy to prepare bidentate ligands. *J. Am. Chem. Soc.*, **126**, 4056–4057.

39 Takacs, J.M., Chaiseeda, K., Moteki, S.A., Reddy, D.S., Wu, D. and Chandra, K. (2006) Rhodium-catalyzed asymmetric hydrogenation using self-assembled chiral bidentate ligands. *Pure Appl. Chem.*, **78**, 501–509; Takacs, J.M., Reddy, D.S., Moteki, S.A., Wu, D. and Palencia, H. (2004) Asymmetric catalysis using self-assembled chiral bidentate P,P-ligands. *J. Am. Chem. Soc.*, **126**, 4494–4495.

40 Slagt, V.F., van Leeuwen, P.W.N.M. and Reek, J.N.H. (2003) Bidentate ligands formed by self-assembly. *Chem. Commun.*, 2474–2475.

41 Kuil, M., Goudriaan, P.E., Kleij, A.W., Tooke, D.M., Spek, A.L., van Leeuwen, P.W.N.M. and Reek, J.N.H. (2007) Rigid bis-zinc(II) salphen building blocks for the formation of template-assisted bidentate ligands and their application in catalysis. *Dalton Trans.*, 2311–2320.

42 Kuil, M., Goudriaan, P.E., Van Leeuwen, P.W.N.M. and Reek, J.N.H. (2006) Template-induced formation of hetero-bidentate ligands and their application in the asymmetric hydroformylation of styrene. *Chem. Commun.*, 4679–4681.

43 Wilson, M.E. and Whitesides, G.M. (1978) Conversion of a protein to a homogeneous asymmetric hydrogenation catalyst by site-specific modification with a diphosphinerhodium(1) moiety. *J. Am. Chem. Soc.*, **100**, 306–307.

44 Lan, J., Li, J., Liu, Z., Li, Y. and Chan, A.S.C. (1999) The first total synthesis of (–)-sinulariol-B and three other cembranoids. *Tetrahedron: Asymmetry*, **10**, 1877–1885.

45 Skander, M., Humbert, N., Collot, J., Gradinaru, J., Klein, G., Loosli, A., Sauser, J., Zocchi, A., Gilardoni, F. and Ward, T.R. (2004) Artificial metalloenzymes: (Strept)avidin as host for enantioselective hydrogenation by achiral biotinylated rhodium-diphosphine complexes. *J. Am. Chem. Soc.*, **126**, 14411–14418; Skander, M., Malan, C., Ivanova, A. and Ward, T.R. (2005) Chemical optimization of artificial metalloenzymes based on the biotin–avidin technology: (S)-selective and solvent-tolerant hydrogenation catalysts via the introduction of chiral amino acid spacers. *Chem. Commun.*, 4815–4817; Ward, T.R. (2005) Artificial metalloenzymes for enantioselective catalysis based on the noncovalent incorporation of organometallic moieties in a host protein. *Chem.–Eur. J.*, **11**, 3798–3804; Letondor, C. and Ward, T.R. (2006) Artificial metalloenzymes for enantioselective catalysis: Recent advances. *Chem. Bio. Chem.*, **7**, 1845–1852.

46 Letondor, C., Humbert, N. and Ward, T.R. (2005) Artificial metalloenzymes based on biotin–avidin technology for the enantioselective reduction of ketones by transfer hydrogenation. *Proc. Natl. Acad. Sci. U.S.A.*, **102**, 4683–4687; Letondor, C., Pordea, A., Humbert, N., Ivanova, A., Mazurek, S., Novic, M. and Ward, T.R. (2006) Artificial transfer hydrogenases based on the biotin-(strept)avidin technology: Fine tuning the selectivity by saturation mutagenesis of the host protein. *J. Am. Chem. Soc.*, **128**, 8320–8328.

47 Roelfes, G. and Feringa, B.L. (2005) DNA-based asymmetric catalysis. *Angew. Chem., Int. Ed.*, **44**, 3230–3232; Kraemer, R. (2006) Supramolecular bioinorganic hybrid catalysts for enantioselective transformations. *Angew. Chem., Int. Ed.*, **45**, 858–860.

48 Ropartz, L., Meeuwenoord, N.J., van der Marel, G.A., van Leeuwen, P.W.N.M., Slawin, A.M.Z. and Kamer, P.C.J. (2007) Phosphine containing oligonucleotides for the development of metallo-deoxyribozymes. *Chem. Commun.*, 1556–1558.

2
Supramolecular Construction of Chelating Bidentate Ligand Libraries through Hydrogen Bonding: Concept and Applications in Homogeneous Metal Complex Catalysis

Bernhard Breit

2.1
Introduction

Modern homogeneous catalysis evolution is driven by the quest for higher efficiency and selectivity. While enzymes are highly substrate specific, their scope is limited. Synthetic catalysts are of particular interest because of their potential for broader substrate range. However, in both fields no general catalyst that gives optimal results for all kinds of substrates is available. Hence, the desired selectivity of a catalytic reaction of interest has to be achieved through catalyst optimization and adjustment to the particular problem. Especially in the field of metal complex catalysis, the choice of the right ligand, which crafts the microenvironment at the catalytically active metal center, is crucial.

However, despite significant progress in the field of theoretical and computational chemistry, there is still no rational way to model from scratch the ligand of choice for a given reaction and selectivity problem. So far, finding the optimal ligand is an unpredictable high-risk endeavor, which is driven to a large extent through various combinations of intuition, experience, hard labor and in many cases serendipity. This is a longstanding problem and has, at least for asymmetric catalysis, a simple physicochemical origin. In 1983 W.S. Knowles noted [1]:

> *Since achieving 95% ee only involves energy differences of about 2 kcal, which is no more than the barrier encountered in a simple rotation of ethane, it is unlikely that before the fact one can predict what kind of ligand structures will be effective.*

Inspired by developments in the pharmaceutical industry to accelerate the lead discovery and lead structure optimization process, methods of combinatorial chemistry have been adapted to homogeneous (and heterogeneous) catalysis to speed up the catalyst discovery and optimization process [2]. Many elegant solutions for high-throughput screening of catalyst libraries are available today, which now allows for testing of large catalyst libraries for optimal activity and selectivity [3].

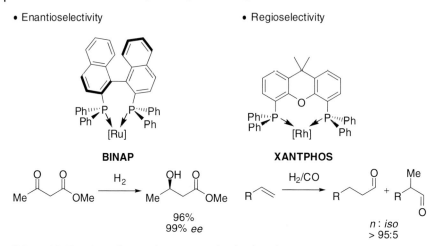

Scheme 2.1 Enantio- and regioselectivity control with selected metal catalysts derived from bidentate ligands.

However, the full potential of the combinatorial approach for catalyst discovery has yet to be reached, due to the difficulty in accessing large and structurally diverse ligand libraries. This problem is particularly valid for the important class of bidentate ligands. Although monodentate ligands have gained increased importance recently, a considerable number of catalytic reactions require a bidentate ligand to achieve maximum selectivity control. A prominent example for enantioselectivity control is the β-ketoester reduction with ruthenium(II)-BINAP catalysts (Scheme 2.1) [4]. Another example is the control of regioselectivity for the industrially important rhodium-catalyzed hydroformylation of terminal alkenes, which requires tailor-made bidentate ligands [5]. Among the few ligands known to achieve efficient regiocontrol is the XANTPHOS system (Scheme 2.1).

The problem of ligand library construction is closely related to the number of synthetic steps required to synthesize a bidentate ligand. These steps are in many cases non-trivial synthetic operations, which renders the ligand synthesis unsuited for automation. A particular challenge is the synthesis of nonsymmetric bidentate ligands equipped with two different donor sites [6,7].

2.2
Emulation of Chelation through Self-Assembly of Monodentate Ligands

Recently, attempts have been made to simplify access to bidentate ligands by a conceptually new approach: The basic idea is to use structurally less complex monodentate ligands, which imitate a bidentate ligand situation at the catalytically active metal center by employing a *noncovalent connection* between the two binding sites (Scheme 2.2). Nature provides a range of noncovalent interactions such as van der Waals interactions, π-stacking, cation–π interactions, charge-transfer

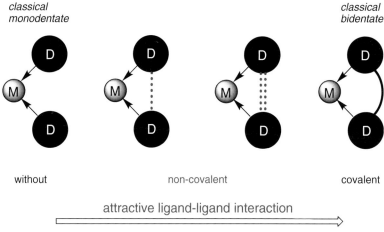

Scheme 2.2 Continuum of attractive ligand–ligand interactions in a ML$_2$ metal complex.

interactions, electrostatic interactions, hydrogen-bonding, and coordinative interactions [8]. Thus, going from two "truly" monodentate ligands with no attractive interaction between the two monodentate ligands bound to one metal center up to a metal–bidentate-ligand complex with two covalently connected binding sites, there is a continuum of two-donor-ligand–metal arrangements that imitate chelation through noncovalent attractive ligand–ligand interactions. A differentiation within this continuum may be made according to the strength and nature of the attractive ligand–ligand interaction [8].

Beyond the benefit of a simplified ligand synthesis, the true potential of this approach now lies in the inherent possibility for combinatorial ligand library generation through simple ligand mixing. Thus, provided that two ligands are bound to one catalytically active metal center, one could obtain from a set of n different monodentate donor ligands a library of $(n^2 + n)/2\,ML^xL^y$-catalysts (Figure 2.1).

If, however, the ligand–ligand interaction between the two ligands bound to one metal center is *non-complementary*, mixing of two different ligands will result in a mixture of the two homodimeric and the heterodimeric ligand metal complex almost irrespective of the nature and strength of the attractive ligand–ligand interaction

Ligand	L^1	L^2	...	L^n
L^1	L^1L^1			
L^2	L^1L^2	L^2L^2		
...	...	...	...	
L^n	L^1L^n	L^2L^n	...	L^nL^n

Figure 2.1 Ligand library through mixing of n monodentate ligands L^x and L^y.

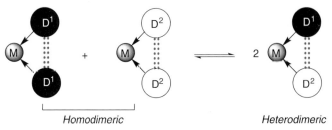

Scheme 2.3 Catalyst mixtures from mixing of two different ligands D^1 and D^2 and a metal source – non-complementary D^1/D^2 interactions.

(Scheme 2.3). In this scenario three potential catalysts are present simultaneously. Only in cases where the heterocombination is more reactive and at the same time more selective than the homodimer is optimization towards a better catalyst possible.

To facilitate the evaluation of structure–activity relations and optimization strategies, it would be desirable to have a situation in which ligand mixing generates the heterodimeric ligand metal complex exclusively. Thus, to shift the equilibrium to the desired heterodimeric species completely, this would require two different sets of monodentate ligands with *complementary* binding sites to generate attractive ligand–ligand interactions (Scheme 2.4).

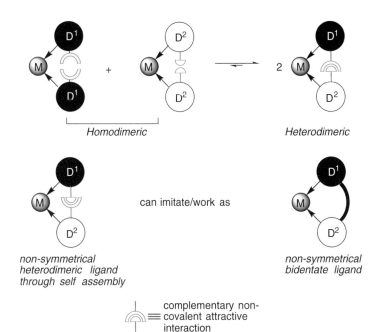

Scheme 2.4 Single defined heterodimeric catalysts from mixing and self-assembly of monodentate ligands with complementary binding sites in the presence of a metal source.

Ligand	m_1	m_2	...	m_i
n_1	m_1n_1	m_2n_1	...	m_in_1
n_2	m_1n_2	m_2n_2	...	m_in_2
...	...	...	...	...
n_j	m_1n_j	m_2n_j	...	m_in_j

Figure 2.2 Ligand library of $m \times n$ different and defined heterodimeric ligands through mixing of two sets of monodentate ligands with complementary binding sites.

Hence, a heterodimeric ligand system formed through complementary self-assembly could imitate a classical non-symmetrical ligand system (Scheme 2.4). Furthermore, mixing of m ligands of set D^1 with n ligands of set D^2 in the presence of a metal species would generate a library of $m \times n$ defined heterodimeric bidentate ligand metal complexes (Figure 2.2).

Herein, we focus on the results of our own research on hydrogen-bonding as a construction element in the process of self-assembly of monodentate to bidentate ligands. Concept development and applications towards combinatorial homogeneous catalysis are included.

2.3
Tautomeric Self-Complementary Interligand Hydrogen Bonding

2.3.1
Hydroformylation

Recently, *in situ* generation of bidentate ligands based on self-assembly through hydrogen bonding of monodentate ligands in the coordination sphere of a metal center has been described [9]. As a platform for hydrogen-bonding the 2-pyridone (**1B**)/2-hydroxypyridine (**1A**) tautomer system was employed. The parent system (D = H) dimerizes in aprotic solvents to form predominantly the symmetrical pyridone dimer **2** (Scheme 2.5) [10,11]. However, if D would be a donor atom capable of binding to a metal center (e.g., PPh_2), one could assume a shift of the equilibrium towards the mixed hydroxypyridine/pyridone dimer **3** [12], stabilized through the chelation effect exhibited through coordinative binding to the metal center (Scheme 2.5).

In fact reaction of 2 equiv. of 6-diphenylphosphanyl-2-pyridone [6-DPPon (**1**, D = PPh_2)] with [$PtCl_2$(1,5-COD)] furnished quantitatively *cis*-[$PtCl_2$(6-DPPon)$_2$], which showed the expected hydrogen bonding network in solution as proven by NMR and in the solid state as proven by X-ray crystal structure analysis (Figure 2.3) [9].

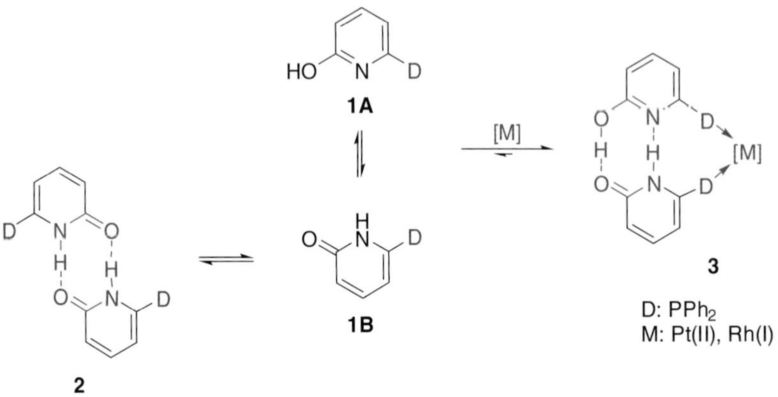

2 Supramolecular Construction of Chelating Bidentate Ligand Libraries

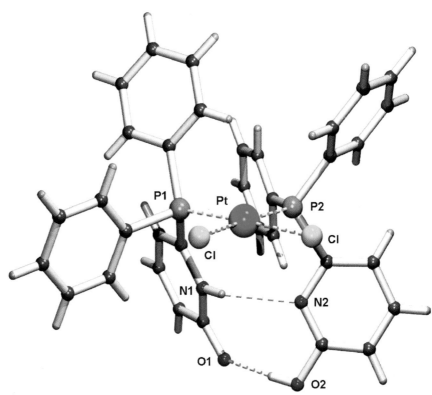

Scheme 2.5 Self-assembly through hydrogen bonding of the 2-pyridone/2-hydroxypyridine system **1** (6-DPPon) to generate bidentate ligand metal complexes **3** for homogeneous catalysis.

Figure 2.3 X-ray plot of *cis*-[PtCl$_2$(6-DPPon)$_2$].

A rhodium catalyst derived from the 6-DPPon ligand **1** displayed behavior typical of a bidentate ligand upon hydroformylation of terminal alkenes [9]. Thus, excellent regioselectivity in favor of the linear aldehyde isomer was noted for hydroformylation of a range of functionalized terminal alkenes (Table 2.1). Among them even those

Table 2.1 Regioselective hydroformylation of functionalized terminal alkenes with the rhodium/6-DPPon (**1**) catalyst in comparison to the standard industrial rhodium/PPh$_3$ catalyst.

Entry[a]	Substrate	n : iso (L = 1)	n : iso (L = PPh$_3$)
1	Br-alkene	97 : 3	72 : 28
2	AcO-alkene	96 : 4	71 : 29
3	MeO$_2$C-alkene	97 : 3	74 : 26
4	Me-C(O)O-alkene	94 : 6	71 : 29
5	PhHN-C(O)O-alkene	96 : 4	69 : 31
6	2-OH-C$_6$H$_4$-C(O)O-alkene	95 : 5	70 : 30
7	HO-alkene (short)	95 : 5	89 : 11
8	HO-alkene	96 : 4	77 : 23
9[b]	HO-alkene	83 : 17	77 : 23
10[c]	HO-alkene	81 : 19	–

[a] Reaction parameters: Rh:L:alkenic substrate (1 : 20 : 1000), c(alkene) = 0.698 M; toluene, 10 bar CO/H$_2$ (1 : 1), 70 °C. Full conversion was reached in every case after 20 h.
[b] MeOH as solvent.
[c] Addition of 0.5 equiv. of AcOH with respect to substrate.

with functional groups capable of hydrogen-bonding, such as carbamates, salicylates and free alcohol functions, either close to or remote from the reaction center were tolerated (Table 2.1, entries 5–8). However, the hydrogen bonding network and thus the chelating binding mode can be disrupted upon employing either temperatures above 110 °C or protic solvents such as methanol and acetic acid, giving rise to low regioselectivity (Table 2.1, entries 9, 10) [9].

Table 2.2 Room temperature/ambient pressure regioselective hydroformylation of functionalized terminal alkenes – substrate scope, a selection from 31 examples.

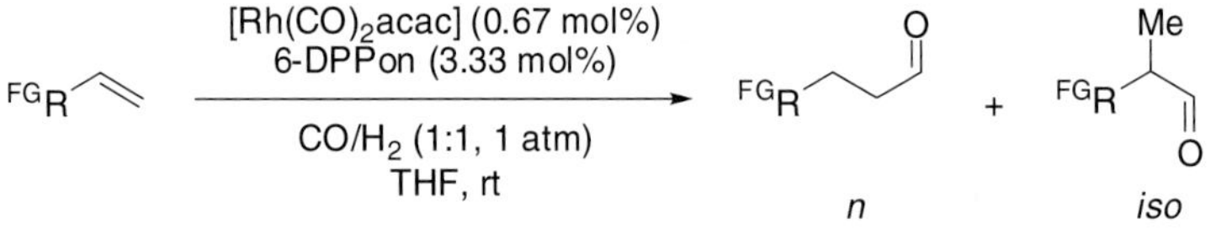

Entry	Substrate	n : iso	Entry	Substrate	n : iso
1	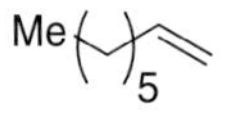 Me(5)=	99 : 1	8	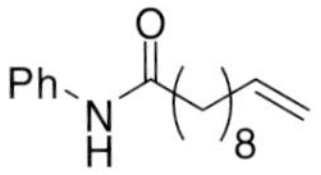 Ph-NH-C(O)-(8)=	99 : 1
2	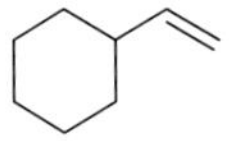 cyclohexyl-CH=CH$_2$	99 : 1	9	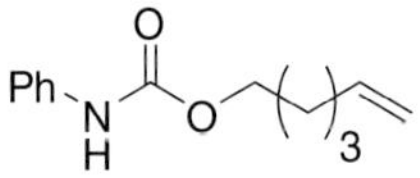 Ph-NH-C(O)-O-(3)=	98 : 2
3	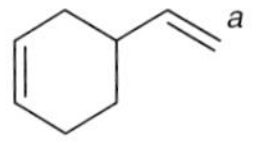 cyclohexenyl-CH$_2$-CH=CH$_2$ [a]	99 : 1	10	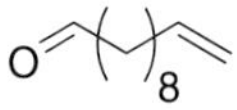 O=(8)=	99 : 1
4	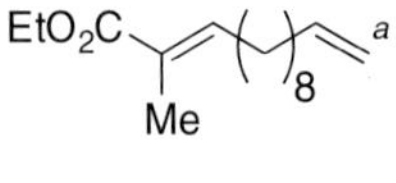 EtO$_2$C-CH(Me)-(8)= [a]	97 : 3	11	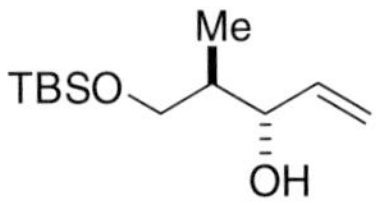 TBSO-CH(Me)-CH(OH)-	95 : 5
5	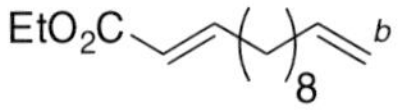 EtO$_2$C-(8)= [b]	91 : 9	12	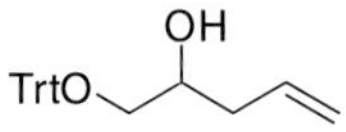 TrtO-CH(OH)-=	99 : 1
6	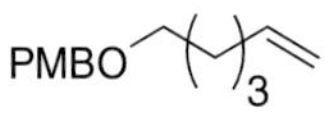 PMBO-(3)=	98 : 2	13	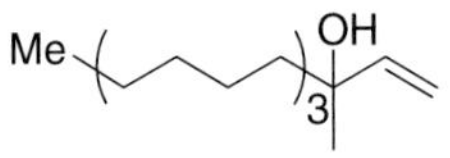 Me-(3)-CH(OH)-=	99 : 1
7	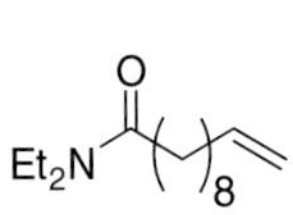 Et$_2$N-C(O)-(8)=	99 : 1	14	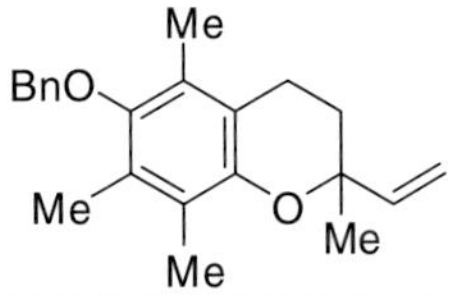 chromanyl derivative	99 : 1

[a] Hydroformylation of the terminal double bond only.
[b] Internal double bond is hydrogenated over time.

2.3.2
Room Temperature/Ambient Pressure Hydroformylation

The 6-DPPon (**1**)/rhodium catalysts are significantly more reactive than classical bidentate diphosphine ligands. Thus, this catalyst allows for the first time a room temperature/ambient pressure regioselective hydroformylation of terminal alkenes with low catalyst loadings in good activity [13]. The generality of this catalyst under these conditions was demonstrated for a wide range of structurally diverse alkenes equipped with many important functional groups (Table 2.2).

This practical and highly selective hydroformylation protocol, which omits the need for special pressure equipment, should find wide application in organic synthesis.

2.3.3
Asymmetric Hydrogenation

Very recently, chiral 6-DPPon derivatives, the phospholanes **4b** and **5b** and the phosphepine **6b**, have been prepared and studied as ligands in asymmetric hydrogenation (Figure 2.4) [14]. For comparison, the corresponding 2-alkoxypyridine systems **4a**–**6a** were studied, too. The latter should behave as truly monodentate ligands while the pyridine systems **4b**–**6b** should allow for complementary hydrogen-bonding.

The coordination behavior as well as the tautomerization equilibria of the new pyridone ligands was studied in detail by NMR and computational methods. For all three the data obtained strongly support a pseudo-bidentate hydroxypyridine–pyridone binding mode.

The new ligands were evaluated in the Rh-catalyzed asymmetric hydrogenation of benchmark substrates methyl α-N-acetamidoacrylate (**7**), methyl α-(Z)-N-acetamidocinnamate (**8**) and dimethyl itaconate (**9**) (Table 2.3). For all three substrates, catalyst performance was superior in CH$_2$Cl$_2$. Differences of up to 83% ee compared to otherwise identical reactions conducted in MeOH could be noted, giving

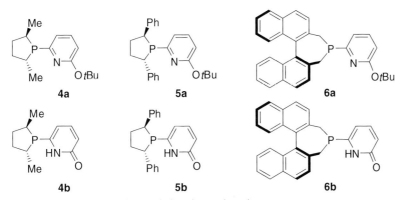

Figure 2.4 Chiral phospholane and phosphepine ligands.

Table 2.3 Rh(I)-catalyzed asymmetric hydrogenation of methyl N-acetamidoacrylate (**7**), methyl α-(Z)-N-acetamidocinnamate (**8**) and dimethyl itaconate (**9**).[a]

			Substrate 7		Substrate 8		Substrate 9	
Entry	Ligand	Solvent	Time (min)[b]	ee (%)	Time (min)[b]	ee (%)	Time (min)[b]	ee (%)
1	4a	CH_2Cl_2	4	32(R)	30	23(R)	1400	11(S)
2	4a	MeOH	10	35(R)	400	22(R)	300	12(S)
3	4b	CH_2Cl_2	80	51(R)	1400	69(R)	1300	83(S)
4	4b	MeOH	1400	35(R)	1300	13(R)	400	0
5	5a	CH_2Cl_2	75	9(R)	60	71(R)	40	49(R)
6	5a	MeOH	17	20(R)	70	77(R)	45	76(R)
7	5b	CH_2Cl_2	960	16(R)	3900	83(R)	2880	51(S)
8	5b	MeOH	400	5(S)	2760	27(R)	2880	47(S)
9	6a	CH_2Cl_2	5	38(R)	7	56(R)	8	87(S)
10	6a	MeOH	2	74(R)	2	90(R)	2	58(S)
11	6b	CH_2Cl_2	10	85(R)	10	94(R)	20	99(S)
12	6b	MeOH	15	91(R)	50	64(R)	60	64(S)

[a] Conditions: precatalyst generated *in situ* by reaction of [Rh(COD)$_2$]BF$_4$ with 2 equiv. of ligand, substrate : catalyst = 100 : 1, H$_2$ (1 bar), 7.5 mL solvent.
[b] Reaction time to achieve full conversion.

impressive evidence for the advantage of the self-assembly through hydrogen bonding (cf. Table 2.3, entries 3 and 4). Noteworthy, is the performance of the phosphepine pyridone ligand **6b**, which enabled the hydrogenation of dimethyl itaconate (**9**) with excellent catalyst activity and enantioselectivity (99% ee, Table 2.3, entry 11).

2.4
A-T Base Pair Analogous Complementary Hydrogen Bonding for the Construction of Heterodimeric Self-Assembling Ligands

2.4.1
Aminopyridine/Isoquinolone Platform

2.4.1.1 Hydroformylation
Since the two tautomers – the hydroxypyridine **1A** and the pyridone **1B** – equilibrate rapidly, mixing of two pyridone ligands with different donor sites would afford

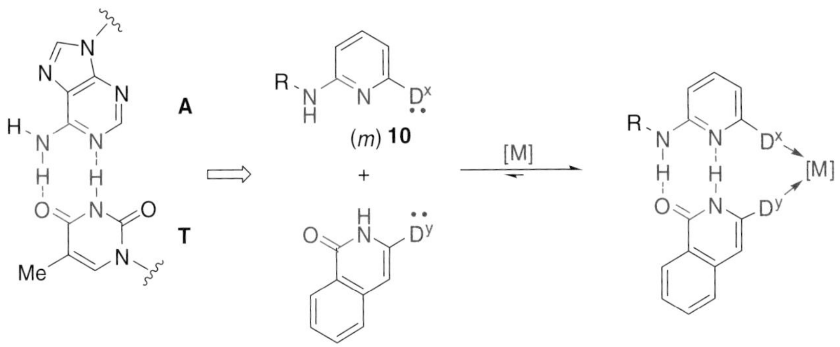

Scheme 2.6 An A-T base pair model (highlighted in red) as a complementary platform for specific self-assembly of heterodimeric bidentate ligands.

mixtures of the heterodimeric and the two homodimeric catalysts (cf. Scheme 2.3). If, however, one would be interested in the formation of the heterodimeric catalyst exclusively, e.g., to allow for a delineation of structure–activity and structure–selectivity relations, one could not rely on the pyridone self-assembly platform. Conversely, this would require self-assembly of two complementary species through hydrogen-bonding – a principle employed by nature in DNA base pairing. Thus an A-T base pair model relying on the aminopyridine/isoquinolone platform was selected to serve for specific heterodimeric ligand assembly (Scheme 2.6) [9,15].

Phosphine ligands based on this platform form, upon mixing in presence of a Pt(II) salt, the heterodimeric complex cis-[(**10**)(**11**)PtCl$_2$] (**12**) exclusively. An X-ray structure analysis of the heterodimeric complex cis-[(**10a**)(**11a**)PtCl$_2$] (**12aa**) (D^x = D^y = PPh$_2$) shows the expected hydrogen-bonding network reminiscent of the Watson–Crick base pairing of A and T in DNA (Figure 2.5). NMR studies provided support the idea that a similar structural situation occurs in solution, too. On this platform the first 4×4 self-assembled ligand library based on hydrogen-bonding was generated and explored for regioselective hydroformylation of terminal alkenes. This study allowed identification of a catalyst that operated with outstanding activity and regioselectivity (Table 2.4) [15]. Interestingly, it is an unsymmetrical ligand combination (**10d**×**11d**) that furnished the optimal catalyst. This result would not be possible to predict on a rational basis, but was reached by employing a combinatorial approach.

2.4.1.2 Hydration of Alkynes

An anti-Markovnikov hydration of terminal alkynes could be a convenient way of preparing aldehydes, but so far only a few ruthenium-complexes have been identified that catalyze this unusual hydration mode [16]. The presence of bidentate phosphine ligands [16b], the coordination of a water molecule stabilized by hydrogen bonding [16e] and the use of phosphinopyridine ligands [16f] seem to be of major importance in these processes.

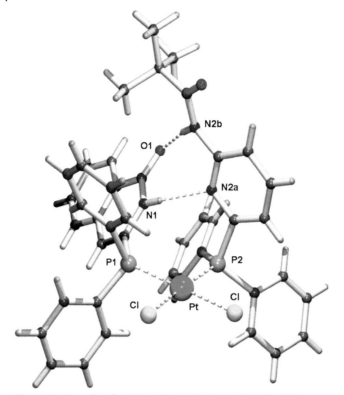

Figure 2.5 X-ray plot of cis-[(10a)(11a)PtCl$_2$] (12aa) (D^x = D^y = PPh$_2$).

When testing various half-sandwich ruthenium complexes bearing monodentate and bidentate phosphine ligands during hydration of 1-nonyne as the test substrate only low catalytic activity and selectivity was noted (Table 2.5) [17]. Using the 6-DPPon (**1**) ligand, which should allow for self-dimerization based hydrogen bonding, between the pyridone form and its hydroxypyridine tautomer [9], a similar unsatisfactory result was obtained. Conversely, employing the self-assembled heterodimeric aminopyridine/isoquinolone bidentate ligands (entry 5) [9b,15] a catalyst was obtained that operated with outstanding activity and perfect regioselectivity. From entries 6 and 7 it is obvious that the self-assembled heterodimer is responsible for this interesting result. Thus, neither 6-DPPAP (**10a**) nor 3-DPICon (**11a**) alone provided an active catalyst.

Furthermore, a small 3 × 3 ligand library with electron donating (p-anisyl) and electron withdrawing (4-F-C$_6$H$_4$) aryl substituents at the phosphine donor were studied, but none of these combinations proved superior to the parent 6-DPPAP (**10a**)/3-DPICon (**11a**) system (Table 2.6).

Single crystals, suitable for X-ray diffraction were obtained from slow diffusion of cyclohexane into a concentrated solution of **13e** in dichloromethane. As proposed, the

Table 2.4 4 × 4 Ligand matrix of aminopyridine (**10a–d**)/isoquinolone(**11a–d**) derived self-assembled bidentate ligands in the [Rh]-catalyzed hydroformylation of 1-octene[a].

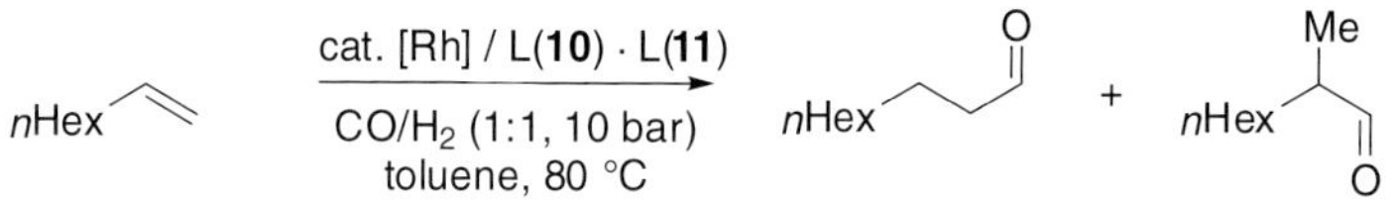

Ligand	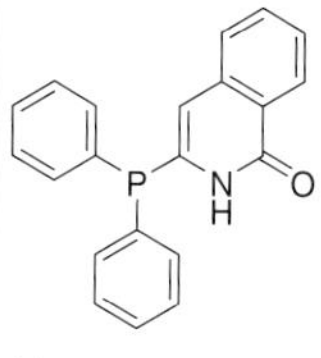 11a	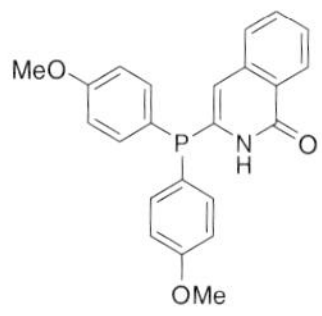 11b	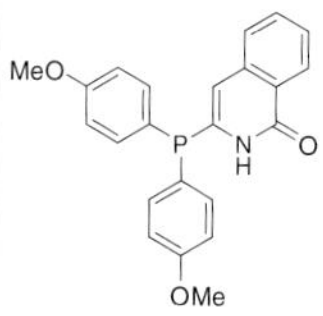 11c	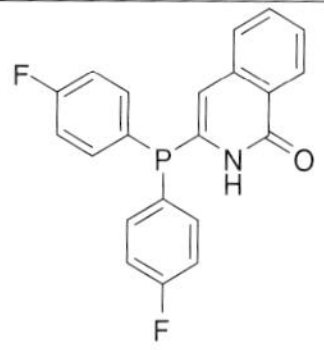 11d
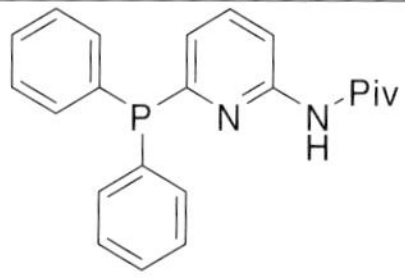 **10a**	2425 h^{-1}[b] 94:6[c]	1040 h^{-1} 94:6	2732 h^{-1} 96:4	2559 h^{-1} 95:5
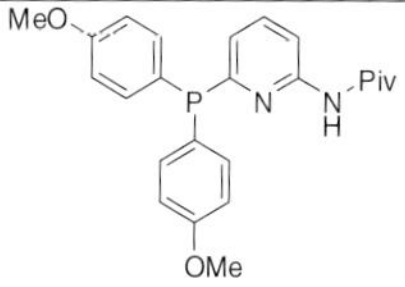 **10b**	2033 h^{-1} 93:7	1058 h^{-1} 92:8	1281 h^{-1} 96:4	1772 h^{-1} 94:6
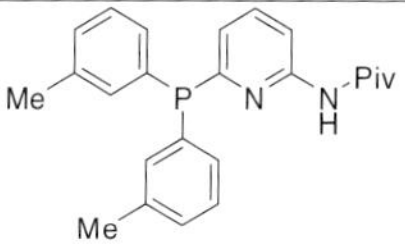 **10c**	3537 h^{-1} 94:6	1842 h^{-1} 93:7	1808 h^{-1} 96:4	2287 h^{-1} 94:6
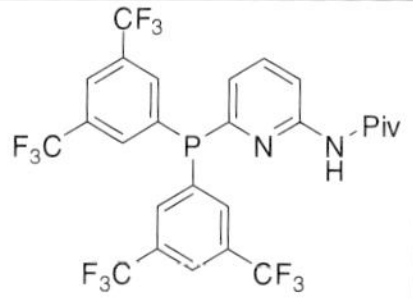 **10d**	7439 h^{-1} 96:4	2695 h^{-1} 95:5	7465 h^{-1} 94:6	**8643 h^{-1}** **96:4**

[a] Reaction conditions: [Rh(CO)$_2$acac], [Rh] : L(**10**) : L(**11**) : 1-octene = 1 : 10 : 10 : 7500, 10 bar CO/H$_2$(1 : 1), toluene [c$_0$(1-octene) = 2.91 M], 5 h. Catalyst preformation: 5 bar CO/H$_2$(1 : 1), 30 min, rt to 80 °C.
[b] Turnover frequency (TOF) = (mol aldehydes) × (mol catalyst)$^{-1}$ × (t/h)$^{-1}$ at 20–30% conversion.
[c] Regioselectivity: n to iso.

Table 2.5 Ruthenium complex-catalyzed hydration of 1-nonyne.

$nC_7H_{15}-\equiv \xrightarrow[H_2O]{13} nC_7H_{15}\text{-CHO} + nC_7H_{15}\text{-C(O)Me}$

Catalyst 13: $[Ru^+(Cp)(MeCN)(L_1)(L_2)]PF_6^-$

aldehyde (anti-Markovnikov product) + ketone (Markovnikov product)

Entry	Cat.	L^{1a}	L^{2a}	Time (h)	a (%)[b]	k (%)[b]
1	13a	PPh₃	PPh₃	140	1.2	18
2	13b[c]	dppy	dppy	168	4.0	2.4
3	13c	dppe	–	168	2.1	20
4	13d[c]	6-DPPon (1)	6-DPPon (1)	168	2.1	25
5	13e	6-DPPAP (10a)	3-DPICon (11a)	26	94	0
6	13f	6-DPPAP (10a)	6-DPPAP (10a)	72	39	3.8
7	13g	3-DPICon (11a)	3-DPICon (11a)	48	1.9	0

[a]dppy: 2-Diphenylphosphinopyridine, dppe: 1,2-bis(diphenylphosphino)ethane, 6-DPPon (1): 6-diphenylphosphino-2-pyridone, 6-DPPAP (10a): 6-diphenylphosphino-N-pivaloyl-2-aminopyridine, 3-DPICon (11a): 3-diphenylphosphinoisoquinolone.
[b]Yield calculated from GC response factors relative to internal hexadecane standard.
[c]η^1-P,η^2-P,N coordination of the phosphinopyridine with replacement of the acetonitrile ligand.

Table 2.6 3 × 3 Ligand matrix of aminopyridine/isoquinolone derived self-assembled bidentate ligands in the Ru-catalyzed hydration of 1-nonyne.

	Aminopyridine[b]		
Isoquinolone[b]	6-DPPAP (10a)	6-D(p-MeO)PPAP (10b)	6-D(p-F)PPAP (10e)
3-DPICon (11a)	94:0[a]	77:0	84:1.8
3-D(p-MeO)PICon (11b)	60:0	60:1.2	66:1.8
3-D(p-F)PICon (11d)	78:1.0	71:2.2	70:4.9

[b]6-D(p-MeO)PPAP (10b): 6-bis(4-methoxyphenyl)phosphino-2-(pivaloylamino)-pyridine, 6-D(p-F)PPAP (10e): 6-bis(4-fluorophenyl)-phosphino-2-(pivaloylamino)pyridine, 3-D(p-MeO)PICon (11b): 3-bis(4-methoxyphenyl)phosphinoisoquinolone, 3-D(p-F)PICon (11d): 3-bis(4-fluorophenyl)phosphinoisoquinolone.
[a]Aldehyde:ketone yields (%) determined by GC after 26 h.

2.4 A-T Base Pair Analogous Complementary Hydrogen Bonding for the Construction | 43

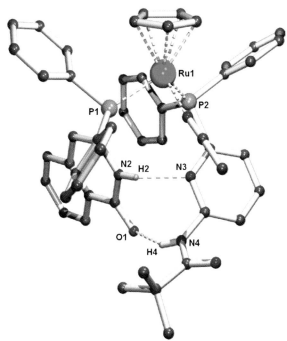

Figure 2.6 Platon plot of **13e** (H atoms bound to carbon and the PF$_6^-$ counterion are omitted for clarity). Selected interatomic distances (Å): Ru1–P1 = 2.3366(7), Ru1–P2 = 2.3193(8), N2···N3 = 2.811(3), O1···N4 = 2.846(3); and angles (°): P1–Ru1–P2 = 98.20(3), N2–H2···N3 = 135.35(3), O1···H4–N4 = 147.50(3).

aminopyridine **10a** and the isoquinolone **11a** ligand form the expected hydrogen-bonding network reminiscent of the Watson–Crick base pairing of A and T in DNA (Figure 2.6) [18].

The scope of the anti-Markovnikov hydration of various 1-alkynes was examined using **13e** as a catalyst (Table 2.7) [17]. Reactions using 2–10 mol.% of **13e** gave the desired aldehydes in good yields. Linear terminal alkynes furnished the corresponding carbonyl products with complete regioselectivity even if the catalyst seems to be partially inhibited by the presence of a nitrile group, which may act as a competing ligand for ruthenium (entries 1–3) [16b]. 1,9-Decadiyne was converted into the corresponding dialdehyde in 78% yield with only traces of the monoaldehyde (entry 4). The regioselectivity is excellent for a range of different functionalized substrates (entries 1–7, 9, 10), with only one exception: a homopropargylic system bearing a benzoate function (entry 8). In this case, the regioselectivity is slightly diminished, which may be due to a chelation effect. Finally, a steroid ring system substituted by a

Table 2.7 Regioselective hydration of functionalized terminal alkynes with **13e** as a catalyst.

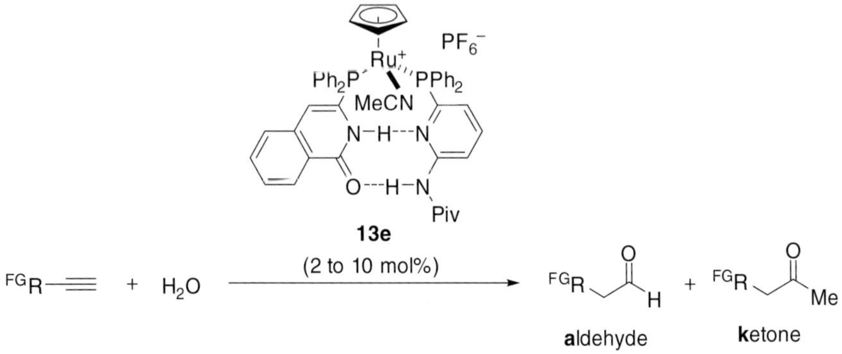

$$^{FG}R-\!\!\!\equiv\;+\;H_2O \xrightarrow[\text{(2 to 10 mol\%)}]{\textbf{13e}} {}^{FG}R\!\!\!\!\!\diagdown\!\!\!\!\!\overset{O}{\underset{H}{\diagup}}\;+\;{}^{FG}R\!\!\!\!\!\diagdown\!\!\!\!\!\overset{O}{\underset{Me}{\diagup}}$$

aldehyde   ketone

Entry	Substrate	13e (%)	Time (h)	a:k (%)[a]	Yield (%)[b]
1	nC₇H₁₅—≡	2	26	>99:1	89
2	Ph—≡	2	26	>99:1	73
3	N≡—(CH₂)₃—≡	10	96	>99:1	78
4	≡—(CH₂)₆—≡	10	78	99:1[c]	82
5	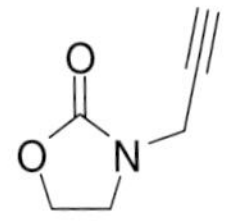 (oxazolidinone-N-propargyl)	5	70	>99:1	87
6	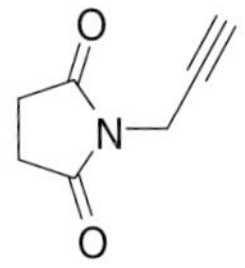 (succinimide-N-propargyl)	5	72	99:1	65
7	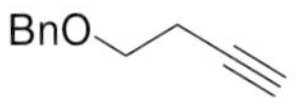 BnO~~~≡	5	48	99:1	83
8	BzO~~~≡	5	50	87:13	74
9	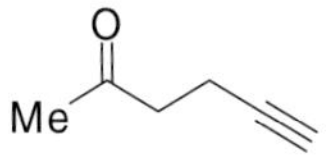 Me-C(O)-CH₂CH₂-≡	5	28	96:4	91
10[d]	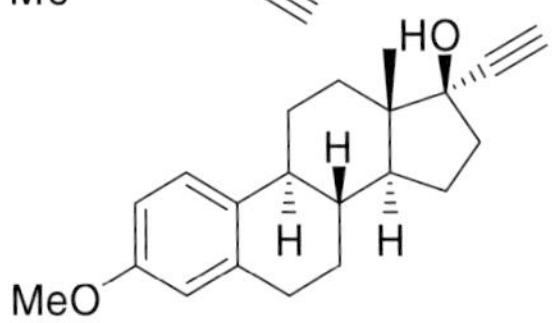 (ethynylestradiol derivative)	10	124	>99:1	61

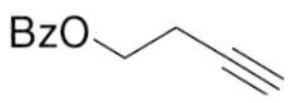

[a] Determined by GC analysis and ¹H-NMR spectroscopy.
[b] Isolated yield.
[c] Decane-1,10-dial : 9-oxo-decanal ratio.
[d] Reaction performed at 70 °C.

propargylic alcohol reacted with water and gave selectively the γ-hydroxy aldehyde in 61% yield (entry 10).

Thus, a new ruthenium catalyst was identified that allows for a highly regioselective anti-Markovnikov hydration of terminal alkynes compatible with a wide range of functional groups. During this reaction the hydrogen-bonding network may serve a dual role: First emulation of a bidentate ligand situation to increase the binding constant of the phosphine ligands, leading to catalyst stabilization, and, second, activation of a water molecule via hydrogen bonding [19].

2.4.1.3 Hydration of Nitriles

To see whether water could be activated and added to π-systems other than alkynes, the metal-catalyzed hydrolysis of nitriles was studied [20]. For this purpose novel homodimeric and heterodimeric bis(acetylacetonato)ruthenium(II) complexes bearing the 6-diphenylphosphino-N-pivaloyl-2-aminopyridine (**10a**) and 3-diphenylphosphinoisoquinolone (**11a**) ligands were prepared. The molecular structures of these precatalyst were studied in solution and also in the solid state and revealed some unusual hydrogen-bonding patterns, in particular for the heterodimeric system in which the acetylacetonato ligand is involved (Scheme 2.7).

On studying the hydrolysis of p-tolylcarbonitrile the highest activity was observed for the isoquinolone homo-complex [100% conversion after 20 h, maxTOF = 20 (mol amide)/(mol catalyst) h^{-1}]. The hetero-complex was less active (90% conversion,

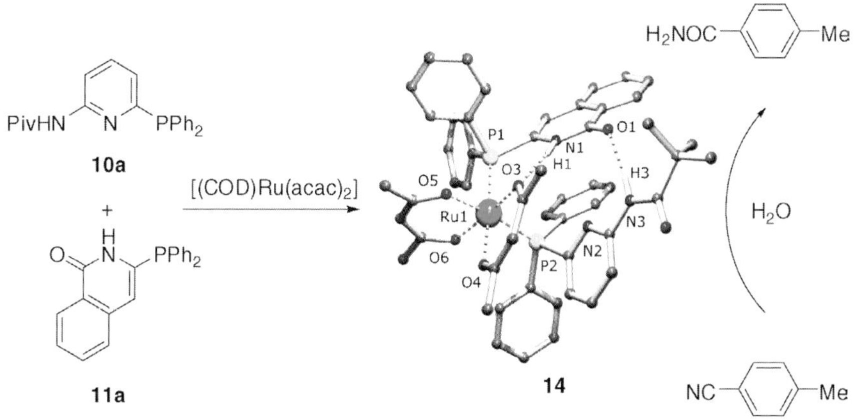

Scheme 2.7 Hydration of nitriles with novel 6-DPPAP/3-DPICon Ru-complexes. Platon plot of [(acac)]$_2$Ru(**10a**)(**11a**)] (**14**); selected interatomic distances (Å): Ru1–P1 = 2.2741(7), Ru1–P2 = 2.2953(7), N1···O3 = 2.749(3), O1···N3 = 2.956(3); and angles (°): P1–Ru1–P2 = 98.40(2), N1–H1···O3 = 161.6, O1···H3–N3 = 173.6.

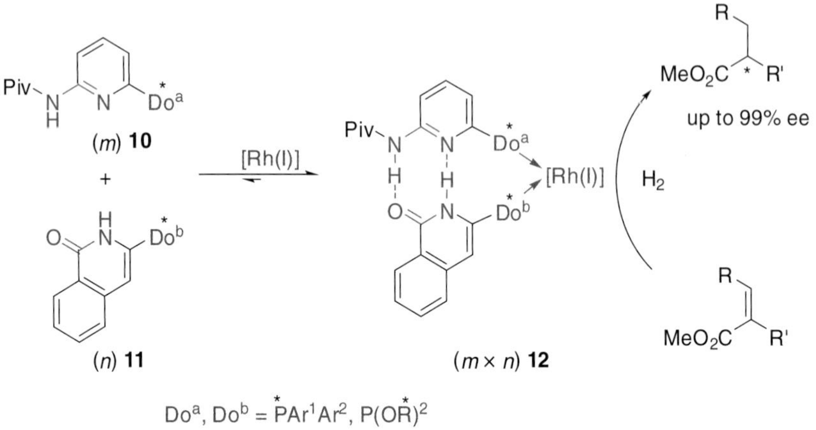
Piv
N
H
N
Do^a
(m) 10
+
O
H
N
Do^b
(n) 11
[Rh(I)]
Piv
N
H
N
H
Do^a
O
N
Do^b
[Rh(I)]
(m × n) 12
H2
R
MeO2C
R'
up to 99% ee
R
MeO2C
R'
Do^a, Do^b = PAr^1Ar^2, P(OR)^2

Scheme 2.8 Self-assembly of chiral monodentate to chiral bidentate ligands through complementary hydrogen-bonding on the basis of an A-T base pair analogue for combinatorial asymmetric catalysis.

maxTOF = $5\,h^{-1}$) and a very low activity (conversion <5%) was detected for the aminopyridine homo-complex.

2.4.1.4 Asymmetric Hydrogenation

To extend the concept to combinatorial asymmetric catalysis a new library of chiral aminopyridine and isoquinolone systems equipped with phosphine and phosphonite donors was prepared and applied to the asymmetric rhodium-catalyzed hydrogenation (Scheme 2.8) [21].

Twelve different chiral aminopyridine and isoquinolone ligands (Figure 2.7) were synthesized and screened against the asymmetric hydrogenation of the standard test

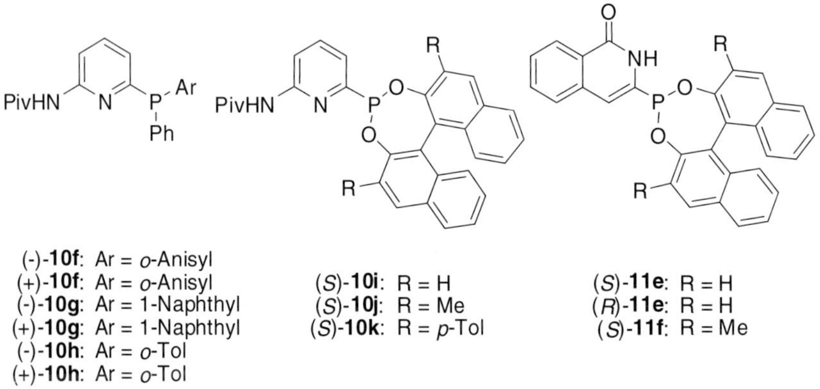
PivHN
N
P
Ar
Ph
PivHN
N
P
O
O
R
R
O
NH
P
O
O
R
R
(-)-10f: Ar = o-Anisyl
(+)-10f: Ar = o-Anisyl
(-)-10g: Ar = 1-Naphthyl
(+)-10g: Ar = 1-Naphthyl
(-)-10h: Ar = o-Tol
(+)-10h: Ar = o-Tol
(S)-10i: R = H
(S)-10j: R = Me
(S)-10k: R = p-Tol
(S)-11e: R = H
(R)-11e: R = H
(S)-11f: R = Me

Figure 2.7 Library of chiral aminopyridine and isoquinolone ligands.

Table 2.8 Results of asymmetric Rh-catalyzed hydrogenation[a] of N-acetamido acrylate (**7**).

Entry	L^a/L^b	Conversion (%)	ee (%)
1[b]	(−)-**10f**/3-DPICon	Quant.	56(R)
2	(−)-**10f**/(S)-**11e**	Quant.	86(R)
3	6-DPPAP (**10a**)/(S)-**11e**	Quant.	82(R)
4	(+)-**10f**/(S)-**11e**	Quant.	92(R)
5[c]	(+)-**10f**/(S)-**11e**	Quant.	94(R)
6	(+)-**10f**/(S)-**11f**	12	70(R)
7	(−)-**10f**/(S)-**11e**	94	81(R)
8	(+)-**10g**/(S)-**11e**	Quant.	89(R)
9	(−)-**10h**/(S)-**11e**	53	79(R)
10	(+)-**10h**/(S)-**11e**	77	82(R)
11[c]	(S)-**10i**/(S)-**11e**	Quant.	99(R)
12[d]	(S)-**10i**/(S)-**11e**	Quant.	99(R)
13	(S)-**10i**/(S)-**10i**	Quant.	98(R)
14	(S)-**11e**/(S)-**11e**	Quant.	94(R)
15	(S)-**10i**/(R)-**11e**	Quant.	80(R)
16	(S)-**10i**/(S)-**11f**	Quant.	92(R)
17	(S)-**10j**/(S)-**11e**	Quant.	88(R)
18	(S)-**10j**/(R)-**11e**	Quant.	63(R)
19	(S)-**10j**/(S)-**11f**	Quant.	59(R)
20	(S)-**10k**/(S)-**11e**	Quant.	33(R)
21	(S)-**10k**/(R)-**11e**	Quant.	83(R)

[a]All reactions in CH$_2$Cl$_2$, Rh : L^a : L^b = 1 : 1.1–1.3 : 1.1–1.3, Rh : olefin = 1 : 100, 1 bar, 24 h.
[b]6 bar, 48 h.
[c]0 °C in ClCH$_2$CH$_2$Cl.
[d]Rh : olefin = 1 : 1000, 1 bar, 24 h.

substrates methyl α-N-acetamidoacrylate (**7**) methyl α-(Z)-N-acetamidocinnamate (**8**) and dimethyl itaconate (**9**).

Original catalyst screening of this 10×4 library was done for the rhodium-catalyzed hydrogenation of acetamidoacrylate [21]. Thus, replacing the achiral 6-DPPAP ligand (**10a** with D^x = PPh$_2$, with the P-chiral (+)-**10f** ligand furnished an active hydrogenation catalyst that performed with moderate enantioselectivity (Table 2.8, entry 1). However, replacing the achiral 3-DPICon (**11a** with D^y = PPh$_2$) with the chiral phosphonite derivative (S)-**11e** led to a catalyst operating with significantly enhanced enantioselectivity (entry 2). Exchanging the aminopyridylphosphine monomer **10** by the corresponding phosphonite system (S)-**10i** gave the best catalyst (entry 11). Even at catalyst loadings of 0.1 mol% complete conversion and perfect enantioselectivity (99% ee, entry 12) was noted. In a control experiment both homocombinations for ligands **10i** and **11e** were tested separately. Both furnished active hydrogenation catalysts which gave slightly lower enantioselectivities (entries 13 and 14) than the heterocombination. This suggests the heterocombination to be not only the prevalent catalyst but also the kinetically competent species. Variation of the ortho-substituents in the BINOL skeleton did not provide any further improvement (entries 16–21).

A small sublibrary based on the bisphosphonite systems was screened against the asymmetric hydrogenation of methyl α-(Z)-N-acetamidocinnamate (**8**) and dimethyl itaconate (**9**) (Table 2.9) [21]. Interestingly, for substrate **8** the optimal ligand combination was found to be (S)-**10i**/(S)-**11f** with 94% ee (entry 2). The same ligand combination had provided only mediocre results with the parent methyl α-N-acetamidoacrylate (**7**) substrate (Table 2.8, entry 16). For dimethyl itaconate (**9**) both (S)-**10i**/(S)-**11e** as well as (S)-**10j**/(S)-**11e** gave best results (94% ee, Table 2.9, entries 5 and 9). In the latter case, control experiments employing the homocombinations clearly showed that the heterocombination provides a catalyst operating with a significantly improved selectivity (entries 10 and 11).

Table 2.9 Rhodium-catalyzed hydrogenation of methyl α-(Z)-N-acetamido cinnamate (**8**) and dimethyl itaconate (**9**).[a]

Entry	Ligands	Substrate	p (bar)	Conversion (%)[b]	ee (%)[c]
1	(S)-**10i**/(S)-**11e**	8	1	Quant.	90(R)
2	(S)-**10i**/(S)-**11f**	8	1	Quant.	94(R)
3	(S)-**10i**/(S)-**10i**	8	1	Quant.	93(R)
4	(S)-**11f**/(S)-**11f**	8	1	33	73(R)
5	(S)-**10i**/(S)-**11e**	9	6	Quant.	94(S)
6	(S)-**10i**/(S)-**11f**	9	6	Quant.	91(S)
7	(S)-**10i**/(R)-**11e**	9	6	Quant.	90(S)
8	(S)-**10j**/(S)-**11e**	9	6	Quant.	70(S)
9	(S)-**10j**/(S)-**11e**	9	30	Quant.	94(S)
10	(S)-**10j**/(S)-**10j**	9	30	33	38(S)
11	(S)-**11e**/(S)-**11e**	9	30	Quant.	89(S)

[a] Rh : L^a : L^b = 1 : 1.1–1.3 : 1.1–1.3, Rh : olefin = 1 : 100, 24 h.
[b] Determined by NMR.
[c] Determined by chiral HPLC or GC analysis.

2.4.2
Platform Variation

2.4.2.1 Hydroformylation

Thus far, it was unclear whether this self-assembly approach is restricted to the aminopyridine (**10**)/isoquinolone (**11**) system or whether indeed a second variation site could stem from variation of the A-T base pair analogous platform (Figure 2.8). It was reasonable to expect that any change of platform geometry as well as the nature of the hydrogen-bonding system should have an immediate impact on the ligand bite angle (θ) and coordination geometry at the metal and, thus, should have an important influence on performance in catalysis [22].

As adenine analogous donor–acceptor ligands the heterocycle functionalized phosphines **10a** and **15–18** were chosen. As T analogous acceptor–donor ligands the known isoquinolone **11a** and the new 7-azaindole system **19** were selected [23].

The coordination properties of all ten possible ligand combinations were studied through NMR spectroscopic investigation of the corresponding platinum complexes [Cl$_2$Pt(L^{DA})(L^{AD})]. ^{31}P as well as ^{1}H NMR spectroscopy showed, in all cases, the selective formation of a defined complex with a heterodimeric ligand. Furthermore, the existence of defined *cis*-heteroleptic complexes in the solid state was confirmed by analysis of the crystal structures of [Cl$_2$Pt(**17**/**11a**)] and [Cl$_2$Pt(**16**/**19**)] by X-ray diffraction (Figure 2.9).

In a first set of experiments all ligands were evaluated independently [23]. Regioselectivities were in the typical order for monodentate triarylphosphine/rhodium catalysts of 75 : 25 to 81 : 19 [24]. Next, all possible ten ligand combinations

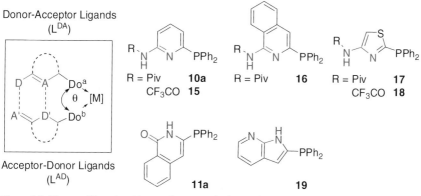

Figure 2.8 Library of ligands with complementary hydrogen bonding motifs.

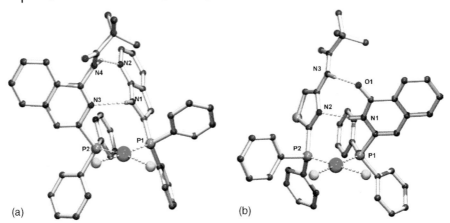

(a) (b)

Figure 2.9 (a) PLATON plot of the structure of cis-[(**16/19**)PtCl$_2$] in the solid state. Selected interatomic distances (Å): Pt–P1 = 2.2417(6), Pt–P2 = 2.2517(6), N1H···N3 = 2.910(3), N4H···N2 = 2.987(3); and angles (°): P1–Pt–P2 = 99.01(2), N1–N1H···N3 = 129.0, N4–N4H···N2 = 153.7. (b) PLATON plot of the structure of cis-[(**17/11a**)PtCl$_2$] in the solid state. Selected interatomic distances (Å): Pt–P1 = 2.2609(5), Pt–P2 = 2.2331(6), N1H···N2 = 2.992(3), N3H···O1 = 3.038(3); and angles (°): P1–Pt–P2 = 97.42(2), N1–N1H···N2 = 159.0, N3–N3H···O = 166.0. H atoms bound to carbon are omitted for clarity. Green Pt, yellow Cl, orange P, blue N, red O, violet S.

Table 2.10 Turnover frequencies[a] and regioselectivities[b] (in parentheses) for a 5 × 2 ligand matrix of **AD**-ligands (**10a** and **15–18**)/**DA**-ligands (**11a** and **19**) derived self-assembled bidentate ligands in the [Rh]-catalyzed hydroformylation of 1-octene.[c]

nHex ⇌ [Rh]/L^{AD}/L^{DA}, CO/H$_2$ (1:1, 10 bar), toluene, 80 °C → nHex–CHO (linear (n)) + nHex–CH(Me)–CHO (branched (iso))

L	10a	15	16	17	18
11a	2394 h^{-1a}	3396 h^{-1}	2452 h^{-1}	3890 h^{-1}	3888 h^{-1}
	(93:7[b])	(96:4)	(95:5)	(98:2)	(>99:1)
19	2679 h^{-1}	4864 h^{-1}	3205 h^{-1}	3233 h^{-1}	2333 h^{-1}
	(89:11)	(96:4)	(95:5)	(95:5)	(99:1)

[a]Turnover frequency (TOF) was calculated as (mol aldehydes) × (mol catalyst)$^{-1}$ × (t/h)$^{-1}$ at 20–30% conversion, determined by GC-analysis.
[b]Regioselectivity: linear to branched, determined by GC analysis.
[c]Reaction conditions: [Rh(CO)$_2$(acac)], [Rh] : L^{AD} : L^{DA} : 1-octene = 1 : 10 : 10 : 7500, 10 bar H$_2$/CO (1:1), toluene, 80 °C, 5 h. Catalyst pre-formation: 5 bar H$_2$/CO (1:1), 30 min, RT → 80 °C.

were studied and the results of these hydroformylation experiments are depicted in Table 2.10. The regioselectivities observed (89 : 11 to >99 : 1) confirm that in all cases bidentate ligand catalysts are the kinetically competent catalyst species.

Interestingly, in all heterocombinations, going from the pivaloyl substituent to a trifluoroacetyl group led to an increase in regioselectivity (**10a/11a** vs. **15/11a**; **10a/19** vs. **15/19**; **17/11a** vs. **18/11a** and **17/19** vs. **18/19**). This goes along with an increase of hydrogen-bond strength.

Among all ligand combinations studied, catalysts derived from thiazole/isoquinolone (**18/11a**) and from thiazole/7-azaindole(**18/19**) were the best. Regioselectivities greater than 99 : 1 in favor of the linear aldehyde were observed even at a reaction temperature of 80 °C. One explanation could be that going from the six-membered aminopyridine systems **10a**, **15** and **16** to the five-membered thiazole heterocycles **17** and **18** might result in stronger hydrogen bonds and hence a more rigid self-assembled system. If this notion is correct, it should be interesting to see whether these new self-assembly platforms employing the thiazole system remain intact in a protic solvent environment (e.g., methanol). Indeed, previous studies had shown that the hydrogen bonding network of first generation self-assembled ligands (e.g., **10a/11a**) is interrupted through the interaction with methanol, which so far limits the application range of self-assembled ligands in homogeneous catalysis.

Thus, the rhodium-catalyzed hydroformylation of 1-octene employing the self-assembled platforms was studied in methanol, and the results were compared to those obtained in toluene (Table 2.11) [23]. As expected, the first-generation platform **10a/11a** showed a significant drop in regioselectivity on going from toluene (94 : 6) to methanol (82 : 18). The drop in regioselectivity is indicative of a disruption of the hydrogen-bonding network between ligands **10a** and **11a**, which forces the system to behave as a monodentate triarylphosphine rhodium catalyst [24]. The same holds for the ligand combination **15/11a** and **18/19**. However,

Table 2.11 Regioselectivities of rhodium-catalyzed hydroformylation of 1-octene using toluene and MeOH as solvents.[a]

Entry	Ligand	l : b[b] in toluene	l : b[b] in MeOH
1	10a/11a	94 : 6	82 : 18
2	15/11a	96 : 4	79 : 21
3	17/11a	98 : 2	97 : 3
4	18/11a	99 : 1	96 : 4

[a] Reaction conditions: [Rh(CO)$_2$(acac)], [Rh] : L^{AD} : **11a** : 1-octene = 1 : 10 : 10 : 7500, 10 bar H$_2$/CO (1 : 1), 80 °C.
[b] Regioselectivity: linear to branched, determined by GC analysis and or ^{1}H NMR spectroscopy.

for the thiazole systems **17/11a** and **18/11a**, high regioselectivities were observed in toluene and in methanol.

Variation of the heterocyclic self-assembly platform has an enormous impact on the resulting catalyst properties. New hydroformylation catalysts operating with excellent activity and outstanding regioselectivities were identified, which can now be applied even in protic solvents such as methanol. This is an important extension for the application range of the self-assembled catalysts based on hydrogen-bonding.

2.5
Conclusion and Outlook

The quest in homogeneous metal catalysis for the ultimate ligand giving rise to a catalyst with optimal activity and selectivity is a difficult task, despite recent progress in rational ligand design. This is why combinatorial methods have gained increased importance in the ligand design and discovery process recently. However, the limiting step so far has been the generation of structurally meaningful and diverse ligand libraries, which has been particularly difficult for the structurally more demanding class of bidentate ligands.

To simplify bidentate ligand synthesis we herein present our concept of ligand self-assembly through complementary hydrogen-bonding. The basic idea is to use structurally less complex monodentate ligands, which imitate a bidentate ligand situation at the catalytically active metal center through complementary hydrogen-bonding interactions between the two ligands. Homodimeric ligands based on the tautomeric self-complementary pyridone/hydroxypyridine platform furnished excellent catalysts for regioselective hydroformylation and asymmetric hydrogenation. Furthermore, combinatorial library generation became possible employing A-T base pair analogous self-assembly platforms. Some of them are even stable in protic solvents such as methanol. From ligand libraries of these systems, excellent catalyst for regioselective hydroformylation, anti-Markovnikov hydration as well as asymmetric hydrogenation have emerged.

Hence, this new field merges the principles of supramolecular chemistry, coordination chemistry, and catalysis to allow for the generation of chelation emulating ligand libraries for homogeneous metal complex catalysis. The proof of principle has been achieved. Thus, the concept is now ready to be used to develop larger ligand libraries to identify new tailor-made ligands for homogeneous catalysis solutions for organic synthesis.

References

1 Knowles, W.S. (1983) Asymmetric hydrogenation. *Acc. Chem. Res.*, **16**, 106–112.

2 Gennari, C. and Piarulli, U. (2003) Combinatorial libraries of chiral ligands

for enantionselective catalysis. *Chem. Rev.*, **103**, 3071–3100.
3 For a review see: Reetz, M.T. (2001) Kombinatorische und evolutionsgesteuerte Methoden zur Bildung enantioselektiver Katalysatoren. *Angew. Chem.*, **113**, 292–310; (2001) Combinatorial and evolution-based methods in the creation of enantioselective catalysts. *Angew. Chem., Int. Ed.*, **40**, 284–310.
4 Ohkuma, T., Kitamura, M. and Noyori, R. (2000) Asymmetric hydrogenation, in *Catalytic Asymmetric Synthesis* (ed. I. Ojima), Wiley-VCH, New York, ch. 1, pp. 1–110. For recent promising results employing monodentate ligands see: Junge, K., Hagemann, B., Enthaler, S., Oehme, G., Michalik, M., Monsees, A., Riermeier, T., Dingerdissen, U. and Beller, M. (2004) Enantioselective hydrogenation of β-ketoesters with monodentate ligands. *Angew. Chem.*, **116**, 5176–5179; (2004) Enantioselective hydrogenation of β-ketoesters with monodentate ligands. *Angew. Chem., Int. Ed.*, **43**, 5066–5096.
5 (a) van Leeuwen, P.W.N.M., Casey, C.P. and Whiteker, G.T. (2000) Bidentate ligands, in *Rhodium Catalyzed Hydroformylation* (eds P.W.N.M. van Leeuwen and C. Claver), Kluwer Academic Publishers, Dordrecht, ch. 4, pp. 76–105; (b) Breit, B. and Seiche, W. (2001) Recent advances on chemo-, regio- and stereoselective hydroformylation. *Synthesis*, 1–36.
6 For reviews on classical bidentate ligand library construction see Ref. [3] and Lavastre, O., Bonnette, F. and Gallard, L. (2004) Parallel and combinatorial approaches for synthesis of ligands. *Curr. Opin. Chem. Biol.*, **8**, 311–318.
7 For an alternative combinatorial approach employing catalytic metal centers bearing two bidentate ligands see: Ding, K., Du, H., Yuan, Y. and Long, J. (2004) Combinatorial chemistry approach to chiral catalyst engineering and screening: Rational design and serendipity. *Chem.–Eur. J.*, **10**, 2872–2884.
8 Schneider, H.-J. and Yatsimirsky, A.K. (2000) *Principles and Methods in Supramolecular Chemistry*, Wiley-VCH, New York.
9 (a) Breit, B. and Seiche, W. (2003) Hydrogen bonding as a construction element for bidentate donor ligands in homogeneous catalysis: Regioselective hydroformylation of terminal alkenes. *J. Am. Chem. Soc.*, **125**, 6608–6609; (b) Breit, B. and Seiche, W. (2006) Self-assembly of bidentate ligands for combinatorial homogeneous catalysis based on an A-T base pair model. *Pure Appl. Chem.*, **78**, 249–256.
10 Beak, P. (1977) Energies and alkylations of tautomeric heterocyclic compounds: Old problems-new answers. *Acc. Chem. Res.*, **10**, 186–192.
11 Chou, P.-T., Wie, C.-Y. and Hung, F.-T. (1997) Conjugated dual hydrogen bonds mediating 2-pyridone/2-hydroxypyridine tautomerism. *J. Phys. Chem. B*, **101**, 9119–9126.
12 Meuwly, M., Müller, A. and Leutwyler, S. (2003) Energetics, dynamics and infrared spectra of the DNA base-pair analogue 2-pyridone 2-hydroxypyridine. *Phys. Chem. Chem. Phys.*, **5**, 2663–2672.
13 Seiche, W., Schuschkowski, A. and Breit, B. (2005) Bidentate ligands by self-assembly through hydrogen bonding: A general room temperature/ambient pressure regioselective hydroformylation of terminal alkenes. *Adv. Synth. Catal.*, **11–13**, 1488–1494.
14 Birkholz, M.-N., Dubrovina, N.V., Jiao, H., Michalik, D., Holz, J., Paciello, R., Breit, B. and Börner, A. (2007) Enantioselective hydrogenation with self-assembling rhodium phosphane catalysts: Influence of ligand structure and solvent. *Chem.–Eur. J.*, **13**, 5896–5907.
15 Breit, B. and Seiche, W. (2005) Selbstorganisation zweizähniger

Liganden für die kombinatorische homogene Katalyse auf der Basis eines AT-Basenpaar-modells. *Angew. Chem.*, **117**, 1666–1669; (2005) Self-assembly of bidentate ligands for combinatorial homogeneous catalysis based on an A-T base-pair model. *Angew. Chem., Int. Ed.*, **44**, 1640–1643.

16 (a) Tokunaga, M. and Wakatsuki, Y. (1998) Die erste anti-Markownikow-hydratisierung terminaler Alkine: Ruthenium(II)/Phosphan-katalysierte Bildung von Aldehyden. *Angew. Chem.*, **110**, 3024–3027; (1998) The first anti-Markovnikov hydration of terminal alkynes: Formation of aldehydes catalyzed by a ruthenium(II)/phosphane mixture. *Angew. Chem., Int. Ed.*, **37**, 2867–2869; (b) Suzuki, T., Tokunaga, M. and Wakatsuki, Y. (2001) Ruthenium complex-catalyzed anti-Markovnikov hydration of terminal alkynes. *Org. Lett.*, **3**, 735–737; (c) Alvarez, P., Bassetti, M., Gimeno, J. and Mancini, G. (2001) Hydration of terminal alkynes to aldehydes in aqueous micellar solutions by ruthenium(II) catalysis; first anti-Markovnikov addition of water to propargylic alcohols. *Tetrahedron Lett.*, **42**, 8467–8470. (d) Tokunaga, M., Suzuki, T., Koga, N., Fukushima, T., Horiuchi, A. and Wakatsuki, Y. (2001) Ruthenium-catalyzed hydration of 1-alkynes to give aldehydes: Insight into anti-Markovnikov regiochemistry. *J. Am. Chem. Soc.*, **123**, 11917–11924; (e) Grotjahn, D.B., Incarvito, C.D. and Rheingold, A.L. (2001) Combined effects of metal and ligand capable of accepting a proton or hydrogen bond catalyze Anti-markovnikov hydration of terminal alkynes. *Angew. Chem.*, **113**, 4002–4005; (2001) Combined effects of metal and ligand capable of accepting a proton or hydrogen bond catalyze Anti-Markovnikov hydration of terminal alkynes. *Angew. Chem., Int. Ed.*, **40**, 3884–3887; (f) Grotjahn, D.B. and Lev, D.A. (2004) A general bifunctional catalyst for the anti-Markovnikov hydration of terminal alkynes to aldehydes gives enzyme-like rate and selectivity enhancements. *J. Am. Chem. Soc.*, **126**, 12232–12233.

17 Chevallier, F. and Breit, B. (2006) Self-assembled bidentate ligands for Ru-catalyzed anti-Markovnikov hydration of terminal alkynes. *Angew. Chem.*, **118**, 1629–1632; (2006) Self-assembled bidentate ligands for Ru-catalyzed anti-Markovnikov hydration of terminal alkynes. *Angew. Chem., Int. Ed.*, **45**, 1599–1602.

18 (a) Watson, J.D. and Crick, F.H.C. (1953) Genetical implications of the structure of deoxyribonucleic acid. *Nature*, **171**, 964–967; (b) Topal, M.D. and Fresco, J.R. (1976) Complementary base pairing and the origin of substitution mutations. *Nature*, **263**, 285–289.

19 A first confirmation for this assumption was obtained from the reaction of 6-DPPAP (1 equiv), 3-DPICon (1 equiv), and *cis*-[PtCl$_2$(cod)] (1 equiv; cod = cyclooctadiene) in toluene in the presence of water to give *cis*-[PtCl$_2$(6-DPPAP)(3-DPICon)·(H$_2$O)] quantitatively.

20 Šmejkal, T. and Breit, B. (2007) Self-assembled bidentate ligands for ruthenium-catalyzed hydration of nitriles. *Organometallics*, **26**, 2461–2464.

21 Weis, M., Waloch, C., Seiche, W. and Breit, B. (2006) Self-assembly of bidentate ligands for combinatorial homogeneous catalysis: Asymmetric rhodium-catalyzed hydrogenation. *J. Am. Chem. Soc.*, **128**, 4188–4189.

22 van Leeuwen, P.W.N.M., Kamer, P.C.J., Reek, J.N.H. and Dierkes, P. (2000) Ligand bite angle effects in metal-catalyzed C–C bond formation. *Chem. Rev.*, **100**, 2741–2769.

23 Waloch, C., Wieland, J., Keller, M. and Breit, B. (2007) Self-assembly of bidentate ligands for combinatorial homogeneous catalysis: Methanol-stable platforms analogous to the adenine-thymine base

pair. *Angew. Chem.*, **119**, 3097–3099; (2007) Self-assembly of bidentate ligands for combinatorial homogeneous catalysis: Methanol-stable platforms analogous to the adenine-thymine base pair. *Angew. Chem., Int. Ed.*, **46**, 3037–3039.

24 Under identical conditions (Table 2.9), a turnover frequency of 1312 h^{-1} and a linear:branched regioselectivity of 76:24 in the rhodium-catalyzed hydroformylation of 1-octene were measured for triphenylphosphine.

3
Bis-Azolylazine Derivatives as Supramolecular Synthons for Copper and Silver [2 × 2] Grids and Coordination Polymers

Félix A. Jalón, Blanca R. Manzano, M. Laura Soriano, and Isabel M. Ortiz

3.1
Introduction

The combination of dynamic coordination chemistry with noncovalent interactions such as hydrogen bonding, π–π stacking or anion–π interactions constitutes a powerful tool for generating complex structures from simple supramolecular synthons [1]. The past few decades have witnessed enormous progress in the self-assembly of elaborately designed building blocks such as metallic anions with different geometric preferences and polytopic organic spacers to give different kinds of supramolecular metal-organic hybrid compounds [2]. Remarkable progress has been made in the synthesis and characterization of infinite one-, two- and three-dimensional (1D, 2D and 3D) coordination compounds [3]. The so-called coordination polymers contain two central components, connectors and linkers that define the framework of the polymer. In addition, there are other auxiliary components, such as blocking ligands, counter-anions, and nonbonding guests or template molecules. Other types of highly organized architectures are grids and these also constitute an important topic in supramolecular chemistry. A molecular grid [$m \times n$] consists of a square ($m = n$) or rectangular matrix array of metal centers (Scheme 3.1) [4].

The grids very frequently consist of tetrahedral metal centers and n-topic rigid, rod-like ligands, although they can also be constructed from octahedral metal centers. In both types of derivative the various topologies of the metal arrays generate interesting optical, redox, magnetic or catalytic properties [5]. There is also increasing interest in the development of nanoscale devices. The generation of molecular motion by interconversion between helical free ligands and rack or grid complexes is also noteworthy [6]. In cases where the material is porous, applications in gas storage, guest inclusion and also in catalysis are possible. Porous materials can be considered as nanospace laboratories where specific transformations involving a guest can occur [3a].

Supramolecular Catalysis. Edited by Piet W. N. M. van Leeuwen
Copyright © 2008 WILEY-VCH Verlag GmbH & Co. KGaA, Weinheim
ISBN: 978-3-527-32191-9

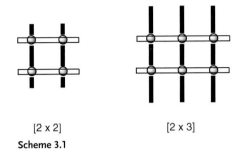

[2 x 2]   [2 x 3]

Scheme 3.1

The supramolecular host–guest compounds involving neutral or cationic entities have received considerable attention [7]. In contrast, the hosting of anions through noncovalent interactions has been less explored [8]. Cation–π interactions have been described in numerous examples of organic and biochemical systems [9] but the expected repulsion between aromatic π-rings and the anions has limited the development of anion–π interactions. Theoretical calculations have recently indicated that noncovalent interactions between anions and π-acidic rings such as hexafluorobenzene, trinitrobenzene, s-triazine and s-tetrazine are energetically favorable [10]. In fact, NMR studies and crystallographic data have provided evidence of their existence in solution and in the solid state. The rings involved are also aromatic fragments with the acidity enhanced by their union to metal centers [11] and different heterocycles such as s-triazines [10c,d,12], pyridazine [12b], tetrazine [12b] and pyrazine [13] rings. Interest in these interactions stems in part from the potential use of aryl hosts in the recognition of anions of biological or environmental importance.

In the present work we describe a series of coordination polymers and grids formed by self-assembly of potentially tetrahedral metal centers like Cu(I) and Ag(I) and bis(azolyl)azine derivatives that can behave as bis-bidentate ligands. We will show that the position of the azolyl groups on the azine ring and, as a consequence, the planarity of the ligand is a decisive factor governing the type of structure formed. On the other hand, the versatility of the coordination number of the silver center will allow the formation of highly varied structures. One aim of the work is to study the anion encapsulation, interaction and interchange with special emphasis on the anion–π interactions. Other types of noncovalent interactions such as π–π stacking and hydrogen bonding will be also analyzed and, in this context, systems containing donor and acceptor groups for hydrogen bonding (diaminotriazines) capable of forming extended 2D or 3D supramolecular structures will be also described.

3.2
"Planar" and "Non-Planar" Azolyl Azines

Scheme 3.2 shows the ligand series used in the first part of this work. In the pyrimidine, pyridazine and triazine derivatives, the azolyl groups are sufficiently

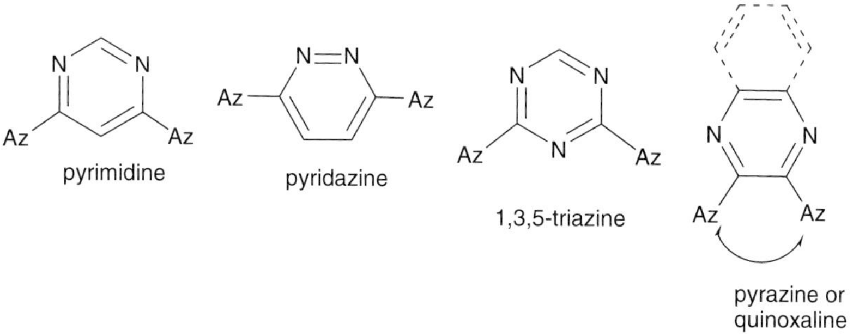

Scheme 3.2

separated in the ligand to exhibit a very stable conformation with coplanar rings, a situation that facilitates an extended delocalization of the π-electronic density. In contrast, in the pyridazine and quinoxaline derivatives at least one of the azine rings is expected to be twisted out of the plane to decrease the steric hindrance between the contiguous substituents of the azine.

As described in the next section, the azole substituents are different N-pyrazolyl or indazolyl rings. These ligands have been much less widely used than the reminiscent bis(2-pyridiyl)azine compounds but they offer advantages in the preparation procedure.

3.2.1
Synthesis

The bis(pyrazol-1-yl)azines depicted in Scheme 3.3 are easily made by the reaction of the sodium pyrazolate salt with the corresponding azine dihalide substituted in the position of the pyrazolate attack. The sodium pyrazolate salts are prepared by reaction of the pyrazole with NaH. If KH is used as the deprotonating agent instead of NaH, the nucleophilic character of the pyrazolate anion is increased and, as a consequence, the reaction time decreases and the yield increases. All of these reactions were carried out under reflux in THF. This procedure can be considered general, and even relatively bulky pyrazoles like (4S,7R)-7,8,8-trimethyl-4,5,6,7-tetrahydro-4,7-methano-2-indazolyl (Camphpz) behave identically. The pyrimidine derivatives bpzpm [14], bpz*pm [15] and related complexes have been prepared previously (see Scheme 3.3 for the structures). The pyridazine derivative bpzpdz [14b,16] has been prepared by a similar procedure whereas bpz*pdz is either prepared by a procedure based on a [4 + 2] cycloaddition [17] or a method similar to our synthesis [18]. The ligand bpz*qnox was prepared by a very different methodology [19].

The procedure based on the reaction of the heterocyclic components offers the possible synthesis of numerous systems considering the diversity of accessible pyrazole and azole compounds, many of which are commercially available.

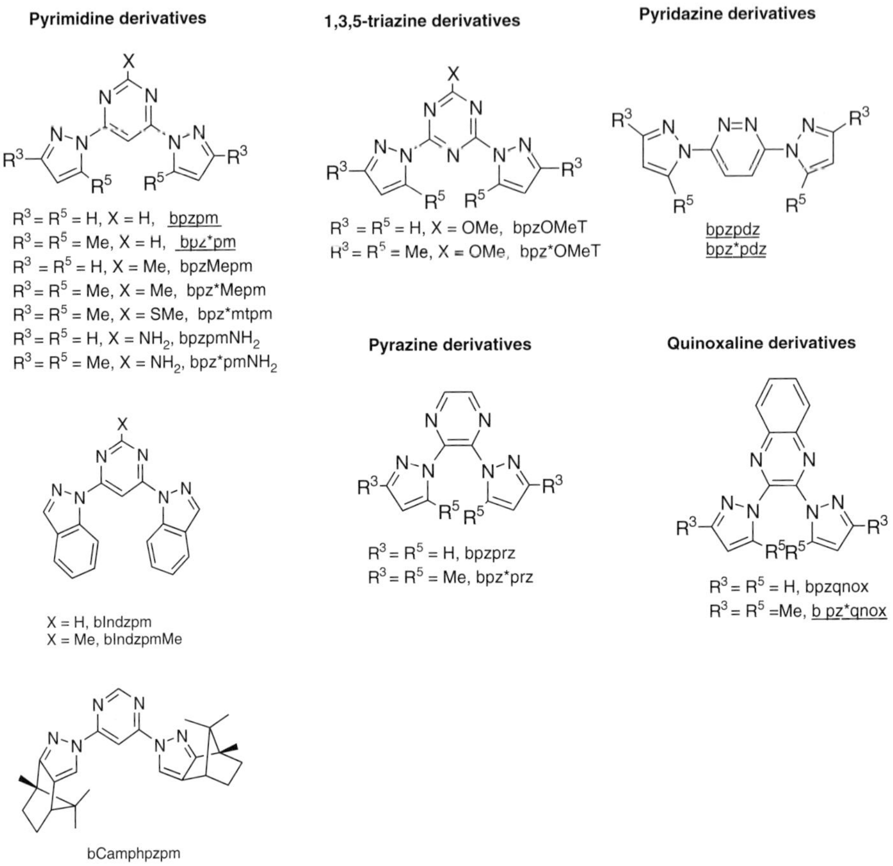

3 Bis-Azolylazine Derivatives as Supramolecular Synthons

Pyrimidine derivatives

$R^3 = R^5 = H$, $X = H$, bpzpm
$R^3 = R^5 = Me$, $X = H$, bpz*pm
$R^3 = R^5 = H$, $X = Me$, bpzMepm
$R^3 = R^5 = Me$, $X = Me$, bpz*Mepm
$R^3 = R^5 = Me$, $X = SMe$, bpz*mtpm
$R^3 = R^5 = H$, $X = NH_2$, bpzpmNH$_2$
$R^3 = R^5 = Me$, $X = NH_2$, bpz*pmNH$_2$

1,3,5-triazine derivatives

$R^3 = R^5 = H$, $X = OMe$, bpzOMeT
$H^3 = R^5 = Me$, $X = OMe$, bpz*OMeT

Pyridazine derivatives

bpzpdz
bpz*pdz

$X = H$, bIndzpm
$X = Me$, bIndzpmMe

Pyrazine derivatives

$R^3 = R^5 = H$, bpzprz
$R^3 = R^5 = Me$, bpz*prz

Quinoxaline derivatives

$R^3 = R^5 = H$, bpzqnox
$R^3 = R^5 = Me$, b pz*qnox

bCamphpzpm

Scheme 3.3 Ligands used for preparing grids or coordination polymers (those previously prepared are underlined).

In comparison with the structurally similar 2-pyridyl derivatives this series of ligands can be more easily prepared and offer chelating positions with a more open bite angle, a feature that could influence both the stability of the derivatives formed and their structural framework.

Ligands not underlined in Scheme 3.3 were prepared in our group for the first time. Notably, the complete regioselectivity in the synthesis of the indazolyl derivatives, bIndzpm and bIndzpmMe, and in the ligand derived from camphor is in contrast with the variable selectivity found in the preparation of 2,6-bis(azolyl) pyridines [20].

The use of the indazolyl ligands could lead to the preparation of luminescent systems, whereas the goal of the synthesis of the chiral ligand bCamphpzpm is the formation of asymmetric systems that could induce enantioselectivity.

All of these ligands can be considered as bis-bidentate (bidentate ditopic) and could coordinate simultaneously to two metal cations.

3.2.2
Crystallographic Evidence for the Planarity

Although there is free rotation around the bonds that link the heterocycles, the coplanar conformation of the three rings for the ligands in Scheme 3.3 is advantageous because it allows extended π-delocalization of the electronic density. However, steric factors not only in the free ligands but also in the resulting complexes could favor conformations with out-of-plane twisting of some rings. This apparently unimportant aspect may be decisive in the formation of discrete or extended structures after coordination to metal centers (see below).

We determined the molecular structures of four bis(pyrazol-1-yl)azine derivatives, three with pyrimidine as the central ring (namely, bpz*pmNH$_2$, bIndzpm and bIndzpmMe) and another with quinoxaline, bis-2,3-(pyrazol-1-yl)-6-Cl-quinoxaline. The expected coplanarity of the three heterocycles in the pyrimidine ligands was confirmed for bpz*pmNH$_2$ and bIndzpm (see Figure 3.1a and b). Dihedral angles between the pyrimidine and azine rings are 4.42° and 3.08° in bpz*pmNH$_2$ and 3.57° and 2.69° in bIndzpm. However, for the bIndzpmMe derivative (Figure 3.1c), although one indazole ring is nearly coplanar with the central ring (dihedral angle 4.63°) the other exhibits a bent disposition, probably induced by the packing. The angle formed by the pyrimidine and indazole heterocycles is of 14.89° in this case. However, the torsion angle N(6)–N(5)–C(11)–N(4) that involves the bent Indz and the central pyrimidine is only of 4.60°, indicating that this distortion is not due to the rotation of the C(11)–N(4) bond between the heterocycles. In contrast, in the quinoxaline derivative the three rings deviate clearly from coplanarity (Figure 3.1d). In this case, the corresponding dihedral angles with the central ring, 29.71° and 43.84°, are clearly higher, with values comparable to the torsion angles of 30.02 and 51.82°, respectively. As explained above, this must be due to the steric hindrance between the two pyrazole fragments that are bonded to contiguous carbon atoms of the quinoxaline ring.

The presence of the NH$_2$ group in position 2 of the pyrimidine ring allows the formation of weak intermolecular hydrogen bonds between the amine hydrogens

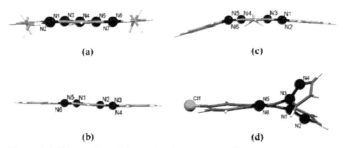

Figure 3.1 Side-on view of the molecular structure of: (a) bpz*pmNH$_2$, (b) bIndzpm, (c) bIndzpmMe and (d) bis-2,3-(pyrazol-1-yl)-6-Cl-quinoxaline.

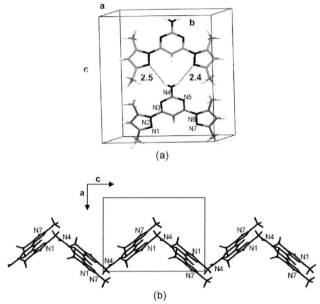

Figure 3.2 (a) Unit cell of the ligand bpz*pmNH$_2$ showing the hydrogen bonds between pyrazole and pyrimidine units (Å) and the labeling of the heteroatoms involved in the hydrogen bonds. (b) Intermolecular arrangement along the c axis.

and the pyrazolic N atoms. The localization and lengths of these bonds are indicated in Figure 3.2(a). To facilitate these interactions the contiguous molecules are disposed in planes that form dihedral angles of approximately 80° (pm rings were considered) and a zigzag arrangement can be observed along the c axis (Figure 3.2b).

Although other 2,3-bis(N-pyrazolyl)quinoxaline, pyrazine or pyridazine heterocycles cannot be found in the Cambridge Crystallographic Data Centre (CCDC), the analogous 2,3-bis(2-pyridyl)quinoxalines [21] and pyrazines [22] do not show coplanarity between the heterocycles. However, as one would expect, coplanar rings were found in the 3,6-bis(2-pyridyl)pyridazine derivative [23].

To evaluate the effect of the coordination on the azolyl-azine dihedral angle of these ligands as bis-bidentate systems, a family of complexes of stoichiometry [{Cu(PPh$_3$)$_2$}$_2$(L)]A$_2$, in which two [Cu(PPh$_3$)$_2$]$^+$ fragments are coordinated, were prepared by the simple reaction of [Cu(MeCN)$_4$]A precursors with the corresponding azolylazine in the presence of PPh$_3$ in a molar ratio Cu : L : PPh$_3$ = 2 : 1 : 4. Detailed discussion of these structures is beyond the scope of this chapter. However, it can be stated that although in some cases coordination to the copper centers leads to an increase in the dihedral angles between the rings, these angles are always clearly smaller (around 25°) for complexes that contain ligands derived from pyrimidine than those containing derivatives with pyrazine or quinoxaline as the central ring (angles around 35–40°).

3.3
Preparation of [2 × 2] Grids with Cu(I) or Ag(I)

The first reported grid architectures were obtained by coordination of multidentate polypyridine or pyridine-polypyridazine ligands to tetrahedral metal centers such as copper(I) [5b,24] or silver(I) [5b,24c,24e,25]. However, other geometries such as octahedral [4,5,26] and square pyramidal [27] have also proven useful for the self-assembly of various building-blocks. In all of the reported examples of grid structures the facing ligands are parallel. However, whereas the use of rigid tri-heterocyclic organic units with two 2-pyridine moieties is very common, the use of pyrazolyl rings in the ligands has hardly been explored in the formation of metal-grids [27].

In a [2×2] metal-grid structure the organic components usually adopt a conformation that is essentially planar and, therefore, heterocycles (Scheme 3.4) that do not have coplanar rings, such as the pyrazine and quinoxaline derivatives, are not very convenient for the preparation of these architectures. In contrast, the pyrimidine, triazine and pyridazine ligands are, in principle, suitable for obtaining grid structures. These ligands differ in the position of the donor atoms and the relative disposition of the two pyrazolyl rings. This gives rise to a different orientation of the "binding vectors" (represented by dotted lines in Scheme 3.4) that represent the projection of the planes where the metal centers and the ligands that complete the coordination sphere should be situated. Consequently, it is expected that the pyridazine ligands could lead to grids in which the facing ligands are parallel, as in other previously described grids. However, the pyrimidine and triazine ligands should give rise to grids in which the facing ligands are divergent (they would ideally define an angle of 60°). This constitutes a new class of grids with intramolecular

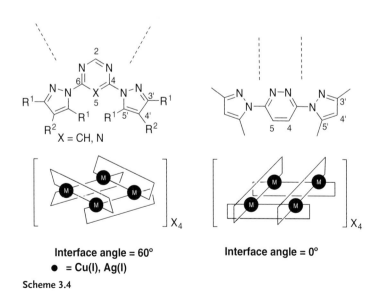

Scheme 3.4

cavities that have a different shape to that encountered in other grid structures. Such cavities could host other types of molecules, thereby offering the potential of broadening the scope for further work in this area. We envisaged that the cavities generated would be able to host anionic guests such as their own counter-anions. One point that could favor this encapsulation is the possible existence of anion–π interactions with the π-acidic rings of pyrimidine or triazine.

Another remarkable difference that can be predicted for these grids is a consequence of the existence of a spacer carbon atom between the N-donor atoms of the pyrimidine or triazine rings that is absent in the pyridazine heterocycle. This leads to a greater distance between the metal centers in the former systems and, consequently, the cavity in the grid should be bigger. All of these envisaged differences are demonstrated in Section 3.3.2, in which X-ray structures are discussed.

3.3.1
Synthesis

The self-assembly of the pyrimidine, triazine and pyridazine derivatives shown in Scheme 3.3 with Cu(I) or Ag(I) metal centers leads to the corresponding [2×2] metal-grids [28]. The organic components were reacted with [Cu(CH$_3$CN)$_4$]X (X = PF$_6^-$, BF$_4^-$, ClO$_4^-$, p-tolSO$_3^-$) or AgX salts (X = BF$_4^-$, PF$_6^-$, ClO$_4^-$, TfO$^-$, p-tolSO$_3^-$) (TfO$^-$ = CF$_3$SO$_3^-$) in a 1:1 molar ratio in acetone [Eqs. (1) and (3)]. When the copper acetonitrile adduct was not accessible, the reaction was performed (in "one pot" or by isolating the chloride intermediate) with CuCl and the chloride was subsequently substituted by the targeted anion through the corresponding silver salt [Eqs. (2a) and (2b)]. In this way the following compounds were prepared:

$$4[Cu(CH_3CN)_4]A + 4NN \rightarrow [Cu(NN)]_4A_4 \qquad (1)$$

NN = bpzpm	A = PF$_6$ (**1**), BF$_4$ (**2**), ClO$_4$ (**3**), p-tolSO$_3$ (**4**)
NN = bpz*pm	A = PF$_6$ (**5**), BF$_4$ (**6**), ClO$_4$ (**7**), p-tolSO$_3$ (**8**)
NN = bpzMepm	A = PF$_6$ (**9**), BF$_4$ (**10**), ClO$_4$ (**11**), p-tolSO$_3$ (**12**)
NN = bpz*Mepm	A = PF$_6$ (**13**), BF$_4$ (**14**)
NN = bpz*mtpm	A = PF$_6$ (**15**), BF$_4$ (**16**), ClO$_4$ (**17**), p-tolSO$_3$ (**18**)
NN = bpzpmNH$_2$	A = PF$_6$ (**19**), BF$_4$ (**20**), ClO$_4$ (**21**), p-tolSO$_3$ (**22**)
NN = bpz*pmNH$_2$	A = PF$_6$ (**23**), BF$_4$ (**24**), ClO$_4$ (**25**), p-tolSO$_3$ (**26**)
NN = bpz*pdz	A = PF$_6$ (**27**), BF$_4$ (**28**), p-tolSO$_3$ (**29**)
NN = bpzOMeT	A = PF$_6$ (**30**), BF$_4$ (**31**), ClO$_4$ (**32**)
NN = bpz*OmeT	A = PF$_6$ (**33**), BF$_4$ (**34**), ClO$_4$ (**35**)
NN = bIndzpm	A = PF$_6$ (**36**), BF$_4$ (**37**)
NN = bIndzpmMe	A = PF$_6$ (**38**), BF$_4$ (**39**)
NN = bcamphorpzpm	A = BF$_4$ (**40**)

3.3 Preparation of [2 × 2] Grids with Cu(I) or Ag(I)

$$n\text{CuCl} + n\text{NN} \rightarrow [\text{Cu(NN)Cl}]_n \quad (2a)$$

NN = bpzpm (**41**), bpz*pm (**42**), bpz*mtpm (**43**), bpz*pmNH$_2$ (**44**)

$$4/n\,[\text{Cu(NN)Cl}]_n + 4\text{AgTfO} \rightarrow 4\text{AgCl} + [\text{Cu(NN)}]_4(\text{TfO})_4 \quad (2b)$$

NN = bpzpm (**45**), bpz*pm (**46**), bpz*Mepm (**47**) "one pot", bpz*mtpm (**48**), bpz*pmNH$_2$ (**49**), bpzpmNH$_2$ (**50**) "one pot"

$$4\text{AgA} + 4\text{NN} \rightarrow [\text{Ag(NN)}]_4\text{A}_4 \quad (3)$$

NN = bpzpm	A = PF$_6$ (**51**), BF$_4$ (**52**), ClO$_4$ (**53**), TfO (**54**)
NN = bpz*pm	A = PF$_6$ (**55**), BF$_4$ (**56**), ClO$_4$ (**57**), TfO (**58**), p-tolSO$_3$ (**59**)
NN = bpzMepm	A = PF$_6$ (**60**), BF$_4$ (**61**), ClO$_4$ (**62**), TfO (**63**), p-tolSO$_3$ (**64**)
NN = bpz*Mepm	A = PF$_6$ (**65**), BF$_4$ (**66**), ClO$_4$ (**67**), TfO(**68**), p-tolSO$_3$ (**69**)
NN = bpz*mtpm	A = PF$_6$ (**70**), BF$_4$ (**71**), ClO$_4$ (**72**), TfO(**73**), p-tolSO$_3$ (**74**)
NN = bpzpmNH$_2$	A = PF$_6$ (**75**), BF$_4$ (**76**), ClO$_4$ (**77**), TfO (**78**), p-tolSO$_3$ (**79**)
NN = bpz*pmNH$_2$	A = PF$_6$ (**80**), BF$_4$ (**81**), ClO$_4$ (**82**), TfO (**83**), p-tolSO$_3$ (**84**)
NN = bpzOMeT	A = PF$_6$ (**85**), BF$_4$ (**86**), Tf O(**87**), p-tolSO$_3$ (**88**)
NN = bpz*OMeT	A = PF$_6$ (**89**), BF$_4$ (**90**), TfO (**91**), p-tolSO$_3$ (**92**)
NN = bpz*pdz	A = PF$_6$ (**93**), BF$_4$ (**94**), TfO (**95**)
NN = bpzpdz	A = TfO (**96**)
NN = bIndzpm	A = PF$_6$ (**97**), BF$_4$ (**98**)
NN = bIndzpmMe	A = PF$_6$ (**99**), BF$_4$ (**100**)

The different ligands were chosen to study the effect of the substituents either on the central ring or in the azolyl fragments. A series of anions were investigated with the aim of analyzing their encapsulation and their interactions with the cationic grid.

3.3.2
X-Ray and other Techniques for Structural Characterization in the Solid State

The structures of the complexes whose number is underlined in the preceding equations [Eqs. (1)–(3)] were studied by X-ray diffraction. All of these complexes are tetranuclear and consist of four ML units with a [2×2] grid structure (see Figure 3.3 below). The cationic charge is neutralized by four counter-anions and different crystallization solvents are trapped in free spaces of the crystal lattices with different solvent-accessible volumes. The Cu or Ag atoms have a distorted tetrahedral environment with each coordinated by two nitrogen atoms of two chelating ligands. A range of N–M–N bond angles can be observed, suggesting a broad tolerance in the coordination sphere of the metal centers, which could lead to structural flexibility – a characteristic favorable for encapsulation properties.

For the complexes of Cu with pyrimidine or triazine ligands (**1**, **2**, **6**, **21**, **24**, **25**, **49**, **33** and **40**) the Cu–Cu distances in the [2×2] grid are about 6.0–6.3 Å, with M–M–M angles close to 90° for most of the complexes or 80/100° for **6** and **75** (Table 3.1). The

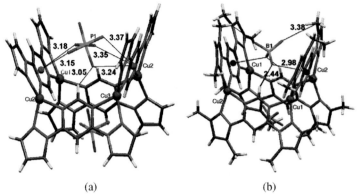

Figure 3.3 Stick representations of complex **1** (a) and **6** (b), including the encapsulated anions. Distances are in Å.

M–M distances are a slightly longer for the silver derivative **75**, probably due to its larger atomic size. The spatial disposition of the two pairs of chelating nitrogen atoms of the organic components in these grids means that the ligands are not perpendicular to the Cu_4 quadrangles but are inclined to them at angles ideally near to 60° (Table 3.1). Thus, the ligand pairs on each side of the complexes adopt a divergent orientation that leads to the formation of two groove-like cavities for each complex and these are occupied by two counter-anions. Complex **40** is an exception and will be discussed separately. Due to the aforementioned flexibility of the Cu coordination, the angles between opposite ligand pairs vary markedly and thus help to host – or even to complex – the two encapsulated anions. Thus, in **6** a low interplanar angle of 34.1° is observed, whereas the corresponding angles in **2** are 40.8° on one side of the Cu_4 quadrangle and 76.5° on the other. Possibly one driving force behind these variations is the interaction of the anions with the "walls" of the $[Cu_4L_4]^{4+}$ cations (see below).

Table 3.1 Selected structural parameters for the grid-shape complexes signaled in the first column.

Complex	Cu–Cu distance (Å)	Cu–Cu–Cu angle (°)	Interplanar angle (°)[a]
1	6.22	90.0	43.6
2	6.02, 6.10, 6.14	89.6, 90.4	40.8, 76.5
6	6.11	79.2, 100.8	34.1
21	6.08, 6.19, 6.24	89.3, 90.7	39.4, 67.2
24	6.11, 6.17	89.8, 90.2	41.4, 50.0
25	6.11, 6.16, 6.17	89.4, 89.5, 90.5	40.6, 48.6
27	3.60, 3.58	81.9, 82.2, 95.2	1.63, 5.96
29	3.56, 3.61, 3.58	84.0, 93.9, 94.9	0.72, 0.95
40	6.30	87.5, 92.5	19.5
49	6.18, 6.30	90.0	40.1, 48.7
75	6.82, 6.62, 6.64, 7.13,	77.8, 79.9, 96.9, 103.8	47.4, 56.5

[a] The pyrimidine or pyridazine ring-planes have been considered.

In complexes with a hydrogen atom in position 2 of the pyrimidine ring (**1**, **2** and **6**) [28] there are contacts between the F atoms of the PF_6^-/BF_4^- anions and the π-electron density of this ring of the ligand (F···π interactions). In **1**, five of the six F atoms of the PF_6 octahedron are remarkably close to the pyrimidine rings, with F···π (centroid) distances in the range 3.15–3.37 Å (from the viewpoint of PF_6^- one face- and one edge-sharing contact, Figure 3.3). A comparable situation is found for BF_4^- in **2** and **6**. Moreover, in **6** the ligand has methylated pyrazolyl groups and two F atoms of each anion interact with the methyl groups (C···F = 3.38 Å). This situation may cause the deviation of the involved pyrazoles from the plane of the rest of the ligand. Noticeably, these are the first reported examples of anion–π interactions with pyrimidine rings.

Ligands with other substituents in position 2 of the pyrimidine or the triazine rings have been designed to orientate these functions towards the inside of the open void created in the grids, with the aim of favoring their interactions with the hosted anions. In the structures determined by X-ray diffraction for complexes **21**, **24**, **25**, **49**, **33** and **75** (containing a NH_2 or OMe substituent) it was demonstrated that this has been achieved. The counter-anions BF_4^- in **24**, ClO_4^- in **21** and **25** and triflate in **49** show practically identical interactions with their respective hosts (see Figure 3.4a for complex **24**). As shown, these interactions consist of F···HN or O···HN moderate hydrogen bonds involving the amino groups and anion–π contacts with the pyrimidine ring. There is, noticeably, no interaction with the Me group of the pyrazoles like that found for **6**. In the structure of **33**, the MeO groups interact with the PF_6^- counter-anion with moderate F···HC hydrogen bonds (see Figure 3.4b). There are also weak F···HC hydrogen bonds involving the pyrazolic methyl groups.

For complexes **27** and **29**, which contain the bpz*dpz ligand, structural architectures that differ markedly from those of the previous complexes were found. The spatial disposition of the two pairs of chelating nitrogen atoms in the pyridazine derivative ligand causes the $[Cu_4L_4]^{4+}$ moieties to have, firstly, much shorter intramolecular Cu–Cu distances (ca. 3.6 Å, see Table 3.1) and, secondly, also favors a nearly parallel alignment of opposite ligands L. As a result of these two features, the

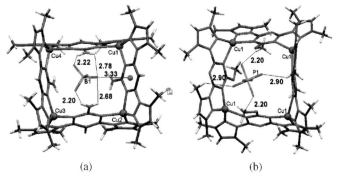

Figure 3.4 Stick representations of complexes **24** (a) and **33** (b), including the encapsulated anions and their interactions. Distances are in Å.

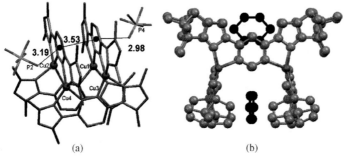

Figure 3.5 (a) Drawing of the cation of complex **27** and two anions showing the anion–π and π – π interactions. Distances are in Å. (b) Molecular structure of the cation of complex **40** showing two molecules of toluene inside the grid.

$[Cu_4L_4]^{4+}$ units of complexes **27** (Figure 3.5a) and **29** are very compact, show extended intramolecular π–π-stacking interactions [29] between adjacent ligands and have no "intramolecular" cavity to complex counter-anions like PF_6^- in **27**. The uneven disphenoidal arrangement of the Cu_4 unit, which is in contrast to the planar Cu_4 units in **1**, **2** and **6**, is advantageous for the intramolecular π–π-stacking interactions because this distortion means that parallel ligand pairs are not superimposed exactly above one another but are instead mutually shifted and laterally rotated, which helps to separate the bulky methyl groups and avoids the superposition of equally charged regions of the π-electron surfaces of the ligands. It is also worth noting the presence of two Cu_4L_4–Cu_4L_4 π–π stacking and anion–π interactions in **27** and two pronounced tosylate–Cu_4L_4 π–π-stacking interactions in **29**.

Complex **40** is exceptional in many respects: (a) sixteen chiral centers are located in the camphorpz groups. (b) Although the organic components are pyrimidine derivatives, they are practically parallel one to the other. (c) None of the counter-anions are hosted inside the internal cavity although it is regular and sufficiently large (see Cu–Cu–Cu angles and Cu–Cu distances in Table 3.1) but toluene molecules are trapped inside. The last two aspects signal the shape flexibility of these grids and a very probable template effect of the guest in the architecture of the macrocycles. Outside the grids other cavities can be visualized and these are formed by four camphor units and two counter-anions. Other toluene molecules of crystallization are hosted within these cavities (Figure 3.6). These molecules show face-to-face π-stacking with the nearest pyrimidine rings (centroid-centroid distance of 3.51 Å)

Other aspects related to the packing of these grid structures, the presence of a multitude of interactions between the tetranuclear cations and the counter-anions not hosted in the cavities, and with the crystallization solvent molecules, are beyond the scope of this discussion. In general all of these solids can be considered as microporous and the anions and solvent molecules located within these pores are normally very disordered. Figure 3.7 shows two selected examples, where channels are formed along the c axis in complexes **49** and **40**. In the former, disordered triflate anions located in the channels are drawn in grey. In the second example, the regular

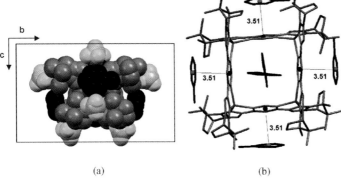

(a) (b)

Figure 3.6 Full space representation of the structure of **40**, showing the cavities of the grid formed by four camphor groups and two counter-anions where the external-grid toluene molecules are trapped. In (b) the face-to-face π stacking (Å) of these toluene molecules with the pyrimidines is emphasized.

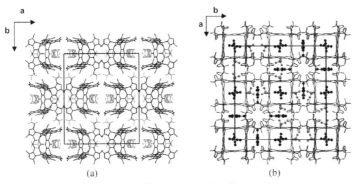

(a) (b)

Figure 3.7 View of the packing of **49** (a) and **40** (b), showing the formation of channels along the c axis. Anions are in grey. For **40** the toluene molecules (in black) are in a ball and stick representation.

disposition of the grids builds hydrophobic channels that are free of counter-anions and solvent molecules.

3.3.3
Structural Characterization in Solution by NMR

The grid compounds described in this chapter exhibit limited solubility in most organic solvents but are sufficiently soluble in acetone-d_6 to record ^{1}H NMR spectra and ^{13}C spectra in the majority of cases. In dmso-d_6 and acetonitrile-d_3 they decompose, releasing the free ligands. Evidence for the stability of the structures in acetone solutions, at least when the counter-anion has a low coordinating ability (PF_6^-, BF_4^-, triflate, ClO_4^-), is obtained from the spectra: (a) The chemical shifts of the carbon and proton resonances are affected by the coordination, particularly the Me^5 of the pz* units, as a result of the conformational change in the ligand. (b) In the complexes containing the ligands bpz*pm, an NOE was observed between H^5 of the pyrimidine ring and Me$^{5'}$ of the pyrazole heterocycle. Similarly, an NOE between Me$^{5'}$ and the H^{4,5} protons of the pyridazine ring is observed. (c) The ^{1}H NMR spectra show sharp resonances in the range -80 to 25 °C, corresponding to two identical halves of the ligands, as one would expect for these structures.

Different behavior was found for the p-toluenesulfonate and chloride derivatives containing ligands derived from pyrimidine. This could be due to a competitive coordination of these anions against the heterocyclic ligands. For the chloride complexes, a clear resonance broadening is observed at room temperature. Considering the chemical shifts it must be accepted that the free ligand is the dominant species.

For the Ag grids a slight broadening of the resonances is observed at low temperature, which can be attributed to a dynamic process, probably due to the marked tendency for decoordination of the silver complexes with N-donor ligands.

A series of complexes containing different counter-anions, BF_4^-, PF_6^-, ClO_4^- and TfO$^-$, show very similar ^{1}H NMR chemical shifts, indicating that the cation–anion interactions are weak in solution.

These aspects have been confirmed by UV/Vis spectroscopy. The corresponding spectra were recorded for the copper derivatives of bpz*pm, bpz*pmNH$_2$ and bpz*pdz. In general, for all the Cu complexes with pyrimidine ligands and with counter-anions of low coordinating ability, the disappearance or a shift in the characteristic absorption bands of the free ligands is observed along with the appearance of a low intensity band at around 407 nm, which is characteristic of an MLCT transition of copper(I) surrounded in a tetrahedral fashion with four nitrogen atoms [24c,30]. However, in the silver complexes and in the copper derivatives with the counter-anions p-tolSO$_3^-$ or Cl$^-$, the MLCT transition is very weak or is not observed, reflecting the aforementioned tendency for decoordination of the heterocyclic ligands. For the bpz*pdz derivatives, the MLCT band is present in all cases, indicating a higher stability for these grids.

A comparative study of the variation in the ^{19}F chemical shift of the PF$_6^-$ resonances of complexes **5** or **27** on addition of increasing amounts of the salt KPF$_6$ was carried out. The resonances were compared with those of free KPF$_6$. The results are represented in Figure 3.8. A similar study was carried out with the BF$_4^-$ complexes **6** and **28** and the salt NBu$_4$BF$_4$. For the complexes, the chemical shift of the anion is different to that of the free salts. When an increasing amount of the salt is added the chemical shift changes and it approaches that of the free anion. This observation points to the existence of some cation–anion interaction in solution that modifies the anion resonances. A ^{19}F, ^{1}H-HOESY experiment confirmed the existence of these contacts in the case of complex **27**. A correlation between the fluorine resonance of the PF$_6^-$ anion and the H^4,H^5 protons of the pyridazine ring indicates that the counter-anion approaches the tetranuclear fragment from the region of these pyridazine protons, a fact that is consistent with the results found in the solid state for the corresponding structure determined by X-ray diffraction (see Figure 3.5a). Unfortunately, the rest of the complexes studied were not sufficiently soluble to record a HOESY spectrum.

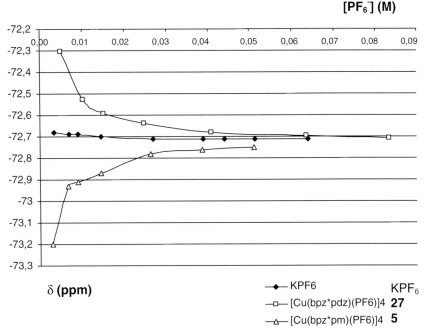

Figure 3.8 ^{19}F NMR chemical shift of the PF$_6^-$ group in KPF$_6$ and complexes **5** and **27** against the PF$_6^-$ concentration.

3.3.4
Anion Exchange in the Solid State

Taking into account the ability of the grids described in this chapter to host counter-anions, the study of anion exchange in the solid state was envisaged. Preliminary studies were carried out in which solid samples of the complexes **1, 5, 19** and **23** were stirred in THF solutions of NBu$_4$BF$_4$ or NBu$_4$TfO salts. The starting materials, grid and ammonium salts were added in an equimolecular ratio. The complexes are initially PF$_6^-$ salts and they are completely insoluble in THF. In this situation the anion substitution must be controlled by diffusion at the surface of the solid samples. The reaction time was 24 h (the data did not change after seven days). The resulting solid products were analyzed by ^{19}F NMR spectroscopy and the ratio of fluorinated counter-anions was determined by integration of the respective resonances. Table 3.2 shows the results obtained.

A relative selectivity toward one of the anions, which depends on the grid-complex used as the starting material, was observed in these experiments. It is interesting to compare the results of tests 2 and 3, on the one hand, and those of tests 4 and 5 on the other. These correspond to experiments in which the same cationic grid is used as the starting material but with a different counter-anion. The data are clearly different, indicating that the solid state structure of the starting material is very important in the selectivity towards a specific anion.

Table 3.2 Anion distribution (%)[a] in the solid phase for complexes [Cu(bpz'pm)]$_4$A$_4$[b] (24 h at room temperature in THF. The anion indicated as A is that of the starting material).

Test		A	PF$_6$	BF$_4$	OTf
1	[Cu(bpzpm)]$_4$A$_4$	PF$_6$	**59.9**	25.2	14.9
2	[Cu(bpz*pm)]$_4$A$_4$	PF$_6$	**40.7**	31.1	28.2
3	[Cu(bpz*pm)]$_4$A$_4$	TfO	38.1	**45.7**	16.3
4	[Cu(bpzpmNH$_2$)]$_4$A$_4$	PF$_6$	15.2	**42.4**	42.4
5	[Cu(bpzpmNH$_2$)]$_4$A$_4$	TfO	**72.5**	27.4	16.1
6	[Cu(bpz*pmNH$_2$)]$_4$A$_4$	PF$_6$	32.0	**51.9**	16.1

[a] % Determined by integration in ^{19}F NMR spectra.
[b] Anion contained in the solid in higher proportion is indicated in bold.

3.4
Preparation of Coordination Polymers with 2,3-Pyrazolylquinoxalines or 2,3-Pyrazolylpyrazines and Cu(I) or Ag(I)

3.4.1
Preparation and Characterization of Dinuclear Building Blocks and Coordination Polymers

As explained previously, the lack of planarity of the pyridazine or quinoxaline derivatives, reflected in Scheme 3.4, make them unsuitable to generate grid-type structures. In this section, we verify that their reaction with Cu(I) or Ag(I) centers gives rise to derivatives with very different structures.

Direct reaction of [Cu(MeCN)$_4$]BF$_4$ or AgA (A = BF$_4^-$, TfO$^-$) salts with the ligands indicated in Eqs. (4) and (5) leads to compounds with a ratio of M:L = 1:1.

$$[Cu(MeCN)_4]BF_4 \longrightarrow 1/n\,[Cu(NN)]_n(BF_4)_n \quad (4)$$

NN = bpzprz (**101**)
NN = bpzqnox (**102**)

$$AgBF_4 \longrightarrow 1/n\,[Ag(NN)]_n(BF_4)_n \quad (5)$$

NN = bpzprz, A = BF$_4$ (**103**)
NN = bpzpqnox, A = BF$_4$ (**104**)

bpzprz, bpzqnox

As the starting materials react, all of these compounds immediately precipitate from the reaction mixture and their metal:ligand stoichiometries are controlled by the lack of solubility. It was only possible to determine the structure of complex **102**, which crystallizes with acetone and has a stoichiometric formula [Cu$_2$(bpzqnx)$_2$(Me$_2$CO)]$_n$(BF$_4$)$_{2n}$(Me$_2$CO)$_{2n}$. As explained below, the structure consists of unsaturated Cu$_2$L$_2$ units, half of which contain the copper center also bonded to acetone molecules.

3.4 Preparation of Coordination Polymers

To gain further information we tried to "cut the polymers" and isolate the dinuclear units. Thus, some of these reactions were performed with a coordinating solvent such as acetonitrile or with pyridine in an M:py ratio of 1:1. We were able to obtain the dinuclear derivatives [see Eqs. (6) and (7)] that, in contrast to the preceding polymeric compounds, are relatively soluble in acetone.

$$\text{bpzprz, bpzqnox} \xrightarrow[S = \text{MeCN, py}]{[\text{Cu}(\text{MeCN})_4]\text{BF}_4} 1/2\ [\text{Cu}(\text{NN})(S)]_2(\text{BF}_4)_2 \quad (6)$$

NN = bpzprz, S = MeCN (**105**)
NN = bpzprz, S = py (**106**)
NN = bpzqnox, S = MeCN (**107**)
NN = bpzqnox, S = py (**108**)

$$\xrightarrow[\text{py}]{\text{AgTfO}} 1/2\ [\text{Ag}(\text{NN})(\text{py})]_2(\text{TfO})_2 \quad \text{NN = bpzqnox (109)} \quad (7)$$

The molecular structures of complexes **107** and **108** (determined by X-ray diffraction, see below) consist of dinuclear units that define a kind of loop, with the metal centers saturated with CH_3CN or py.

We propose that derivatives **101**, **103** and **104** are probably polymers formed by such {M_2L_2} units.

These loops of Cu or Ag can be used as building blocks for 1D coordination polymers with two different architectures: (a) Loops and chains (A) and (b) condensed loops (B).

Polymers of the loop and chain type (A) can be obtained after crystallization by slow diffusion of the metallic salts and the quinoxaline or pyrazine derivatives in two different phases in the presence of organic spacers such as pyrazine (prz) or 4,4'-bipyridine (bpy) [see Eq. (8)].

$$2\ \text{Cu or Ag salt} + 2\ \text{NN} + \text{spacer} \rightarrow 1/n[M_2(\text{NN})_2(\text{spacer})]_n(\text{BF}_4)_{2n}$$

M = Cu, spacer = prz, NN = bpzqnox (**110**)
M = Cu, spacer = bpy, NN = bpzqnox (**111**)
M = Cu, spacer = prz, NN = bpzprz (**112**)
M = Cu, spacer = bpy, NN = bpzprz (**113**)
M = Ag, spacer = prz, NN = bpzqnox (**114**)
M = Ag, spacer = bpy, NN = bpzqnox (**115**) (8)

Other polymers with structures of type A or B can be obtained by the slow diffusion of the starting materials indicated in Eqs. (4) and (5) in two separate phases. This process leads in some cases to monocrystals suitable for X-ray structure determination. It was found that the M:L ratio is different to 1:1, according to Eqs. (9)–(11):

$$3[\text{Cu}(\text{MeCN})_4](\text{BF}_4) + 4\text{bpzprz} \rightarrow 1/n[\text{Cu}_3(\text{bpzprz})_4]_n(\text{BF}_4)_{3n}\ (\underline{\textbf{116}}) \quad (9)$$

$$\text{AgBF}_4 + 2\text{bpzprz} \rightarrow 1/n[\text{Ag}(\text{bpzprz})_2]_n(\text{BF}_4)_n\ (\underline{\textbf{117}}) \quad (10)$$

$$2\text{AgBF}_4 + 3\text{bpzqnox} \rightarrow 1/n[\text{Ag}_2(\text{bpzqnox})_3]_n(\text{BF}_4)_{2n}\ (\underline{\textbf{118}}) \quad (11)$$

Finally, when the methylated ligands bpz*prz and bpz*qnox are self-assembled with Cu or Ag ions, metallopolymers with an M:ligand ratio of 1:1 are formed. In one case [Eq. (12), (122)], the structure was determined by X-ray diffraction (see below) and consists of a sequential arrangement of metal and organic components.

$$\text{Cu or Ag salt} + \text{NN} \rightarrow 1/n\,[\text{M(NN)}]_n(\text{A})_n \tag{12}$$

M=Cu, NN=bpz*qnox, A=BF$_4$ (119)   M=Ag, NN=bpz*qnox, A=TfO (121)
M=Cu, NN=bpz*prz, A=BF$_4$ (120)   M=Ag, NN=bpz*prz, A=TfO (**122**)

3.4.2
X-Ray and other Techniques for Structural Characterization

The structures of the underlined compounds in the equations were determined by X-ray diffraction. This discussion begins with dinuclear derivatives **107** and **108**. The molecular structures are reflected in Figure 3.9.

This type of dinuclear unit has been reported elsewhere and has been characterized with bis-2-pyridyl-pyrazine or quinoxaline both in Cu [31] and in Ag [32] compounds. However, until now such units have not been used as building blocks (loops) in the construction of coordination polymers, an aspect that is discussed below. Each ligand acts in a bidentate manner with one metal center and as a monodentate system with the other. The two rings that are coordinated to the same metal center are nearly coplanar but the other deviates clearly from this plane. In this way, the structure if this dinuclear unit is very well adapted to the preferred conformation of the free ligand. The relative disposition of the ligands allows the existence of face-to-face π–π-stacking interactions between the quinoxaline and

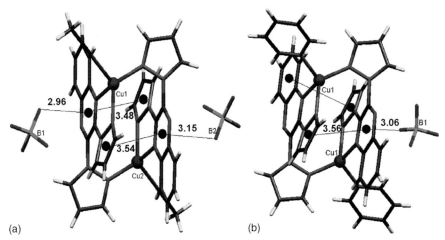

Figure 3.9 Molecular structures of compounds **107** (a) and **108** (b), showing the anion–π interactions. The π – π stacking (the centroids are indicated) between the organic components is reflected. Distances are in Å.

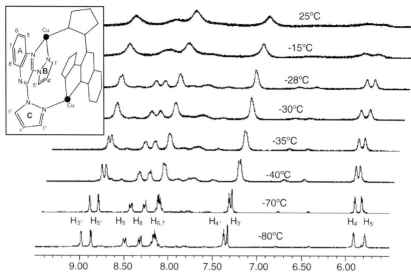

Figure 3.10 Variable-temperature ^{1}H NMR spectra of complex **107** in acetone-d_6 solution.

pyrazole rings. These noncovalent interactions probably make a clear contribution to the stability of the units. In the unit {Cu$_2$L$_2$}, each copper atom has a coordinate position saturated with CH$_3$CN (**107**) or py (**108**) and, in addition, each ligand has a nitrogen atom that is not involved in coordination. These two facts enable, as we will see below, the formation of polymers through different possibilities of using these positions to form new bonds.

In both structures anion–π interactions between one or two of the BF$_4^-$ counteranions and the electronically poor nitrogenated ring of the quinoxaline are observed in the solid state (Figure 3.9).

The dinuclear structure is probably retained in solution and this is likely to be in equilibrium with dissociation products in a process that is slowed down at low temperature – as deduced from the ^{1}H NMR data (see Figure 3.10 for a collection of ^{1}H NMR spectra of complex **107** at different temperatures). The broadening of the resonances at room temperature reflects the existence of a fluxional process in solution that is slowed down as the temperature decreases. A free energy value of 54.30 kJ mol^{-1} was determined at the coalescence temperature (266 K) of the H^5 and H^8 protons of the quinoxaline ring. At −80 °C the shift to high field of the resonances of pyrazole B can be ascribed to the electronic shielding created by the π stacking and by the pyrazole C. The shift to low field of the resonances of the pyrazole C could be a consequence of the perpendicular arrangement of this ring with regard to the rest of the heterocycle.

The structure of derivative **102** is striking (Figure 3.11). It consists of an alternating array of two types of loop, {Cu$_2$L$_2$} and {Cu$_2$L$_2$(acetone)$_2$}, giving rise to a polymer extended along the c axis. In the {Cu$_2$L$_2$} units the copper centers are saturated by coordination of the quinoxaline nitrogen of the adjacent units and this allows the

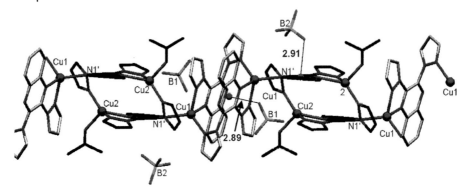

Figure 3.11 Molecular view of the structure of complex **102** showing the alternating array of loops and the anion–π interactions. Distances are in Å.

propagation of the polymer. The quinoxaline nitrogen of these units is not used because in the other units the metal centers are bonded to acetone molecules. As in the preceding cases, anion–π interactions were found between the nitrogenated ring of the quinoxaline and the BF_4^- counter-anions.

The structures of the coordination polymers **114** and **115** are represented in Figure 3.12. In both derivatives a sequential arrangement of the dinuclear loops of Ag and the spacers (prz or bpy) was found. The structure of the loops is similar to that described for the copper derivatives. The silver centers exhibit a distorted tetrahedral geometry. The metallopolymers grow along one of the crystallographic axes (b and c, respectively) and the BF_4^- anions are packed between them with F–qnox$_{centroid}$ distances that suggest the existence of anion–π interactions. Noticeably, a disordered anion links two vicinal polymers in **115** through these interactions.

Polymers **116–118**, obtained by slow diffusion of the reactants, also contain {M$_2$L$_2$} units but these are connected in different ways (Figure 3.13). As with the previous derivatives, complexes **116** and **117** also consist of a sequence of loops and chains, although in these cases the spacers are {Cu(bpzprz)$_2$} (Figure 3.13a) or the ligand bpzqnox (Figure 3.13b), respectively. In **116**, the Cu centers are tetrahedrally

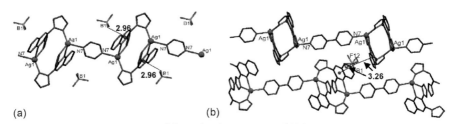

(a)   (b)

Figure 3.12 Schematic views of the molecular structures of **114** (a) and **115** (b). The anion–π interactions found are indicated. Distances are in Å.

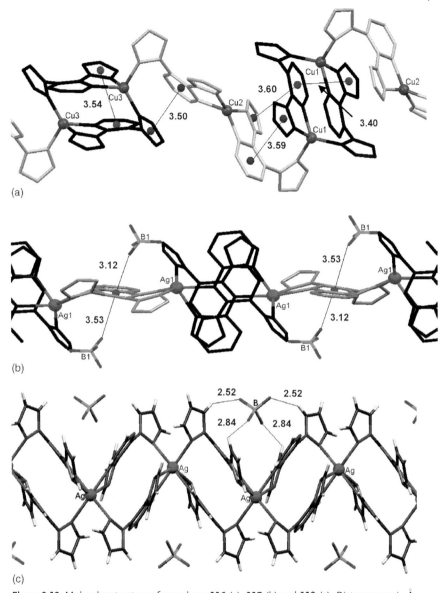

Figure 3.13 Molecular structure of complexes **116** (a), **117** (b) and **118** (c). Distances are in Å.

coordinated while in the silver structure **117** the metal has a coordination number of five because the spacer ligand behaves in a bis-bidentate manner. In **116**, face-to-face π–π-stacking interactions are present not only inside the loops but also between these units and the CuL$_2$ spacer (Figure 3.13a). In both **116** and **117** anion–π interactions were found, although only those in the second example are represented in Figure 3.13.

The structure of complex **118** (Figure 3.13c) involves a different arrangement of the structural units of these metallopolymers. Each silver center can be considered as being part of two contiguous loops. As in other examples, each of these loops consists of two Ag atoms and two ligands. These bpzqnox ligands are coordinated in a bidentate chelate manner to one of the silver atoms and as monodentate systems to the second. However, the difference is that each silver atom is bonded to four bpzqnox ligands, two bidentate and two monodentate, to achieve a coordination number of six. This high coordination number is unusual in silver [33]. The counteranions are packed close to the coordination polymer and adapt to the holes that form the condensed loops. F···HC hydrogen bonds between the anions and pyrazolic hydrogens are observed in this structure.

All of the structures described so far in this section exhibit dinuclear loops as building blocks and it is probably not by chance that the organic component always has unmethylated pyrazole rings. The only structure containing methylated pyrazoles that was determined has a completely different sequence of components and the aforementioned loops are not present. Figure 3.14 shows a schematic view of the structure of **122**. It consists of a sequential arrangement of silver atoms and ligands along the crystallographic direction [−101]. Two types of silver atoms can be distinguished as a consequence of the presence of two types of ligand coordination. All of the ligands bridge two silver centers but half of them chelate one of the silver atoms and act in a monodentate fashion with the other metal center, while the other half of the ligands are coordinated as monodentate systems with both silver centers. The silver atoms are also bonded to the TfO$^-$ counter-anion. In this way, alternating tri- and tetracoordinated silver centers are present in the structure. The conformation of the pyrazole and the pyridazine rings in the structure define a rotational axis in the polymeric chain (P in the figure) and therefore this structure can be considered as a helix. Helixes of both rotation senses (P and M) are present in the crystal. As emphasized in the figure, together with the covalent bonds, anion–π

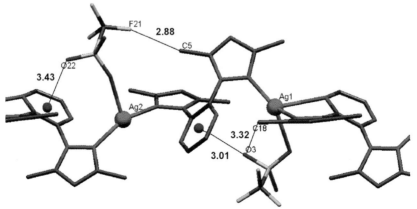

Figure 3.14 Schematic view of the structure of **122** with indication of anion–π and F···H hydrogen bonds (Å).

interactions with the pyrazine rings and F···HC hydrogen bonds involving the triflate anions are present.

3.5
Preparation of Supramolecular Structures with 2,4-Diamino-6-R-1,3,5-triazines and Ag(I)

In this section we discuss the formation of supramolecular structures using 2,4-diamino-6-R-1,3,5-tiazines as the organic components (Scheme 3.5). The triazine ring possesses nitrogen atoms that can coordinate to metal centers and the amino groups could give rise to hydrogen bonds either with the nitrogen atoms of the ring or with other atoms such as, for example, those of the counter-anions. The ligands have an additional substituent in the 6-position of the triazine ring and this could play different roles in the structure and packing of the resulting compounds (face-to-face π–π stacking, C–H···π interactions, weak hydrogen bonds, hydrophobic interactions, etc.). We envisaged that 2D or 3D networks could be formed after the reaction with metal centers. We chose the silver center on the basis that the high versatility that this cation exhibits could give rise to different and potentially interesting structures. Examination of the influence of the counter-anion on the final structure was also a goal of the study.

These ligands exhibit interesting solid-state structures [34] but discussion of these is beyond the scope of this chapter. This and other similar ligands [35] have been used in the preparation of interesting supramolecular arrays combined with organic substrates that offer complementary donor–acceptor functions and lead to the formation of strong hydrogen bonds.

In these three ligands the N^1 and N^5 atoms are suitable for covalent coordination with the metal center (see Scheme 3.5 for labeling). Although N^3 also possesses an electron pair, this will not be the preferred position for metal coordination because normally it is involved in the formation of donor–acceptor hydrogen bonds together with one or two hydrogen atoms of the amino groups. Very few examples of the coordination of an N atom with two vicinal NH_2 functions have been reported

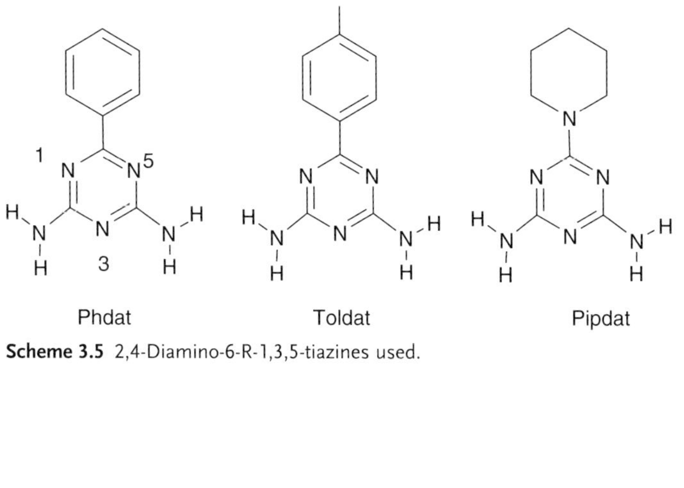

Phdat   Toldat   Pipdat

Scheme 3.5 2,4-Diamino-6-R-1,3,5-tiazines used.

for diaminotriazines (melamine has not been considered) [36]. Examples of the coordination of one NH$_2$ group of a triazine heterocycle to a metal atom are not known, probably because of delocalization of the electron pair in the electron-poor triazine ring.

3.5.1
Synthesis

Complexes **123–131** were obtained by the direct reaction of the corresponding triazine and different silver salts according to Eq. (13). These products are sparingly soluble in acetone and insoluble in other conventional organic solvents.

$$\text{triazine} \xrightarrow{\text{AgX}} [\text{Ag}(L)]_n(X)_n \tag{13}$$

R = Phenyl, Phdat
Tolyl, Toldat
Pyperidino, Pipdat

L = Phdat, X = PF$_6$, **123**; X = BF$_4$, **124**; X = TfO, **125**
L = Toldat, X = PF$_6$, **126**; X = BF$_4$, **127**; X = TfO, **128**
L = Pipdat, X = PF$_6$, **129**; X = BF$_4$, **130**; X = TfO, **131**

3.5.2
X-Ray Structure Determination

Slow diffusion of the triazine and the silver salt solutions afforded monocrystals suitable for X-ray diffraction in the cases of complexes **124**, **125** and **127**. Compounds **124** and **127** have structures that are practically identical and only the latter will be discussed. Figure 3.15 gives a representation of complex **125**.

A dimeric unit of formula [Ag$_2$(Phdat)$_2$(CF$_3$SO$_3$)$_2$] with two triflate bridges can be considered as the elemental building block for this structure (Figure 3.16a). Clearly, the two rings of each triazine ligand are not situated in the same plane and the corresponding dihedral angle is 41.86° (N12–C11–C21–C22; see Figure 3.15 for labeling). The silver atoms are also not in the same plane as the triazine rings, with distances between the silver centers and these planes being between 0.64 and 0.92 Å.

This dimeric unit is bonded to four other units through the N^5 of the triazines and the silver atoms to form a 2D sheet of silver and triazine ligands. These four units are rotated through 180° with respect to the central unit around the crystallographic *b* axis (Figure 3.16b and c). In this way a sheet is formed that extends along the [001] plane. The silver atoms exhibit a distorted tetrahedral arrangement and are bonded to two triflate groups and two triazine nitrogen atoms of different ligands.

As depicted in Figures 3.15 and 3.16(c), there is another noticeable motif for assembly between the elemental building blocks, namely triazine-phenyl π–π stacking. This face-to-face stacking is extended along the crystallographic *b* axis and leads to columns of alternate electronically poor and rich aromatic rings (the columns are indicated by arrows in Figure 3.16c). The inter-centroid ring distances have

3.5 Preparation of Supramolecular Structures with 2,4-Diamino-6-R-1,3,5-triazines and Ag(I)

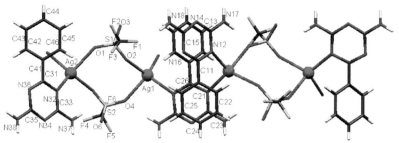

Figure 3.15 Graphic representation of two asymmetric units of complex **125**.

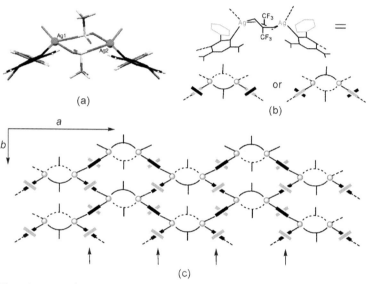

Figure 3.16 (a) Schematic view of the building unit of **125**. The two orientations of these units are represented in (b).
(c) Representation of the 2D network of the sheets in the complex.

alternate values of 3.78 and 3.52 Å and the centroid-plane distances are in the range 3.36–3.40 Å. Figure 3.17 shows another perspective of these sheets.

As can be seen in Figure 3.18, the asymmetric unit exhibits an arrangement of four ADAD centers consisting of the uncoordinated oxygen atom of the triflate (O3), an amino group (N18), the triazine N^2 atom (N14) and the second amino group (N17). The second acceptor center provided by the TfO⁻ bridging equals the number of donor and acceptor centers. The numbering included represents an example of the ADAD groups that exist along both sides of the sheets. All of these acceptor and donor centers (ADAD/DADA system) form lines of hydrogen bonds that connect the 2D-networks along the [010] plane, forming a three-dimensional superstructure (Figure 3.18). The N–O (NH···O) and N–N (NH···N) distances are 2.93–3.36 and

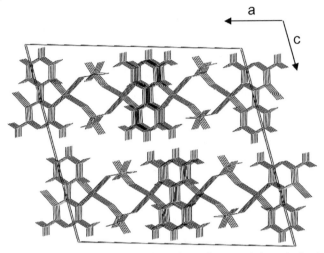

Figure 3.17 Complex **125**. View of two sheets that extend along the *b* axis.

3.00–3.02 Å, respectively. There are also intralayer N–H···O hydrogen bonds between NH$_2$ and triflate groups (Table 3.3).

Complex **127** [Ag(Toldat)]$_n$(BF$_4$)$_n$ crystallizes in the orthorhombic space group. In contrast to the situation found in complex **125**, the BF$_4^-$ counter-anion is not bonded to the silver center. This is not unexpected considering its low coordinating ability. In this case, a polymeric zigzag chain is formed that is extended along the *a* axis and consists of fragments {Ag(Toldat)} (Figure 3.19). The metal center exhibits a linear geometry (N1–Ag1–N1A angle of 171.87°) and is bonded to two triazine nitrogen atoms. The Ag1–N1 distance of 2.151 Å is shorter than those found in complex **125**.

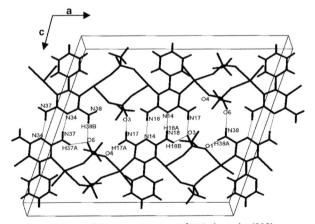

Figure 3.18 View of the X-ray structure of **125** along the [010] plane. The NH···N and NH···O hydrogen bonds are indicated.

3.5 Preparation of Supramolecular Structures with 2,4-Diamino-6-R-1,3,5-triazines and Ag(I)

Table 3.3 Hydrogen bond distances (Å) and angles (°) in the structure of **125**.

D–H···A	d(D–H) (Å)	d(H···A) (Å)	∠DHA (°)	d(D···A) (Å)
N17–H17B···O3	0.900	2.144	158.29	2.998
N18–H18A···N14	0.900	2.145	158.69	3.002
N18–H18B···O3	0.900	2.437	131.24	3.105
N37–H37A···O6	0.900	2.209	136.40	2.928
N37–H37B···N34	0.900	2.165	157.39	3.016
N38–H38A···O1	0.900	2.183	152.33	3.009
N17–H17A···O4	0.900	2.111	165.83	2.992
N38–H38B···O6	0.900	2.603	141.71	3.356

As in complex **125**, the two rings of the ligand are not coplanar (dihedral angle N1–C2–C3–C4 = 43.72°) and the silver atoms are not in the same plane as the triazine rings (distance to the plane of 0.84 Å). The angle formed by two consecutive triazine rings is 59.58°. The triazine molecules alternate the orientation of their symmetric C_2 axis along the crystallographic b axis. Evidence for π–π–stacking is not observed between the aromatic rings and the relative rotation of the tolyl and triazine moieties is probably due to steric requirements. The BF_4^- counter-anions are located in the inter-polymeric positions and short N–H···F–B distances of 2.12 Å are observed. These weak interactions mean that the Ag metal centers approach the BF_4^- anions (F···Ag distances of 2.9 Å were measured). As depicted in Figure 3.20 (a), these contacts could influence the sinusoidal disposition of the elemental units in the polymer since the maximum deviations from linearity along the polymer fit with the point where these contacts are located. As a consequence of these structural aspects, channels of ca. 8.5 Å in diameter (inter-anionic F–F distances were considered) are formed in the solid along the b axis (Figure 3.20b). These channels are full of disordered solvent, the nature of which could not been established. Another possibility to understand the formation of these channels and the consequent sinusoidal arrangement of units is their template effect in the crystal formation.

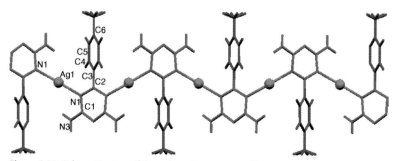

Figure 3.19 Schematic view of the polymeric structure of **127** along the b axis. Labeling for the independent atoms is indicated.

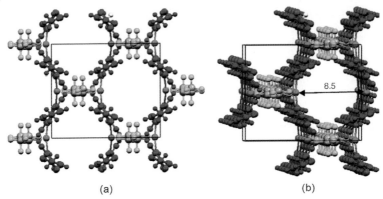

Figure 3.20 X-Ray structure of **127** in the [010] plane, showing the zigzag disposition of the elemental units (a) and the formation of channels (b). Distance is in Å.

Although complexes **125** and **127** differ in the R group of the triazine ring (phenyl versus p-tolyl, respectively) we believe that the dramatic differences found in both structures originate from the different coordinating ability of the corresponding counter-anions. In fact the structure of **124**, with Ph groups in the triazine, was also determined by X-ray diffraction and it is essentially as same as that described for **127**. The coordination of the triflate group makes the silver adopt a tetrahedral geometry instead of the linear one exhibited by the BF_4^- complexes. In the tetrahedral arrangement it is possible for the two nitrogenated ligands to be relatively near in space and interact through π–π stacking of the aromatic rings. This could be the driving force for the formation of the sheets, with the columns exhibiting the π–π stacking.

3.5.3
Structural Characterization in Solution by NMR

^{1}H NMR spectra at room temperature reveal the presence of the R groups in position 6 of the triazine ring and these are seen as symmetric. A single broad resonance assigned to the NH_2 groups is also observed. With the exception of the ortho protons of the aromatic rings, deshielding of the signals is observed with respect to those in the free ligands, $\Delta\delta = 0.4$–1.0, a fact due to the donation of electron density to the metal center. In the corresponding ^{19}F NMR spectra the counter-anion resonances appear as sharp signals. The ^{13}C {^{1}H} NMR spectra were only recorded for the more soluble derivatives and this also reflects the presence of symmetric ligands.

To obtain more information about the behavior of the derivatives in solution, the spectra of several complexes were also recorded at low temperature ($-80\,°C$, acetone-d_6). For complex **123**, at low temperature a broadening of the ortho resonance was observed and also a splitting of the amine signal (7.07 at r.t.) into two resonances shifted to higher frequency (7.8 and 8.05 ppm). This could be due to the existence of

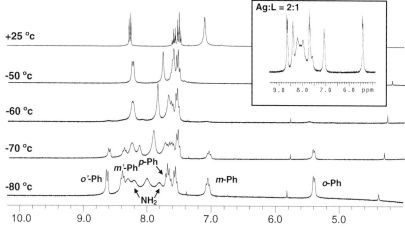

Figure 3.21 Variable-temperature ^{1}H NMR spectra of **124** in acetone-d_6 solution. The resonances assigned to the new derivative are indicated. The inset corresponds to the spectrum at −80 °C for a solution with an AgBF$_4$: Phdat ratio of 2.

hydrogen bonds that are more fixed at low temperature. This splitting has also been observed for the free ligand Phdat [34]. The spectra recorded at different temperatures for complexes **124** and **125** are quite similar, and resonances corresponding to a new species appear at low temperature (Figure 3.21). It contains an asymmetric phenyl group with a particularly shielded *ortho* proton (5.43 ppm for **124**) that exhibit a reduction in the $J_{Hortho-Hmeta}$ value. The amount of the new species increased when the Ag:Phdat ratio was higher than 1 (see insert of Figure 3.21). Although other options can not be conclusively discarded we propose that the new specie contain C−H...Ag agostic interactions.

3.6
Conclusions

The combination of polytopic organic components with Cu(I) and Ag(I) centers can give rise to an enormous variety of supramolecular structures. With bis(azolyl)azine derivatives the relative position of the two azolyl rings is a very important factor in influencing the type of structure obtained. For the ligands in which the azolyl (pyrazole or indazole) rings are not located at contiguous carbons of the central ring, i.e., pyrimidine, triazine or pyridazine heterocycles, the reaction with Cu(I) or Ag(I) yields [2×2] metal-grids. The position of the donor atoms determines the size and shape (parallel or divergent facing ligands) of the grid and in some cases cavities capable of hosting the counter-anions are formed. The hosting process is probably favored by the existence of anion−π interactions with acidic rings such as triazine, pyrimidine and pyridazine. Some of these interactions have not been described before. Hydrogen bonds are also established between the anions and the cation "walls". Interchange of the anions in the solid state has been observed to have a certain degree of selectivity.

When the azolyl rings (pyrazole) are bonded to contiguous carbon atoms of the central ring (pyrazine or quinoxaline), grid structures are not formed. In this case, polymeric species are more favorable. If the pyrazolyl rings are not methylated, $\{Cu_2L_2\}$ units with internal π–π stacking interactions are formed. These have the form of a loop and are the building block for the formation of polymers. The loops can be bonded directly or connected through spacers that can be the ligand itself, $\{ML_2\}$ units or other ditopic ligands. If the pyrazolyl rings are methylated the $\{Cu_2L_2\}$ units are not formed, probably for steric reasons. Polymers with alternating metal and organic fragments are formed. Interactions between the counter-anions and the polymers have been observed.

The introduction of amino groups into a triazine ring (and consequently the possible formation of strong hydrogen bonds) allows the expansion of the type of derivatives obtained. The self-assembly between 2,4-diamino-6-R-1,3,5-triazines and different Ag(I) salts gives rise to networks, the structure of which depends strongly on the coordinating ability of the counter-anion. With TfO^-, sheets with tetrahedral silver atoms and with columns exhibiting π–π stacking interactions are formed. The sheets are connected through quadruple hydrogen bonds with the participation of the triflate groups. When the anion is BF_4^-, the silver center is linear and zigzag chains are formed. These are connected through interactions with the anions and the formation of channels is observed in which the solvent is encapsulated.

Acknowledgments

Financial support from the Junta de Castilla-La Mancha-FEDER Funds (PBI-05-003) and the Spanish DGES/MCyT (CTQ2005-01430/BQU) is acknowledged.

References

1 (a) Beatty, A.M. (2003) Open-framework coordination complexes from hydrogen-bonded networks: Toward host/guest complexes. *Coord. Chem.*, **1–2**, 2751. (b) Lehn, J.-M. (1995) *Supramolecular Chemistry: Concepts and Perspectives*, VCH, Weinheim.

2 (a) Swiegers, G.F. and Malefetse, T.J. (2000) New self-assembled structural motifs in coordination chemistry. *Chem. Rev.*, **100**, 3483. (b) Eddaoudi, M., Moler, D.B., Li, H., Chen, B., Reineke, T.M., O'Keefe, M. and Yaghi, O.M. (2001) Modular chemistry: Secondary building units as a basis for the design of highly porous and robust metal-organic carboxylate frameworks. *Acc. Chem. Res.*, **34**, 319. (c) Ockwig, N.W., Delagao-Friedrichs, O., O'Keefe, M. and Yaghi, O.M. (2005) Reticular chemistry: Occurrence and taxonomy of nets and grammar for the design of frameworks.

Acc. Chem. Res., **38**, 176. (d) Leininger, S., Olenyuk, B. and Stang, P.J. (2000) Self-assembly of discrete cyclic nanostructures mediated by transition metals. *Chem. Rev.*, **100**, 853.

3 (a) Kitagawa, S., Kitaura, R. and Noro, S.-I. (2004) Functional porous coordination polymers. *Angew. Chem., Int. Ed.*, **43**, 2334–2375. (b) Chen, C.-L., Kang, B.-S. and Su, C.-Y. (2006) Recent advances in supramolecular design and assembly of silver(I) coordination polymers. *Aust. J. Chem.* **59**, 3–18.

4 (a) Andruh, M. *Encyclopedia of Supramolecular Chemistry* (eds J.L. Atwood and J.W. Steed), Marcel Dekker, Inc., New York, pp. 1186–1193. (b) Waldmann, O., Hassmann, J., Müller, P., Hanan, G.S., Volkmer, D., Schubert, U.S. and Lehn, J.-M. (1997) Intramolecular antiferromagnetic coupling in supramolecular grid structures with Co^{2+} metal centers. *Phys. Rev. Lett.*, **78**, 3390.

5 (a) Ruben, M., Breuning, E., Lehn, J.-M., Ksenofontov, V., Renz, F., Gütlich, P. and Vaughan, B.M. (2003) Supramolecular spintronic devices: Spin transitions and magnetostructural correlations in $[Fe_4^{II}L_4]^{8+}$ [2 × 2]-grid-type complexes. *Chem.–Eur. J.*, **9**, 4422.
(b) Ruben, M., Rojo, J., Romero-Salguero, F.J., Uppadine, L.H. and Lehn, J.-M. (2004) Grid-type metal ion architectures: Functional metallosupramolecular arrays. *Angew. Chem., Int. Ed.*, **43**, 3644–3662.
(c) Waldmann, O., Ruben, M., Zeiner, U. and Lehn, J.-M. (2006) Supramolecular Co (II)-[2 × 2] Grids: Metamagnetic behavior in a single molecule. *Inorg. Chem.*, **45**, 6535.

6 Stadler, A.-M., Kyritsakas, N., Graff, R. and Lehn, J.-M. (2006) Formation of rack- and grid-type metallosupramolecular architectures and generation of molecular motion by reversible uncoiling of helical ligand strands. *Chem.–Eur. J.*, **12**, 4503.

7 (a) Steed, J.W. and Atwood, J.L. (2001) *Supramolecular Chemistry*, John Wiley, New York. (b) Hof, F., Craig, S.L., Nuckolls, C. and Rebek, J., Jr (2002) Molecular encapsulation. *Angew. Chem., Int. Ed.*, **41**, 1488–1508.

8 (a) Atwood, J.L. (1997) Structural and topological aspects of anion coordination, in: *Supramolecular Chemistry of Anions* (eds A. Bianchi, K. Bowman-James and E. García-España), Wiley-VCH, New York. p. 147.(b) Beer, P.D. and Gale, P.A. (2001) Anion recognition and sensing: The state of the art and future perspectives. *Angew. Chem., Int. Ed.*, **40**, 486.

9 (a) Meyer, E.A., Castellano, R.K. and Diederich, F. (2003) Interactions with aromatic rings in chemical and biological recognition. *Angew. Chem., Int. Ed.*, **42**, 1210. (b) Ma, J.C. and Dougherty, D.A. (1997) The cation-π interaction. *Chem. Rev.*, **97**, 1303.

10 (a) Quiñonero, D., Garau, C., Rotger, C., Frontera, A., Ballester, P., Costa, A. and Deyà, P.M. (2002) Anion-π interactions: Do they exist?. *Angew. Chem., Int. Ed.*, **41**, 3389. (b) Garau, C., Quiñonero, D., Frontera, A., Ballester, P., Costa, A. and Deyà, P.M. (2005) Approximate additivity of anion-π interactions: An ab initio study on anion-π anion-$π_2$ and anion-$π_3$ complexes. *J. Phys. Chem. A*, **109**, 9341. (c) Frontera, A., Saczewski, F., Gdaniec, M., Dziemidowicz-Borys, E., Kurland, A., Deyà, P.M., Quiñonero, D. and Garau, C. (2005) Anion-π interactions in cyanuric acids: A combined crystallographic and computational study. *Chem.–Eur. J.*, **11**, 6560. (d) Mascal, M., Armstrong, A. and Bartberger, M.D. (2002) Anion-aromatic bonding: A case for anion recognition by π-acidic rings. *J. Am. Chem. Soc.*, **124**, 6274.

11 (a) Fairchild, R.M. and Holman, K.T. (2005) Selective anion encapsulation by a metalated cryptophane with a π-acidic interior. *J. Am. Chem. Soc.*, **127**, 16364. (b) Holman, K.T., Halihan, M.M., Jurisson, S.S., Atwood, J.L., Burkhalter, R.S., Mitchell, A.R. and Steed, J.W. (1996) Inclusion of neutral and anionic guests within the cavity of π-metalated cyclotriveratrylenes. *J. Am. Chem. Soc.*,

118, 9567. (c) Staffilani, M., Hancock, K.S.B., Steed, J.W., Holman, K.T., Atwood, J.L., Juneja, R.K. and Burkhalter, R.S. (1997) Anion binding within the cavity of π metalated calixarenes. *J. Am. Chem. Soc.*, **119**, 6324.

12 (a) Demeshko, S., Dechert, S. and Meyer, F. (2004) Anion-π interactions in a carousel copper(II)-triazine complex. *J. Am. Chem. Soc.*, **126**, 4508. (b) Schottel, B.L., Chifotides, H.T., Shatruck, M., Chouai, A., Pérez, L.M., Bacsa, J. and Dunbar, K.R. (2006) Anion-π interactions as controlling elements in self-assembly reactions of Ag(I) complexes with π-acidic aromatic rings. *J. Am. Chem. Soc.*, **128**, 5895.

13 Black, C.A., Hanton, L.R. and Spicer, M.D. (2007) Probing anion–π interactions in 1-D Co(II), Ni(II), and Cd(II) coordination polymers containing flexible pyrazine ligands. *Inorg. Chem.*, **46**, 3669.

14 (a) Yamada, T. and Onishi, M. (1992) Preparation of bisazolylpyrimidine derivatives as antiulcer agents *Jpn. Kokai Tokkyo Koho*, 9 pp, Patent. (b) Steel, P.J. and Constable, E.C. (1989) Synthesis of new pyrazole-derived chelating ligands *J. Chem. Res. (S).*, **7**, 189. (c) Gómez-de, la Torre, F., de la Hoz, A., Jalón, F.A., Manzano, B.R., Rodríguez, A.M., Elguero, J., and Martínez-Ripoll, M. (2000) Pd(II) complexes with polydentate nitrogen ligands. Molecular recognition and dynamic behavior involving Pd-N bond rupture. X-ray molecular structures of [{Pd (C$_6$HF$_4$)$_2$}(bpzpm)] and [{Pd(η^3-C$_4$H$_7$)}$_2$ (bpzpm)] (CF$_3$SO$_3$)$_2$ [bpzpm = 4,6-Bis (pyrazol-1-yl)pyrimidine]. *Inorg. Chem.*, **39**, 1152.

15 Uson, R., Oro, L.A., Esteban, M., Carmona, D., Claramunt, R.M. and Elguero, J. (1984) Rhodium(I) complexes with pyridazine, 4,6-dimethyl-pyrimidine, 4,6-bis(3,5-dimethylpyrazol-1-yl)pyrimidine, 3,6-bis (3,5-dimethylpyrazol-1-yl)pyridazine and 3-(3,5-dimethylpyrazol-1-yl)-6-chloropyridazine. *Polyhedron*, **3**, 213–221.

16 Addison, A.W. and Burke, Ph. J. (1981) Synthesis of some imidazole- and pyrazole-derived chelating agents. *J. Heterocycl. Chem.*, **18**, 803–805.

17 Rusinov, G.L., Ishmetova, R.I., Latosh, N.I., Ganebnych, I.N., Chupakhin, O.N. and Potemkin, V.A. (2000) [4 + 2]-cycloaddition of 3,6-bis(3,5-dimethyl-4-R-pyrazol-1-yl)-1,2,4,5-tetrazines with alkenes. *Russ. Chem. Bull.*, **49**, 355–362.

18 Elguero, J., Jacquier, R. and Mondon, S. (1970) NMR studies in the heterocyclic series, IV conformation of N-aryl azoles; influence of sp^2 hybridized nitrogen atoms on the deshielding of o-protons of the neighboring aromatic nucleus. *Bull. Soc. Chim. Fr.* 1346–1351.

19 Zayed, S.E., Taha, M.M. and Mohammed, A.E. (1995) Synthesis of some azaheterocycles condensed to and fused with quinoxaline. *Mansoura J. Pharm. Sci.* **11**, 266–284.

20 (a) Jameson, D.L. and Goldsby, K.A. (1990) 2,6-bis(N-pyrazolyl)pyridines: The convenient synthesis of a family of planar tridentate N3 ligands that are terpyridine analogues. *J. Org. Chem.*, **55**, 4992. (b) Holland, J.M., Kilner, C.A., Thornton-Pett, M. and Halcrow, M.A. (2001) Steric effects on the electronic and molecular structures of nickel(II) and cobalt(II) 2,6-dipyrazol-1-ylpyridine complexes. *Polyhedron*, **20**, 2829. (c) Watson, A.A., House, D.A. and Steel, P.J. (1991) Chiral heterocyclic ligands. 7. Syntheses of some chiral 2,6-di-N-pyrazolylpyridines. *J. Org. Chem.*, **56**, 4072.

21 Rasmussen, S.C., Richter, M.M., Yi, E., Place, H. and Brewer, K.J. (1990) Synthesis and characterization of a series of novel rhodium and iridium complexes containing polypyridyl bridging ligands: Potential uses in the development of multimetal catalysts for carbon dioxide reduction. *Inorg. Chem.*, **29**, 3926–3932.

22 (a) Huang, N.T., Pennington, W.T. and Petersen, J.D. (1991) Structure of 2,3-bis(2-pyridyl)pyrazine. *Acta Crystallogr. Sect C: Cryst. Struct. Commun.*, **47**, 2011. (b) Robertson, K.N., Bakshi, P.K., Lantos, S.D., Cameron, T.S., and Knop, O. (1998)

Crystal chemistry of tetraradial species. Part 9. The versatile BPh_4^- anion, or how organoammonium H(N) atoms compete for hydrogen bonding. *Can. J. Chem.*, **73**, 583.

23 Ghumaan, S., Sarkar, B., Patra, S., Parimal, K., van Slageren, J., Fiedler, J., Kai, W. and Kumar Lahiri, G. (2005) 3,6-bis(2'-pyridyl) pyridazine (L) and its deprotonated form $(L-H^+)^-$ as ligands for $\{(acac)_2Ru^{n+}\}$ or $\{(bpy)_2Ru^{m+}\}$: Investigation of mixed valency in $[\{(acac)_2Ru\}_2(\mu\text{-}L\text{-}H^+)]^0$ and $[\{(bpy)_2Ru\}_2(\mu\text{-}L\text{-}H^+)]^{4+}$ by spectroelectrochemistry and EPR. *Dalton Trans.*, 706–712.

24 (a) Recent and selected papers: Price, J.R., Lan, Y. and Brooker, S. (2007) Pyridazine-bridged copper(I) complexes of bis-bidentate ligands: Tetranuclear [2 × 2] grid versus dinuclear side-by-side architectures as a function of ligand substituents. *Dalton Trans.*, 1807–1820. Hoogenboom, R., Wouters, D. and Schubert, U.S. (2003) L-Lactide polymerization utilizing a hydroxy-functionalized 3,6-bis(2-pyridyl) pyridazine as supramolecular (Co) initiator: Construction of polymeric [2 × 2] grids. *Macromolecules*, **36**, 4743.
(b) Hoogenboom, R., Moore, B.C. and Schubert, U.S. (2006) Synthesis of star-shaped poly(e-caprolactone) via "click" chemistry and "supramolecular click" chemistry. *Chem. Commun.*, 4010. (c) Patroniak, V., Lehn, J.-M., Kubicki, M., Ciesielski, A. and Wałesa, M. (2006) Chameleonic ligand in self-assembly and synthesis of polymeric manganese(II), and grid-type copper(I) and silver(I) complexes. *Polyhedron*, **25**, 2643–2649. (d) Barboiu, M., Petit, E., van der Lee, A. and Vaughan, G. (2006) Constitutional self-selection of [2 × 2] homonuclear grids from a dynamic mixture of copper(I) and silver(I) metal complexes. *Inorg. Chem.*, **45**, 484–486. (e) Patroniak, V., Stefankiewicz, A.R., Lehn, J.-M. and Kubicki, M. (2005) Self-assembly and characterization of grid-type copper(I), silver(I), and zinc(II) complexes. *Eur. J. Inorg. Chem.*, 4168–4173.
(f) Nitschke, J.R., Hutin, M. and Bernardinelli, G. (2004) The hydrophobic effect as a driving force in the self-assembly of a [2 × 2] copper(I) grid. *Angew. Chem., Int. Ed.*, **43**, 6724–6727. (g) Youinou, M.-T., Rahmouni, N., Fisher, J. and Osborn, J.A. (1992) Self-assembly of a Cu_4 complex with coplanar copper(I) ions: Synthesis, structure, and electrochemical properties. *Angew. Chem., Int. Ed.*, **31**, 733.

25 (a) Weissbuch, I., Baxter, P.N.W., Kuzmenko, I., Cohen, H., Cohen, S., Kjaer, K., Howes, P.B., Als-Nielsen, J. and Lehn, J.-M. (2000) Oriented crystalline monolayers and bilayers of 2 × 2 silver(i) grid architectures at the air-solution interface: Their assembly and crystal structure elucidation. *Chem.–Eur. J.*, **6**, 725. (b) Baxter, P.N.W., Lehn, J.-M., Baum, G. and Fenske, D. (2000) Self-assembly and structure of interconverting multinuclear inorganic arrays: A [4 × 5]-Ag-20(I) grid and an Ag-10(I) quadruple helicate. *Chem.–Eur. J.*, **6**, 4510–4517.
(c) Marquis, A., Kintzinger, J.-P., Graff, R., Baxter, P.N.W. and Lehn, J.-M. (2002) Mechanistic features, cooperativity, and robustness in the self-assembly of multicomponent silver(I) grid-type metalloarchitectures. *Angew. Chem., Int. Ed.*, **41**, 2760. (d) Baxter, P.N.W., Lehn, J.-M., Baum, G. and Fenske, D. (2000) Self-assembly and structure of interconverting multinuclear inorganic arrays: A [4 × 5]-AgI20 grid and an AgI10 quadruple helicate. *Chem.–Eur. J.*, **6**, 4510.
(e) Brooker, S., Davidson, T.C., Hay, S.J., Kelly, R.J., Kennepohl, D.K., Plieger, P.G., Moubaraki, B., Murray, K.S., Bill, E. and Bothe, E. (2001) Doubly pyridazine-bridged macrocyclic complexes of copper in +1, +2 and mixed valent oxidation states. *Coord. Chem. Rev.*, **216–217**, 3–30.

26 (a) Rojo, J., Romero-Salguero, F.J., Lehn, J.-M., Baum, G. and Fenske, D. (1999) Self-assembly, structure, and physical properties of tetranuclear Zn^{II} and Co^{II} complexes of [2 × 2] grid-type. *Eur. J. Inorg. Chem.*, 1421–1428. (b) Milway, V.A.,

Abedin, S.M.T., Niel, V., Nelly, T.L., Dawe, L.N., Dey, S.K., Thompson, D.W., Miller, D.O., Alam, M.S., Müller, P. and Thompson, L.K. (2006) Supramolecular 'flat' Mn$_9$ grid complexes – towards functional molecular platforms. *Dalton Trans*, 2835–2851. (c) Ruben, M., Breuning, E., Gisselbrecht, J.-P. and Lehn, J.-M. (2000) Multilevel molecular electronic species: Electrochemical reduction of a [2 × 2] Co$_4^{II}$ grid-type complex by 11 electrons in 10 reversible steps. *Angew. Chem., Int. Ed.*, **39**, 4139–4142. (d) Uppadine, L.H. and Lehn, J.-M. (2004) Three-level synthetic strategy towards mixed-valence and heterometallic [2 × 2] gridlike arrays. *Angew. Chem., Int. Ed.*, **43**, 240–243. (e) Bark, T., Düggeli, M., Stoeckli-Evans, H. and von Zelewsky, A. (2001) Designed molecules for self-assembly: The controlled formation of two chiral self-assembled polynuclear species with predetermined configuration. *Angew. Chem., Int. Ed.*, **40**, 2848–2851.

27 (a) Selected recent references: Klingele, J., Prikhod'ko, A.I., Leibeling, G., Demeshko, S., Dechert, S. and Meyer, F. (2007) Pyrazolate-based copper(II) and nickel(II) [2 × 2] grid complexes: Protonation-dependent self-assembly, structures and properties. *Dalton Trans*. 2003–2013. (b) Dawe, L.N., Abedin, T.S.M., Kelly, T.L., Thompson, L.K., Miller, D.O., Zhao, L., Wilson, C., Leech, M.A. and Howard, J.A.K. (2006) Self-assembled polymetallic square grids ([2 × 2] M$_4$, [3 × 3] M$_9$) and trigonal bipyramidal clusters (M$_5$) – structural and magnetic properties. *J. Mater. Chem.*, **16**, 2645–2659.

28 Manzano, B.R., Jalón, F.A., Ortiz, I.M., Soriano, M.L., Gómez de la Torre, F., Elguero, J., Maestro, M.A., Mereiter, K. and Claridge, T.D.W. *Inorg. Chem.*, accepted for publication. DOI: 10.1021/ic 701117a

29 (a) Janiak, C. (2000) A critical account on π–π stacking in metal complexes with aromatic nitrogen-containing ligands. *Dalton Trans*, 3885. (b) Mukhopadhyay, U., Choquesillo-Lazarte, D., Niclós-Gutierrez, J. and Bernal, I. (2004) A critical look on the nature of the intra-molecular interligand π,π-stacking interaction in mixed-ligand copper(II) complexes of aromatic side-chain amino acidates and α,α'-diimines. *Cryst Eng Comm*, **6**, 627.

30 Ziessel, R., Charbonnière, L., Cesario, M., Prangé, T. and Nierengarten, H. (2002) Assembly of a face-to-face tetranuclear copper(I) complex as a host for an anthracene guest. *Angew. Chem., Int. Ed.*, **41**, 975.

31 Chesnut, D.J., Kusnetzow, A., Birge, R.R. and Zubieta, J. (1999) Solid state coordination chemistry of the copper halide- and pseudo-halide-organoamine system, Cu-X-[(bis-2,3-(2-pyridyl) pyrazine)] (X = Cl, Br, CN): Hydrothermal synthesis and structural characterization. *Inorg. Chem.*, **38**, 2663–2671.

32 (a) Jung, O.-S., Park, S.H., Kim, Y.J., Lee, Y.-A., Jang, H.G. and Lee, U. (2001) Synthesis, structure, and thermal behavior of discrete Co(II), Ag(I), and Pd(II) complexes with 2,3-bis(2-pyridyl) quinoxaline. Insight into coordination modes. *Inorg. Chim. Acta*, **312**, 93–99. (b) Bu, X.-H., Liu, H., Du, M., Wong, K.M.-Ch., Yam, V.W.-W. and Shionoya, M. (2001) Novel boxlike dinuclear or chain polymeric silver(I) complexes with polypyridyl bridging ligands: Syntheses, crystal structures, and spectroscopic and electrochemical properties. *Inorg. Chem.*, **40**, 4143–4149.

33 (a) Some selected examples: Brandi, L., Schottel, J.B. and Dunbar, K.R. (2005) Anion dependence of Ag(I) reactions with 3,6-bis(2-pyridyl)-1,2,4,5-tetrazine (bptz): Isolation of the molecular propeller compound [Ag$_2$(bptz)$_3$][AsF$_6$]$_2$. *Chem. Commun.*, 46–47. (b) Reger, D.L., Collins, J.E., Rheingold, A.L., Liable-Sands, L.M. and Yap, G.P.A. (1997) Syntheses and characterization of cationic (tris(pyrazolyl) methane)silver(I) complexes. Solid-state structures of {[HC(3,5-Me$_2$pz)$_3$]$_2$Ag} (O$_3$SCF$_3$), {[HC(3-Bu^{t}pz)$_3$]Ag}(O$_3$SCF$_3$) and {[HC(3-Bu^{t}pz)$_3$]Ag(CNBu^t)}(O$_3$SCF$_3$).

Organometallics, **16**, 349–353. (c) Jørgensen, M. and Krebs, F.C. (2001) A new prearranged tripodant ligand N,N′,N″-trimethyl-N,N′,N″-tris(3-pyridyl)-1,3,5-benzene tricarboxamide is easily obtained via the N-methyl amide effect. *Tetrahedron Lett.*, **42**, 4717–4720. (d) Dong, G., Ke-Liang, P., Chun-Ying, D., Yong-Gang, Z. and Qing-Jin, M. (2002) Solid-state stabilization of discrete triple-stranded silver(I) helicates. *Chem. Lett.*, **31**, 1014.

34 Díaz-Ortiz, A., Elguero, J., Foces-Foces, C., de la Hoz, A., Moreno, A., Mateo, M.C., Sánchez-Mingallón, A. and Valiente, G. (2004) Green synthesis and self-association of 2,4-diamino-1,3,5-triazine derivatives. *New J. Chem.*, **28**, 952.

35 (a) Selected references are: Vázquez-Campos, S., Crego-Calama, M. and Reinhoudt, D.N. (2007) Supramolecular chirality of hydrogen-bonded rosette assemblies. *Supramol. Chem.*, **19**, 95. (b) Whitesides, G.M., Simanek, E.E., Mathias, J.P., Seto, C.T., Chin, D.N., Mammen, M. and Gordon, D.M. (1995) Noncovalent synthesis - using physical-organic chemistry to make aggregates. *Acc. Chem. Res.*, **28**, 37. (c) Sherrington, D.C. and Taskinen, K.A. (2001) Self-assembly in synthetic macromolecular systems *via* multiple hydrogen bonding interactions. *Chem. Soc. Rev.*, **30**, 83. (d) Prins, L.J., Huskens, J., de Jong, F., Timmerman, F.P. and Reinhoudt, D.N. (1999) Complete asymmetric induction of supramolecular chirality in a hydrogen-bonded assembly. *Nature*, **398**, 498. (e) Kerckhoffs, J.M.C.A., van Leeuwen, F.W.R., Spek, A.L. Kooijman, H., Crego-Calama, M. and Reinhoudt, D.N. (2003) Regulatory strategies in the complexation and release of a noncovalent guest trimer by a self-assembled molecular cage. *Angew. Chem., Int. Ed.*, **42**, 5717. (f) Kimizuka, N., Kawasaki, T., Hirata, K. and Kunitake, T. (1995) Tube-like nanostructures composed of networks of complementary hydrogen bonds. *J. Am. Chem. Soc.*, **117**, 6360. (g) Lange, R.F.M., Beijer, F.H., Sijbesma, R.P., Hooft, R.W.W., Kooijman, H., Spek, A.L., Kroon, J. and Meijer, E.W. (1997) Crystal engineering of melamine-imide complexes; tuning the stoichiometry by steric hindrance of the imide carbonyl groups. *Angew. Chem., Int. Ed. Engl.*, **36**, 969.

36 (a) Deak, A., Kalman, A., Parkamyi, L. and Haiduc, I. (2001) Hydrogen-bonded hexagonal and pseudo-hexagonal grid motifs in supramolecular cobalt(II) and nickel(II) cupferronato complexes incorporating neutral N-donors with intermolecular NH_2 connectors and solvent molecules. *Acta Crystallogr., Sect. B: Struct. Sci.*, **57**, 303. (b) Aoki, K., Inaba, M., Teratani, S., Yamazaki, H. and Miyashita, Y. (1994) Interligand interactions affecting site-specific metal bonding. X-ray crystal structures of $[Rh_2(acetato)_4(L)_2]$, where L = 2,4-diamino-6-methyl-s-triazine, 2,6-dimethyl-4-aminopyrimidine, and 2,6-dimethylpyridine. *Inorg. Chem.*, **33**, 3018.

4
Chiral Metallocycles for Asymmetric Catalysis
Wenbin Lin

4.1
Introduction

Mimicking the form and function of Nature's highly efficient biochemical machineries is one of the most important goals of contemporary synthetic chemistry [1,2]. Macrocyclic compounds are the simplest form of supramolecular system and have been extensively studied over the past three decades [3,4]. As covalent macrocycles are typically synthesized under high-dilution conditions to avoid the formation of undesired linear oligomers, large amounts of solvents are typically required for their synthesis, which presents a major burden to the environment. Large-scale synthesis of covalent macrocycles thus presents a significant challenge. Many applications of macrocyclic compounds have been hindered by their typically tedious, low-yield, stepwise synthesis based on covalent bond formation.

Seminal work by Fujita [5,6] and Stang [7,8] in the 1990s clearly demonstrated the ability to efficiently construct small molecular polygons based on metal–ligand coordination. The synthesis of metallocycles typically utilizes appropriate well-defined rigid and directional building blocks, which can be classified into two types – linear and angular subunits, to construct molecular polygons with high efficiency. Each subunit has two active functional end groups to interact with other building blocks. By using appropriate angular and linear building blocks, numerous small molecular polygons have been constructed via self-assembly processes. Figure 4.1 lists some of the combinations of angular and linear building blocks that can lead to different sizes and shapes of molecular polygons [9]. For example, a planar triangle can be assembled by combining three linear building subunits and three 60° angular ones. A molecular square can be assembled in several different ways, including a combination of four linear and four 90° angular building units and a combination of two different 90° angular subunits. Combining five linear subunits with five angular ones that possess a 109.5° angle between their binding sites will generate a molecular pentagon. These predictions of metallocycle sizes assume total conformational rigidity of subunits, but some distortion of the binding angle can occur. Metallocycles

Supramolecular Catalysis. Edited by Piet W. N. M. van Leeuwen
Copyright © 2008 WILEY-VCH Verlag GmbH & Co. KGaA, Weinheim
ISBN: 978-3-527-32191-9

4.2
Thermodynamically-Controlled Metallocycles

Most known metallocycles were synthesized under thermodynamic control and their structures could be readily rationalized as shown in Figure 4.1 [9]. For a metallocycle to be synthesized under thermodynamic control, three conditions must be met: (a) coordination bonds must form between the donor and acceptor elements involved, (b) the bonds must be kinetically labile to allow self-correction, and (c) the desired assembly must be thermodynamically favorable over competing species [10]. Thermodynamic factors in the self-assembly of macrocyclic compounds have been widely studied [11,12]. It is well-established that cyclic structures are preferred over linear ones for enthalpic reasons, while small cycles are favored over large cycles for entropic reasons. The enthalpic preference arises from the fact that an increased number of bonds per subunit are present in a cyclic arrangement relative to a linear one. For example, a square consisting of two angular units and two linear units contains four coordinate bonds, i.e., one bond per building block (Figure 4.2a). Its

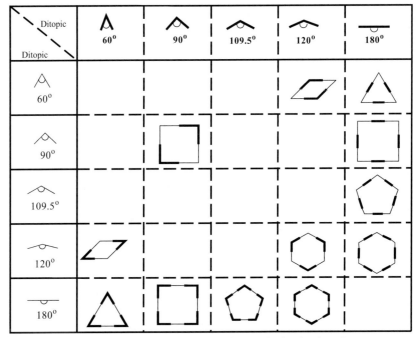

Figure 4.1 Schematic depiction of routes for the self-assembly of molecular polygons.

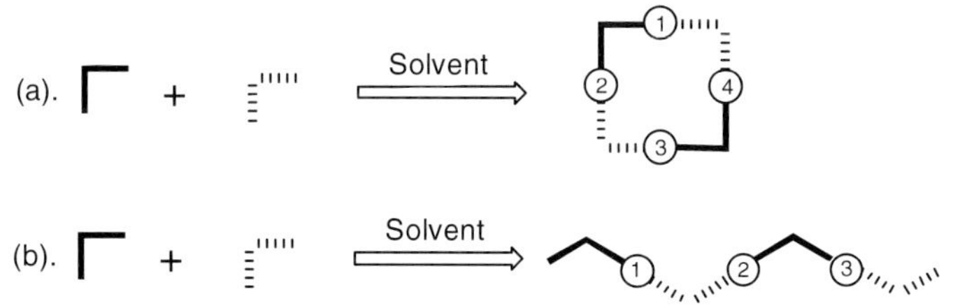

Figure 4.2 Schematic illustrating different number of bonds formed in cyclic vs. acyclic structures.

equivalent acyclic coordination oligomer is coordinatively unsaturated in one end and therefore contains only three bonds, i.e., 0.75 bonds per building block. (Figure 4.2b).

Acyclic oligomers can polymerize to increase the number of bonds formed. However, the donor and acceptor sites at each end of the polymer will always remain uncoordinated. Thus, the ideal scenario of one bond per building block cannot be achieved unless cyclization occurs. When cyclization is unfavorable, it becomes likely that the oligomers will polymerize until higher oligomers precipitate as kinetic products.

Small metallocycles are favored over larger metallocycles due to entropic factors. Considering the equilibrium between smaller cycles and larger cycles, the enthalpy ΔH is 0 when the steric effect and ring strains are ignored. Thus, the free energy of the equilibrium is $\Delta G = -T\Delta S$, which indicates that the equilibrium is dominated by the entropy. When the equilibrium shifts from smaller cycles to larger cycles, the entropy change is negative ($\Delta S < 0$). Thus the free energy in such an equilibrium shift is positive ($\Delta G > 0$), and the equilibrium shift from smaller cycles to larger cycles is disfavored. This is why there are many molecular triangles and squares reported in the literature and yet very few large metallocycles are known.

4.3
Kinetically-Controlled Metallocycles

The above discussion is based on the reversible reactions that produce the most thermodynamically stable products. In contrast, irreversible reactions produce kinetically-controlled products. In such cases, the self-correction mechanism is not operative. Well-defined building blocks with appropriate geometrical features are even more important for the synthesis of kinetically-controlled macrocycles. It is possible to obtain larger cycles because different sizes of linear species form in the reaction, which will cyclize to give a mixture of different sizes of cycles. Smaller cycles will still be favored over larger cycles in such kinetically-controlled reactions owing to kinetic factors.

Assuming a fundamental reaction, shown in Eq. (4.1), to generate the cyclic species $[LM]_n$ from ligand L and metal M, the reaction rate is $r = k[L]^n[M]^n$.

$$nL + nM \rightarrow [LM]_n \tag{4.1}$$

$$r = k[L]^n[M]^n$$

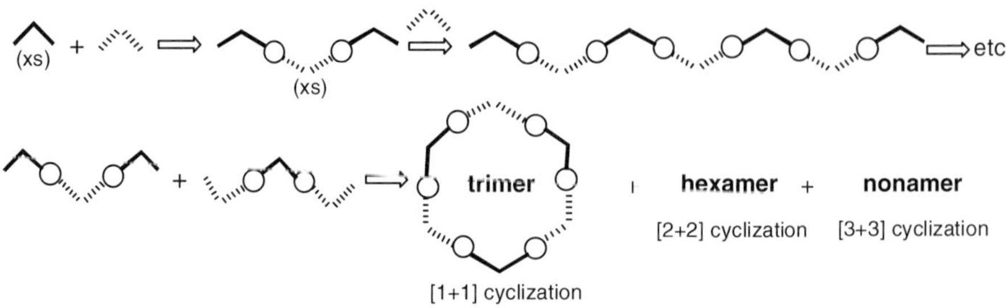

Figure 4.3 Schematic for directed assembly of large kinetically-controlled metallocycles.

where r is reaction rate, k is rate constant, [L] is the concentration of ligand L, and [M] is the concentration of metal M.

Since most cyclization reactions are carried out in highly dilute conditions, [L] and [M] are usually much smaller than 1 M. The reaction rate thus is faster for smaller cycles than larger cycles, and, as a result, higher yields of smaller cycles are typically obtained.

The preference for forming smaller metallocycles during kinetically-controlled syntheses can be overcome by using the so-called directed assembly strategy. As schematically shown in Figure 4.3, ligand-terminated acyclic oligomers can be synthesized by using a large excess of the ligand relative to the metal connecting unit via an iterative stepwise growth process. Under optimal conditions, the acyclic ligand-terminated acyclic oligomers can be purified by chromatography or crystallization. Metal-terminated acyclic oligomers can be synthesized using the same strategy. Combination of both ligand- and metal-terminated acyclic oligomers in a 1 : 1 molar ratio will lead to much larger metallocycles than those possible from the self-assembly of simple ligand and metal connecting units. In such directed assembly processes, metallocycles can only result from [1 + 1], [2 + 2], [3 + 3], or higher-order cyclization reactions. Our recent work has shown that metallocycles of unprecedentedly large sizes can be synthesized using this directed assembly strategy [13].

4.4
General Synthetic Strategies for Chiral Metallocycles

Compared to covalently-bonded organic counterparts, metallocycles can be assembled with much ease and higher efficiency by mixing appropriately-designed multitopic ligands and metal connectors with suitable binding geometries. Although numerous metallocycles have been reported over the past decade, few of them were designed with functionality (applications) in mind. There has been tremendous progress in the catalytic asymmetric synthesis of chiral compounds (both chemically and enzymatically) over the last two decades [14,15]. Chiral metallocycles are thus an interesting synthetic target owing to the prospect that well designed chiral metallocycles can possess enzyme-like chiral pockets and functionalities for applications in chiral sensing and asymmetric catalysis.

Chiral metallocycles have been constructed using three distinct strategies: (1) Introduction of metallocorners that are coordinated to chiral capping groups; (2) use of metal-based chirality owing to specific coordination arrangements; and (3) introduction of chiral bridging ligands. In the first approach, chiral metallocycles were prepared by attaching a chiral capping ligand to the metal centers. Stang et al. [16] used optically active metal complexes [M{(R)-BINAP}(OTf)$_2$] (M = Pd or Pt, BINAP = 2,2′-bis(diphenylphosphino)-1,1′-binaphthalene) as a chiral building block to construct chiral molecular squares (Figure 4.4). When combined with bis[4-(4-pyridyl)phenyl]iodonium triflate, heteronuclear optically active cyclic species **1a** (M = Pd) and **1b** (M = Pt) were obtained. These molecular squares are chiral due only to the chiral transition-metal auxiliary (BINAP) in the assembly. Both **1a** and **1b** possess D_2 symmetry, with one C_2 axis passing through the center of the binaphthyl rings and the other C_2 axis passing through the two iodine atoms.

Chiral tetranuclear molecular squares were readily synthesized when [M{(R)-BINAP}(OTf)$_2$] was treated with C_{2h}-symmetrical diaza ligands 2,6-diazaanthracene (DAA) in acetone. Only a single diastereomer was obtained, as evidenced by a single signal in the ^{31}P NMR spectrum. In contrast, when 2,6-diazaanthracene-9,10-dione (DAAD) was used in place of DAA (Figure 4.4), the reaction mixture consisted of a dominant diastereomer of **2a** (M = Pd) and **2b** (M = Pt) in a de of 81% and minor amounts of other diastereomers. These chiral molecular squares have also been characterized by electrospray mass spectrometry (ESI-MS). Stang et al. have also synthesized a family of interesting chiral molecular squares containing porphyrins by using trans-5,15-di(4-pyridyl)porphyrin (trans-DPyDPP) as the linear modules, and (R)-(+)-, (S)-(−)-BINAP-Pd(II) as the angular units [17]. Rotation of the metal—pyridyl bonds in **3** is restricted at room temperature, but becomes unrestricted at elevated temperatures. The chirality of the metallocorners thus promotes the formation of enantiomeric macrocycles with a puckered geometry.

In the second strategy, chiral metallocycles were assembled by taking advantage of metal-based chirality owing to specific coordination arrangements. Wang and coworkers investigated the assembly of chiral molecular squares using achiral building blocks [18]. Unlike the square-planar Pt(II) or Pd(II) complexes, octahedral Co(II) or Mn(II) complexes with two achiral chelating ligands can be made chiral. No chiral auxiliary ligands are required for the construction of chiral molecular squares using this synthetic strategy. For example, when the D_{2d} symmetric bridging ligand, tetraacetylethylene dianion (tae), was coordinated to octahedral Co(II) metallocorners that are capped with two chelating di-2-pyridylamine (dpa) ligands, the D_{2d} symmetry is reduced to pure rotational symmetries, and chiral molecular squares were obtained. Chiral molecular square Co$_4$(tae)$_4$(dpa)$_4$ (**4**) was obtained in 15% yield and characterized by single crystal X-ray crystallography (Figure 4.5). The synthesis of **4** demonstrates that ligands with a ∼90° twist between the binding sites can facilitate the formation of chiral molecular squares when octahedral metal ions are employed. Based on a related strategy, MacDonnell et al. synthesized chiral molecular hexagons by using octahedral complexes with propeller-like arrangements of three chelating ligands around a metal center. Chiral building blocks [(bpy)Ru(tpphz)$_2$]$^{2+}$ and [(bpy)Os(tpphz)$_2$]$^{2+}$ (where tpphz is tetrapyrido[3,2-a:2′,3′-c:3″,2″-h:2‴,3‴-j]phenazine)

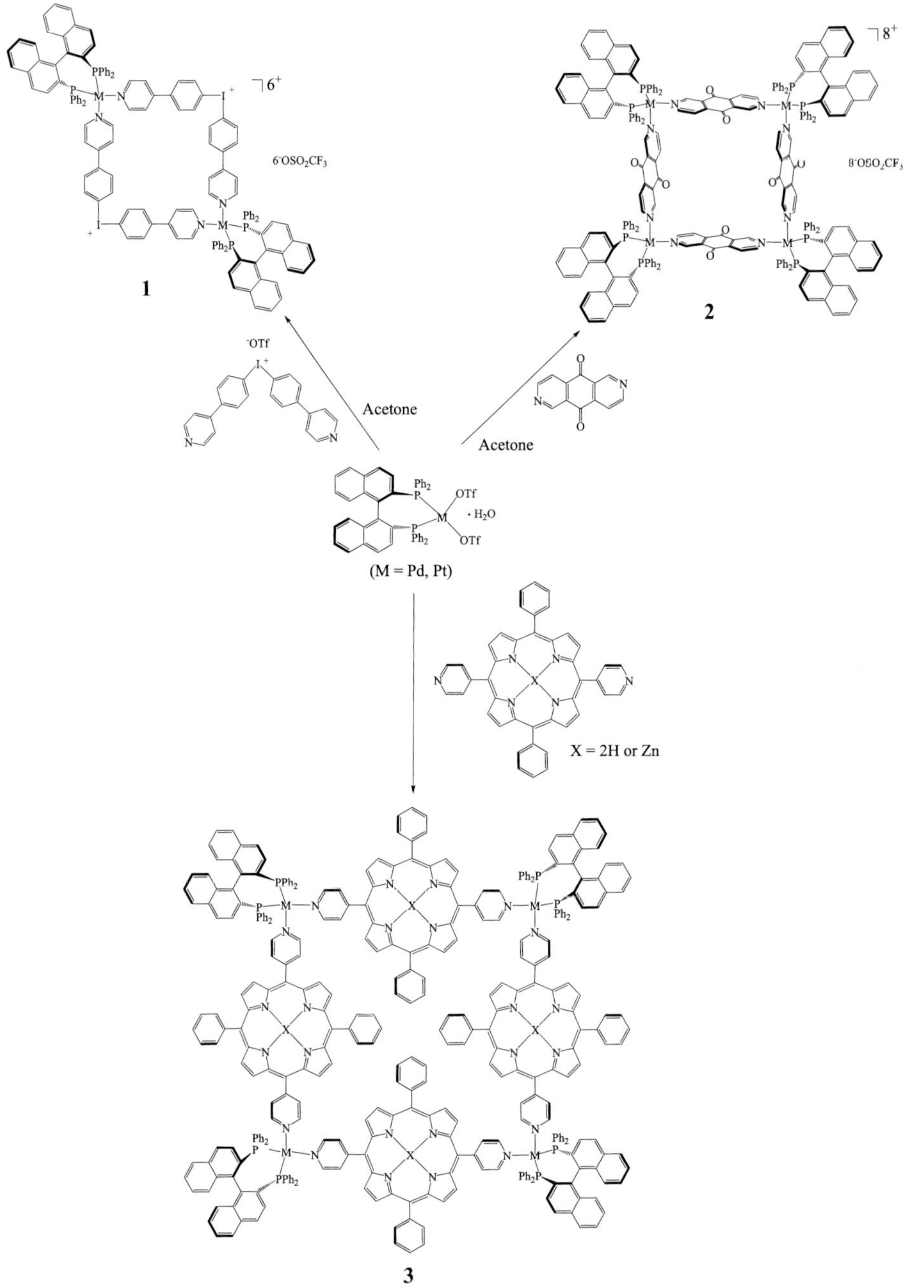

Figure 4.4 Synthesis of chiral metallocycles by attaching chiral capping ligands to the metal centers.

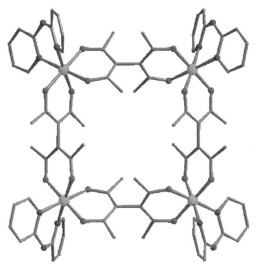

Figure 4.5 Single-crystal X-ray structure of $Co_4(tae)_4(dpa)_4$, **4**.

were connected by palladium metal centers to afford hexagonal molecules with cavities as large as 5.5 nm in diameter [19].

In the third approach, metallocorners are simply linked by chiral bridging ligands to form chiral metallocycles. Although this approach requires the most synthetic manipulations, it offers the versatility of fine-tuning the chiral pockets or functionalities for potential applications. Lin et al. assembled a family of novel chiral molecular squares using enantiopure linear 1,1′-binaphthyl-derived bipyridyl bridging ligands (**5a–d**) and *fac*-Re(CO)$_3$Cl corners (Figure 4.6). Based on their spectroscopic data, one single enantiomer formed during the assembly, possessing an approximate D_4 symmetry [20]. IR spectra of metallocycles **6a–d** exhibit three carbonyl stretches, which is consistent with the formation of the *fac*-[Cl(CO)$_3$Re] metallocorners that have local C_s symmetry. Available spectroscopic data cannot pinpoint the position of Cl atoms on the *fac*-Re(CO)$_3$Cl corners. Metallocycle **6d** exhibits interesting enantioselective luminescence quenching by chiral amino alcohols. (R)-**6d** has an enantioselectivity factor k_{sv}(R-S)/k_{sv}(R-R) of 1.22 for fluorescence quenching in favor of (S)-2-amino-1-propanol.

Lin et al. have also assembled D_4-symmetric chiral molecular squares based on [M(dppe)]$^{2+}$ metallocorners (M = Pd or Pt) and enantiopure angular 1,1′-binaphthyl-derived bipyridine bridging ligands (Figure 4.7). These chiral metallocycles were characterized by various analytical techniques, including IR, UV/Vis, circular dichroism (CD), and NMR spectroscopy, and ESI mass spectrometry [21]. All these chiral metallocycles are highly luminescent in solution at room temperature with quantum efficiencies of 0.06–0.63. Interestingly, mixing equal molar enantiopure molecular squares of opposite handedness in solution led to a new a meso dimeric metallocycle with C_2 symmetry. This indicates the lability of the M–pyridyl bonds in these metallocycles, which may hinder their applications in several areas that require integrity of the metallocycles.

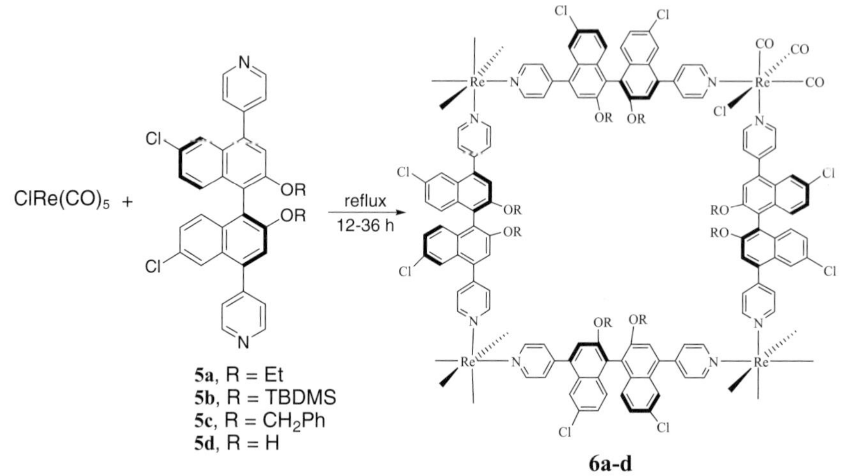

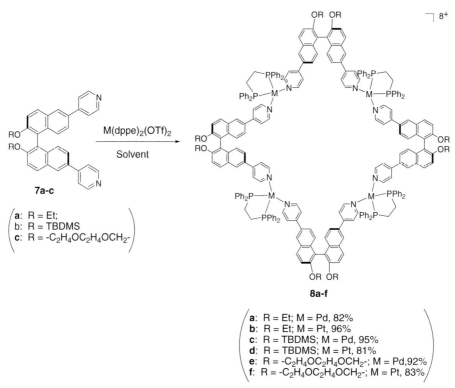

Figure 4.6 Synthesis of chiral molecular squares **6a–d** based on enantiopure linear 1,1′-binaphthyl-derived bipyridyl bridging ligands and *fac*-Re(CO)$_3$Cl corners.

Figure 4.7 Synthesis of chiral molecular squares **8a–f** based on enantiopure angular 1,1′-binaphthyl-derived bipyridyl bridging ligands and [M(dppe)]$^{2+}$ metallocorners.

4.5
Self- and Directed-Assembly of Chiral Pt-Alkynyl Metallocycles

To design chiral macrocycles exhibiting better stability and favorable solubility in nonpolar solvents, which are required for many asymmetric catalytic reactions, Lin et al. has taken advantage of the robust Pt-alkynyl linkage to construct a family of chiral metallocycles by both self- and directed-assembly routes. As shown in Figure 4.8, treatment of ligands **9a–c** with an equal molar equivalent of cis-PtCl$_2$(PEt$_3$)$_2$ in the presence of CuI catalyst at room temperature generated chiral molecular triangles **10a–c** in modest yields [22]. Chiral cycle **10d** was obtained quantitatively by treating **10c** with TBAF in THF. Compounds **10a–d** were characterized by ^{1}H, ^{13}C{^{1}H}, and ^{31}P{^{1}H} NMR spectroscopy, mass spectrometry, elemental analysis, and IR, UV/Vis, and circular dichroism (CD) spectroscopies. In particular, CD spectra of **10a–d** exhibit an intense band at ∼202 nm, in addition to three lower-energy bands corresponding to naphthyl $\pi \rightarrow \pi^*$ transitions and two acetylenic $\pi \rightarrow \pi^*$ transitions. The new CD band at ∼202 nm can be assigned to the transitions associated with cis-Pt(PEt$_3$)$_2$ moieties. This result suggests that triethylphosphines on the Pt centers adopt a propeller-type arrangement (relative to the naphthyl groups), apparently steered by chiral binaphthyl moieties. Cycles **10a–d** exhibit enhanced lower energy CD signals over the free ligands, which is consistent with the presence of multiple ligands in each metallocycle.

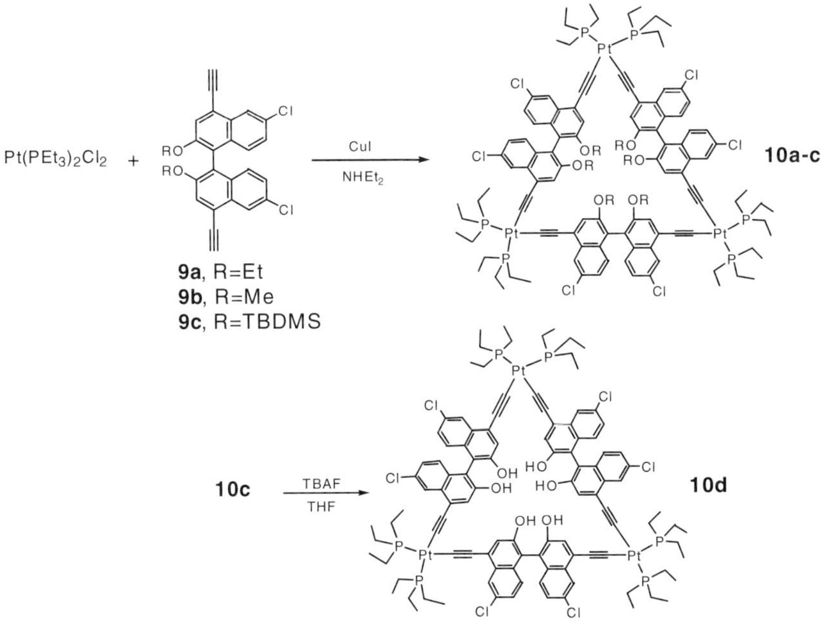

Figure 4.8 Self-assembly of Pt-alkynyl chiral molecular triangles.

Molecular triangles were obtained from the self-assembly linear ligands **9a–c** and the *cis*-[PtCl$_2$(PEt$_3$)$_2$] metallocorner, although molecular squares were the expected products based on the geometrical predictions shown in Figure 4.1. This result illustrates the kinetic preference of smaller metallocycles. Lin *et al.* were able to alter the reaction course and synthesized chiral molecular squares based on the same building blocks via stepwise directed-assembly processes [23]. The requisite monoprotected enantiopure bis(alkynes) **11a–c** were synthesized in ∼50% yield by treatment of their corresponding bis(alkynes) with one equiv of *n*-BuLi followed by bromotrimethylsilane at −78 °C (Figure 4.9). Treatment of ligands **11a–c** with 0.5 equiv of *cis*-[PtCl$_2$(PEt$_3$)$_2$] in the presence of CuCl catalyst in diethylamine at room temperature gave the Pt-containing intermediates, **12a–c**, in relatively high yield (74–89%), which were further deprotected by K$_2$CO$_3$ to afford the Pt-containing intermediates with terminal alkynes (**13a–c**). Chiral molecular squares **14a–c** were then prepared in 34–46% yield by treating **13a–c** with one equiv of *cis*-[PtCl$_2$(PEt$_3$)$_2$] in the presence of CuCl catalyst and diethylamine at low temperatures (0 to −20 °C). All these intermediates and chiral molecular squares were characterized by ^{1}H, ^{13}C{^{1}H}, and ^{31}P{^{1}H} NMR and IR spectroscopy and FAB mass spectrometry.

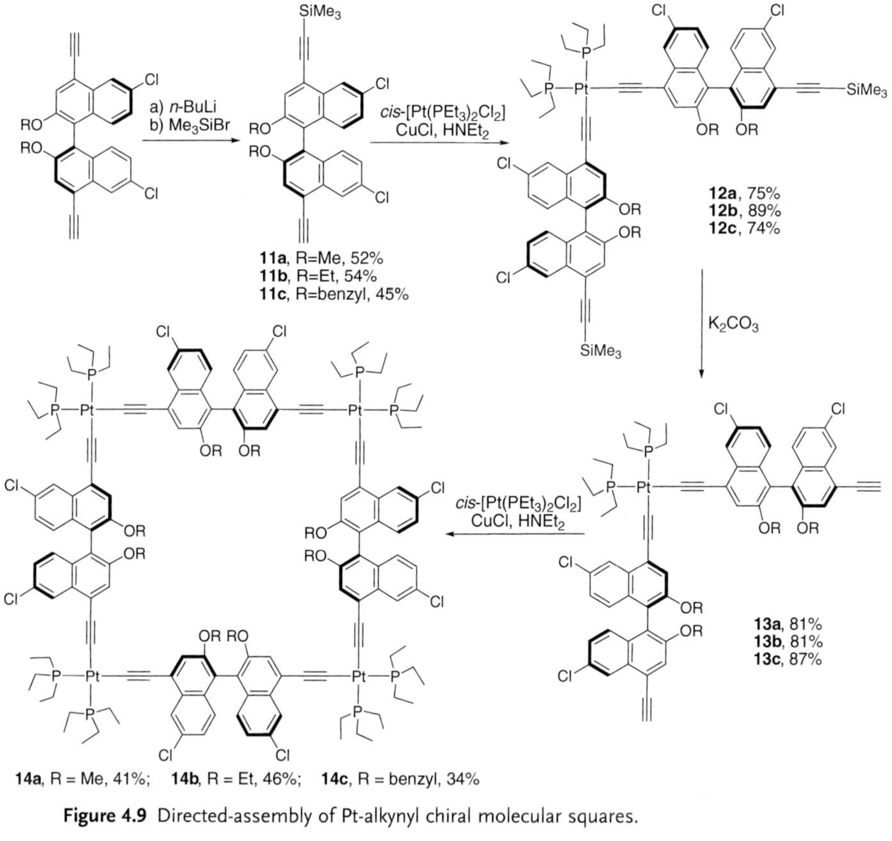

Figure 4.9 Directed-assembly of Pt-alkynyl chiral molecular squares.

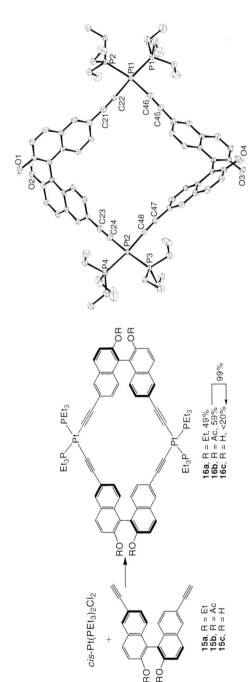

Figure 4.10 Self-assembly of Pt-alkynyl chiral metallacyclophanes.

Chiral dinuclear metallacyclophanes **16a–c** were obtained when enantiomerically pure angular bis(acetylenes) **15a–c** were treated with one equiv of cis-[Pt(PEt$_3$)$_2$Cl$_2$] [24]. Single-crystal X-ray diffraction studies showed that the two cis-Pt(PEt$_3$)$_2$ units were linked by two angular bis(acetylene) ligands to form a cyclic dinuclear structure **16c** (Figure 4.10). Both Pt centers adopted slightly distorted square planar geometry with the cis angles around the Pt1 center ranging from 82.4(2) to 101.3(1)° and the cis angles around the Pt2 center ranging from 84.3(2) to 100.3(1)°. The rigid metallacyclophane structure of **16c** was characterized by small dihedral angles between the naphthyl rings within each **15c** ligand (62.18 and 73.45°).

Interestingly, when topologically different 3,3'-bis(alkynyl)-1,1'-binaphthalene bridging ligands (**17a–c**) were used in place of their geometric isomers of **15a–c**, Lin et al. obtained chiral dinuclear metallacyclophanes **18a–c** that are supramolecular isomers of **16a–c** (Figure 4.11) [25]. Single-crystal X-ray structure determination showed that the asymmetric unit of **16c** contained two molecules of **16c** and one ethyl acetate solvent molecule. No crystallographic symmetry is present in molecules of **16c** in the solid state. Close proximity of the 1,1'-binaphthalene units is clearly evident

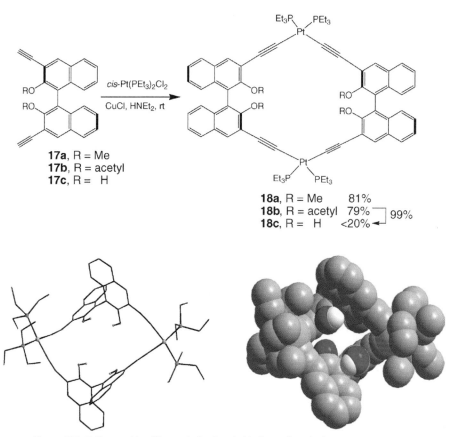

Figure 4.11 Self-assembly of isomeric Pt-alkynyl chiral metallacyclophanes.

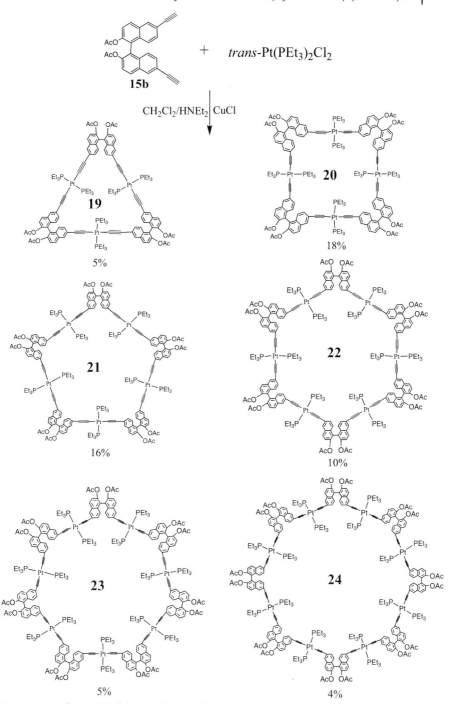

Figure 4.12 Self-assembly of chiral Pt-alkynyl molecular polygons.

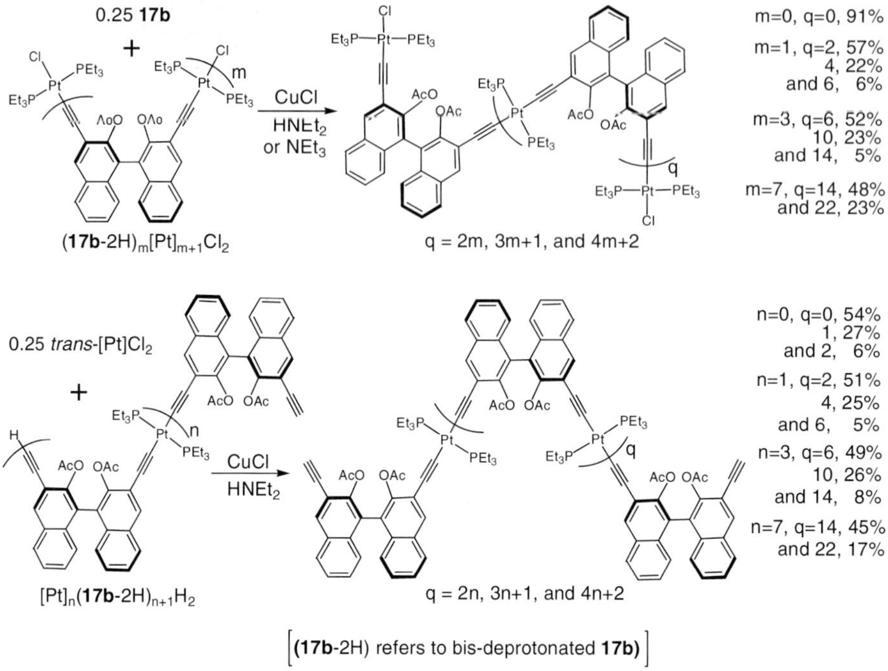

Figure 4.13 Directed-assembly of acyclic Pt-alkynyl building blocks.

in X-ray structure of **16c**, which prevents the reaction of **16c** with Ti(O^{i}Pr)$_4$ to generate the active catalyst for the addition of diethylzinc to aromatic aldehydes (see below).

When the linear trans-[Pt(PEt$_3$)$_2$Cl$_2$] was used to connect the bis(alkyne) ligand **15a–c**, a mixture of different sizes of chiral metallocycles [trans-(PEt$_3$)$_2$Pt(**15a–c**)]$_n$, ($n = 3$–8, **19–24**) was obtained [26]. Each of the chiral molecular polygons **19–24** was purified by silica-gel column chromatography and analytically pure **19–24** was obtained in yields (%) of 5, 18, 16, 10, 5, and 4, respectively (Figure 4.12). Compounds **19–24** have been characterized by ^{1}H, ^{13}C{^{1}H}, and ^{31}P{^{1}H} NMR spectroscopy, FAB and MALDI-TOF MS, and IR, UV/Vis, and circular dichroism (CD) spectroscopies, and microanalysis. The limited conformational flexibility of the bridging ligand is key to the facile one-pot self-assembly of chiral molecular polygons **19–24**.

This one-pot self-assembly strategy represents a rare example in which multiple products can be readily isolated from a coordination-directed self-assembly process, but is not amenable for the synthesis of even larger metallocycles. Lin et al. has developed a directed-assembly strategy for expeditious stepwise synthesis of chiral metallocycles containing as many as 47 metal centers by cyclization of metal- and ligand-terminated oligomers [27]. The requisite metal- and ligand-terminated oligomers were synthesized via an iterative process (Figure 4.13). These chiral metallocycles were unambiguously characterized by MALDI-TOF MS and resulted from [1 + 1], [2 + 2], and [3 + 3] cyclization process, respectively (Figure 4.14). Exceptionally large metallocycles have been efficiently synthesized by this process (Table 4.1). The largest metallocycle contains 47 [Pt] and 47 **17b** units, with an expected molecular

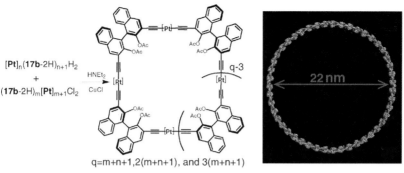

$[Pt]_n(17b-2H)_{n+1}H_2$
+
$(17b-2H)_m[Pt]_{m+1}Cl_2$

$\xrightarrow[CuCl]{HNEt_2}$

q-3

q=m+n+1, 2(m+n+1), and 3(m+n+1)

Figure 4.14 Directed assembly of unprecedentedly large chiral metallocycles.

Table 4.1 Efficient directed-assembly of exceptionally large chiral metallocycles (see Figure 4.14).

Reagents	Products (yield %)			Total yield (%)
	$(m+n+1)$	$2(m+n+1)$	$3(m+n+1)$	
$m=1, n=1$	25	45	15	85
$m=1, n=2$	59	20	8	87
$m=1, n=3$	77	9	2	88
$m=3, n=2$	74	15	/	89
$m=3, n=3$	66	9	/	75
$m=7, n=7$	62	7	/	69
$m=11, n=11$	63	/	/	63
$m=15, n=15$	57	/	/	57
$m=23, n=23$	10	/	/	10

weight of 39847.5 Da. Molecular mechanics simulations indicated that the internal cavities of these molecular polygons range from 0.9 to 22 nm. Supramolecular isomers of these exceptionally large metallocycles were similarly prepared when topologically different bis(alkynyl) bridging ligands **15a–c** were used in place of **17a–c**), illustrating the generality of this directed-assembly synthetic methodology [13,28].

4.6
Chiral Pt-Alkynyl Metallocycles for Asymmetric Catalysis

Molecular triangle **10d** contains chiral dihydroxy functionalities and has been used for highly enantioselective catalytic diethylzinc additions to aromatic aldehydes, affording chiral secondary alcohols upon hydrolytic work-up, as shown in Eq. (4.2) (Table 4.2) [22]. With Ti(IV) complexes of **10d** as catalyst, chiral secondary alcohols were obtained in >95% yield and 89–92% ee for a wide range of aromatic aldehydes with varying steric demands and electronic properties (Table 4.2). In comparison, when the free ligand 6,6'-dichloro-4,4'-diethynyl-2,2'-binaphthol was used instead of

4 Chiral Metallocycles for Asymmetric Catalysis

Table 4.2 Diethylzinc additions to aldehydes catalyzed by Ti(IV) complexes of **10d**.

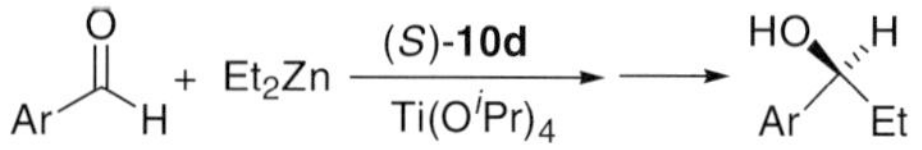

$$\text{Ar-CHO} + \text{Et}_2\text{Zn} \xrightarrow[\text{Ti(O}^i\text{Pr)}_4]{(S)\text{-10d}} \text{Ar-CH(OH)-Et} \quad \text{(Eq. 2)}$$

Entry	Aldehyde	Temp	Conversion	ee (%)
1	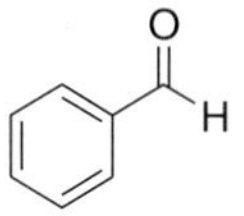 benzaldehyde	rt	>95%	91
2	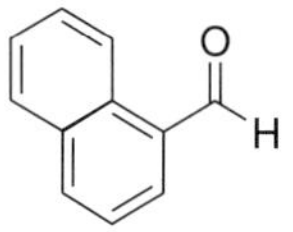 1-naphthaldehyde	rt 0 °C	>95% >95%	91 92
3	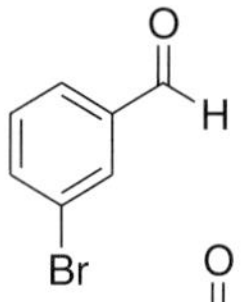 3-bromobenzaldehyde	rt	>95%	90
4	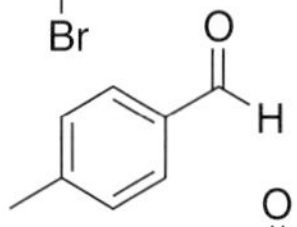 4-methylbenzaldehyde	rt	>95%	91
5	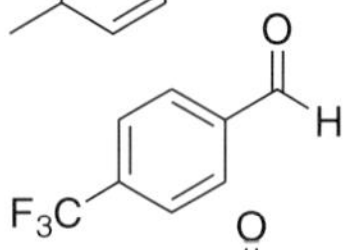 4-(trifluoromethyl)benzaldehyde	rt	>95%	89
6	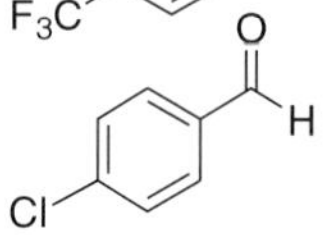 4-chlorobenzaldehyde	rt	>95%	90

10d, a lower ee (80%) was obtained for the addition of diethylzinc to 1-naphthaldehyde. The broad substrate scope for catalytic diethylzinc additions using **10d** and Ti(O^{i}Pr)$_4$ suggests that there is significant flexibility in the dihydroxy groups to accommodate aldehydes of various sizes.

As shown in Table 4.3, the Ti(IV) complexes of **16c** are excellent catalysts for the additions of diethylzinc to 1-naphthaldehyde, with 94% ee and >95% conversion at 0 °C [24]. The enantioselectivity, however, dropped significantly when other smaller aromatic aldehydes were used as the substrates. This result differs from the performance of **10d**, which has a very broad substrate scope. Such a difference is a direct consequence of the much more rigid structure of **16c**; the dihedral angles of naphthyl rings in the Ti(IV) catalyst can not vary to accommodate aldehydes of various sizes to give high enantioselectivity. The chiral dihydroxy groups in **16c** thus differ from those of BINOL, and may prove useful for mechanistic work owing to their rigid structure.

Interestingly, the steric congestion around the chiral dihydroxy groups in isomeric metallacyclophane **18c** prevented its reaction with Ti(O^{i}Pr)$_4$ to form active catalysts

Table 4.3 Diethylzinc additions to aldehydes catalyzed by Ti(IV) complexes of **16c** at 0 °C.

Aldehyde	Time (h)	Conversion	ee (%)
benzaldehyde	16	>95%	84
1-naphthaldehyde	16	>95%	94
3-bromobenzaldehyde	16	>95%	78
4-methylbenzaldehyde	40	~40%	78
4-(trifluoromethyl)benzaldehyde	40	~80%	77
4-chlorobenzaldehyde	16	>95%	78

for enantioselective diethylzinc additions to aromatic aldehydes. This result highlights the influence of supramolecular arrangement on not only stereoselectivity but also activity of asymmetric catalysts derived from metallocycles.

4.7
Concluding Remarks

The examples illustrated in this chapter testify to chemists' ability to create large chiral metallocycles that can potentially serve as mimics of natural enzyme systems. By judicious combination of multitopic bridging ligands and unsaturated metal centers, chiral metallocycles can be constructed with unprecedented predictability and ease. Preliminary results have demonstrated the influence of supramolecular arrangement on both activity and stereoselectivity of asymmetric catalysts derived from metallocycles. Future work will be directed towards the exploitation of supramolecular functions of chiral metallocycles in many areas of applications, including molecular recognition, host–guest interaction, chiral recognition, and catalysis. Such research efforts will lead to new materials with desirable tunable properties and ultimately to the demonstration of nanoscale devices and molecular machinery.

References

1 Swiegers, G.F. and Malefetse, T.J. (2000) New self-assembled structural motifs in coordination chemistry. *Chem. Rev.*, **100**, (9) 3483–3537.
2 Riley, D.P. (1999) Functional mimics of superoxide dismutase enzymes as therapeutic agents. *Chem. Rev.*, **99** (9), 2573–2588.
3 Peterson, C.J. (1967) Cyclic polyethers and their complexes with metal salts. *J. Am. Chem. Soc.*, **89** (26), 7017–7036.
4 Helgeson, R.C., Weisman, G.R., Toner, J.L., Tarnowski, T.L., Chao, Y., Mayer, J.M. and Cram, D.J. (1979) Host–guest complexation. 18. Effects on cation binding of convergent ligand sites appended to macrocyclic polyethers. *J. Am. Chem. Soc.*, **101** (17), 4928–4941.
5 Fujita, M. and Ogura, K. (1996) Transition-metal-directed assembly of well-defined organic architectures possessing large voids: from macrocycles to [2]catenanes. *Coord. Chem. Rev.*, **148**, 249–264.
6 Fujita, M. (1998) Metal-directed self-assembly of two- and three-dimensional synthetic receptors. *Chem. Soc. Rev.*, **27** (6), 417–425.
7 Stang, P.J. and Olenyuk, B. (1997) Self-assembly, symmetry, and molecular architecture: Coordination as the motif in the rational design of supramolecular metallacyclic polygons and polyhedra. *Acc. Chem. Res.*, **30** (12), 502–518.
8 Olenyuk, B. Fechtenköter, A. and Stang, P.J. (1998) Molecular architecture of cyclic nanostructures: use of coordination chemistry in the building of supermolecules with predefined geometric shapes. *J. Chem. Soc., Dalton Trans.*, 1707–1728.
9 Leininger, S., Olenyuk, B. and Stang, P.J. (2000) Self-assembly of discrete cyclic nanostructures mediated by transition metals. *Chem. Rev.*, **100** (3), 853–907.
10 Lawrence, D.S., Jiang, T. and Levett, M. (1995) *Chem. Rev.*, **95** (6), 2229–2260.
11 Chi, X., Guerin, A.J., Haycock, R.A., Hunter, C.A. and Sarson, L.D. (1995) The thermodynamics of self-assembly. *J. Chem. Soc., Chem. Commun.*, 2563–2565.
12 Chi, X., Guerin, A.J., Haycock, R.A., Hunter, C.A. and Sarson, L.D. (1995) Self-assembly of macrocyclic porphyrin oligomers. *J. Chem. Soc., Chem. Commun.*, 2567–2569.
13 Jiang, H. and Lin, W. (2006) Directed assembly of mesoscopic metallocycles with controllable size, chirality, and functionality based on the robust Pt-alkynyl linkage. *J. Am. Chem. Soc.*, **128** (34), 11286–11297.
14 Noyori, R. (2002) Catalysis: Science and opportunities (Nobel Lecture), *Angew. Chem., Int. Edn.*, **41** (12), 2008–2022.
15 Patel, R.N. (1999) *Stereoselective Biocatalysis*, (ed.) Marcel Dekker, New York.
16 Oenyuk, B., Whiteford, J.A. and Stang, P.J. (1996) Design and study of synthetic chiral nanoscopic assemblies. Preparation and characterization of optically active hybrid, iodonium-transition-metal and all-transition-metal macrocyclic molecular squares. *J. Am. Chem. Soc.*, **118** (35), 8221–8230.
17 Fan, J., Whitehold, J.A., Olenyuk, B., Levin, M.D., Stang, P.J. and Fleischer, E.B. (1999) Self-assembly of porphyrin arrays via coordination to transition metal bisphosphine complexes and the unique spectral properties of the product metallacyclic ensembles. *J. Am. Chem. Soc.*, **121** (12), 2741–2752.
18 Zhang, Y., Wang, S., Enright, G.D. and Breeze, S.R. (1998) Tetraacetylethane dianion (Tae) as a bridging ligand for molecular square complexes: $Co^{II}_4(Tae)_4(Dpa)_4$, Dpa = Di-2-pyridylamine, a chiral molecular square in the solid state. *J. Am. Chem. Soc.*, (36) **120**, 9398–9399.
19 Ali, M.M. and MacDonnell, F.M. (2000) Topospecific self-assembly of mixed-metal molecular hexagons with diameters of 5.5 nm using chiral control. *J. Am. Chem. Soc.*, **122** (46), 11527–11528.

20 Lee, S.J. and Lin, W. (2002) A chiral molecular square with metallo-corners for enantioselective sensing. *J. Am. Chem. Soc.*, **124** (17), 4554–4555.

21 Lee, S.J., Kim, J.S. and Lin, W. (2004) Chiral molecular squares based on angular bipyridines: Self-assembly, characterization, and photophysical properties. *Inorg. Chem.*, **43** (21), 6579–6588.

22 Lee, S.J., Hu, A. and Lin, W. (2002) The first chiral organometallic triangle for asymmetric catalysis. *J. Am. Chem. Soc.*, **124**, 12948–12949.

23 Lee, S.J., Luman, C.R., Castellano, F.N. and Lin, W. (2003) Directed assembly of chiral organometallic squares that exhibit dual luminescence. *Chem. Commun.*, 2124–2125.

24 Jiang, H., Hu, A. and Lin, W. (2003) A chiral metallacyclophane for asymmetric catalysis. *Chem. Commun.*, 96–97.

25 Jiang, H. and Lin, W. (2004) Chiral metallacyclophanes: Self-assembly, characterization, and application in asymmetric catalysis. *Org. Lett.*, **6** (6), 861–864.

26 Jiang, H. and Lin, W. (2003) Self-assembly of chiral molecular polygons. *J. Am. Chem. Soc.*, **125** (27), 8084–8085.

27 Jiang, H. and Lin, W. (2004) Expeditious assembly of mesoscopic metallocycles. *J. Am. Chem. Soc.*, **126** (24), 7426–7427.

28 Jiang, H. and Lin, W. (2005) Chiral molecular polygons based on the Pt-alkynyl linkage: Self-assembly, characterization, and functionalization. *J. Organomet. Chem.*, **690** (23), 5159–5169.

5
Catalysis of Acyl Transfer Processes by Crown-Ether Supported Alkaline-Earth Metal Ions

Roberta Cacciapaglia, Stefano Di Stefano, and Luigi Mandolini

5.1
Introduction

Supramolecular control of reactivity and catalysis is among the most important functions in supramolecular chemistry. Since catalysis arises from a differential binding between transition and reactant states, a supramolecular catalyst is, in essence, chemical machinery in which a fraction of the available binding energy arising from noncovalent interactions is utilized for specific stabilization of the transition state or, in other words, is transformed into catalysis.

This chapter describes the various ways that the rates of acyl transfer reactions can be enhanced in supramolecular complexes or by supramolecular catalysts in which the crucial ingredients are such simple and relatively featureless chemical species as alkaline-earth metal ions, mainly Sr^{2+} and Ba^{2+} ions, and occasionally Ca^{2+}.

Binding interactions available to these ions are essentially electrostatic in nature, namely, cation–anion attractions responsible for cation-pairing to counter-anions, and ion–dipole interactions that provide the basis for the well-known complexation with crown ethers.

5.2
Basic Facts and Concepts

Earlier work in this field has been thoroughly reviewed [1,2]. However, to illustrate in a sensible and logical way the evolution from simple metal ion promotion of acyl transfer in supramolecular complexes to supramolecular catalysts capable of turn-over catalysis, an account of earlier work is appropriate. The following sections present a brief overview of our earlier observations related to the influence of alkaline-earth metal ions and their complexes with crown ethers on the alcoholysis of esters and of activated amides under basic conditions.

Supramolecular Catalysis. Edited by Piet W. N. M. van Leeuwen
Copyright © 2008 WILEY-VCH Verlag GmbH & Co. KGaA, Weinheim
ISBN: 978-3-527-32191-9

5.2.1
Reactivity of Alkaline-Earth Metal Alkoxides

MeONMe$_4$-catalyzed methanolysis of phenyl acetate is accelerated by the addition of alkaline-earth metal salts. The kinetics are consistent with a reaction scheme [Eq. (1)] involving preassociation of MeO$^-$ ion with the metal cation (K_M) and independent contributions to the overall rate from free (k_o) and cation-paired (k_M) MeO$^-$ [3,4].

$$\text{MeO}^- + \text{M}^{2+} \xrightleftharpoons{K_M} (\text{MeOM})^+ \tag{1a}$$

$$\text{CH}_3\text{CO}_2\text{C}_6\text{H}_5 + \text{MeOH} \xrightarrow{\text{MeO}^-, k_o} \text{CH}_3\text{CO}_2\text{Me} + \text{C}_6\text{H}_5\text{OH} \tag{1b}$$

$$\text{CH}_3\text{CO}_2\text{C}_6\text{H}_5 + \text{MeOH} \xrightarrow{(\text{MeOM})^+, k_M} \text{CH}_3\text{CO}_2\text{Me} + \text{C}_6\text{H}_5\text{OH} \tag{1c}$$

Treatment of rate data according to a standard binding isotherm [4] gave $K_{Sr} = 60$ M^{-1}, $K_{Ba} = 44$ M^{-1}, $k_{Sr}/k_o = 4.7$, $k_{Ba}/k_o = 3.7$. Larger rate enhancements were observed in the EtONMe$_4$/EtOH base–solvent system [5,6]. The binding of EtO$^-$ to Sr^{2+} and Ba^{2+} is so strong ($K > 10^4$ M^{-1}) [5] that upon mixing equimolar amounts of EtONMe$_4$ and MBr$_2$ in the concentration range 1–10 mM a metal-bound ethoxide species is formed quantitatively [Eq. (2)].

$$\text{EtONMe}_4 + \text{MBr}_2 \rightleftharpoons \text{EtOMBr} + \text{Me}_4\text{NBr} \tag{2}$$

It is unknown whether and to what an extent the metal-ethoxide species contains a bromide counterion, but its kinetic behavior is that of a single species. Quite remarkably, cleavage of phenyl acetate is 62 and 45 times faster with EtOSrBr and EtOBaBr, respectively, than with EtONMe$_4$ [6]. The corresponding figures in the cleavage of p-nitrophenyl acetate are 8.0 and 7.0, respectively [5].

In view of the widespread notion that ion pairs are less reactive than free anions [7], the very finding that metal-bound alkoxides cleave esters more rapidly than free alkoxides do came as a surprise. It seems likely that transition state stabilization takes place via chelate structures having the form of a four-membered contact ion pair **I**, or of the kinetically equivalent six-membered solvent shared ion pairs **II** and **III**.

Whatever the detailed structure of the transition state, the important conclusion is reached that such a simple species as an alkaline-earth metal ion can provide the reacting carbonyl with electrophilic assistance, which is tantamount to saying that interaction of the metal ion with the dispersed negative charge in the transition state is stronger than interaction with the localized negative charge of the alkoxide reactant. Thus, cation–anion electrostatic attraction is an effective force for catalysis. Consistently, rate enhancements in the less polar EtOH are much larger than in the more polar MeOH, and become either much smaller or even negligible in the presence of alkali metal ions [4,8]. Furthermore, the presence of a strong electron-withdrawing group such as in p-nitrophenyl acetate strongly decreases the need for electrophilic assistance.

5.2.2
The Influence of Crown Ethers

A selection of the results obtained in an extensive investigation of the influence of cation-complexing agents on the barium-assisted basic ethanolysis of phenyl acetate [9] are shown in the upper part of Table 5.1. Interestingly, the statistically corrected reactivity of $(EtO)_2Ba$ is very similar to that of EtOBaSCN, showing that the behavior of the metal bound ethoxide in EtOMX is not dramatically affected by the nature of X. The most surprising result is, however, that the reactivity of the 18C6-complexed ion pair is significantly higher than that of the uncomplexed ion pair. This is clearly at variance with the common notion that addition of a metal ion complexing agent is expected to cause a rate decrease whenever the metal ion plays a role as electrophilic catalyst [10–12].

Even more surprising is the finding that the reactivity picture exhibited by the various ethoxide species in the ethanolysis of phenyl acetate is very similar to that found in the ethanolysis of the activated amide N-methyl-2,2,2-trifluoroacetanilide [Eq. (3)], despite the different mechanisms of the two reactions [13]. Schowen et al. [14] showed that C–N bond breaking in the rate-limiting decomposition of the tetrahedral intermediate is assisted by proton transfer from a general acid to the leaving group.

$$CF_3C(O)-N(CH_3)-C_6H_5 + EtOH \xrightarrow{EtONMe_4} CF_3CO_2Et + C_6H_5NHCH_3 \quad (3)$$

Table 5.1 Ethanolysis of phenyl acetate and N-methyltrifluoroacetanilide at 25 °C; relative reactivity of various metal-bound species.[a]

	EtO⁻	⁻OEt⋯Ba²⁺	⁻OEt⋯Ba²⁺⋯OEt⁻	18C6⋅Ba²⁺(OEt)₂
$CH_3CO_2C_6H_5$	1[b]	33	26[c]	60
$CF_3C(O)N(CH_3)C_6H_5$	1[d]	15[e]	11[c]	55[e]

[a] Source of barium ion is $Ba(SCN)_2$.
[b] Absolute reactivity: $k_2 = 1.44\ M^{-1} s^{-1}$.
[c] Statistically corrected.
[d] Absolute reactivity: $k_2 = 0.051\ M^{-1} s^{-1}$.
[e] Relative reactivity of $(EtOSr)^+$ is 19, and that of its 18C6 complex is 150.

Accordingly, the simplest mechanism that accounts for the enhanced ethanolysis rates in the presence of metal ions is shown in **IV**, where a metal-bound ethanol molecule, made acidic by metal coordination, serves as a general acid catalyst for C–N bond cleavage. Here again enhanced reactivity is exhibited by the 18C6-complexed metal ethoxide species [13].

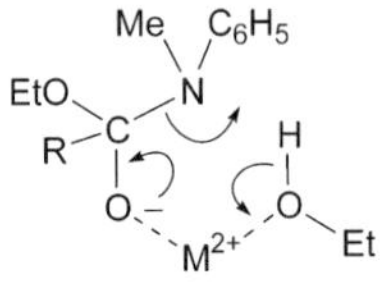

IV

The findings that, both in ester and amide cleavage, an alkaline-earth metal ion is still catalytically active when complexed with a crown ether, and that a fraction of the binding energy made available by coordinative interactions with the polyether chain can be translated into catalysis, provide the basis for the construction of supramolecular catalysts capable of esterase and amidase activity.

5.2.3
Preorganized Systems

Scheme 5.1 depicts the various strategies followed in the design of supramolecular complexes endowed with enhanced reactivity in transacylation processes.

It was Swain [15] who pointed out more than half a century ago that when a reaction between a nucleophile N and a substrate S is catalyzed by an electrophile E (case a), enhanced catalysis is expected if two of the components, either N and S (case b), S and E (case c), or N and E (case d) are bound together in the same molecule or in the same complex by covalent bonding or noncovalent interactions, respectively. An additional possibility is given by case e, where the two reactants and catalyst are held together in a ternary complex.

5.2.3.1 Selected Examples
Several earlier studies [3–6,8] clearly showed that enhanced metal ion effects on rates of ester cleavage are obtained in model substrates in which a covalently linked polyether chain holds the metal ion in close proximity to the ester function undergoing nucleophilic attack.

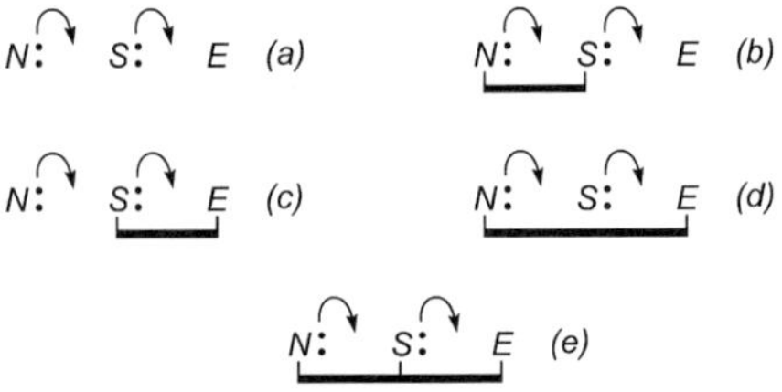

Scheme 5.1

Table 5.2 Metal-ion promoted basic alcoholysis of aryl acetates at 25 °C. Rate enhancements (k_M/k_o).

			k_{Ba}/k_o	k_{Sr}/k_o
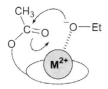 n = 4 2-AcO-15C4 n = 5 2-AcO-18C5 n = 6 2-AcO-21C6	2-AcO-15C4	(MeOH)	93	40
	2-AcO-15C4	(EtOH)	7200	2300
	2-AcO-18C5	(MeOH)	760	230
	2-AcO-18C5	(EtOH)	46 000	48 000
	2-AcO-21C6	(MeOH)	250	190
	2-AcO-21C6	(EtOH)	500 000	41 000

The sample kinetic data listed in Table 5.2 shows that the size of rate enhancement critically depends on the substrate–metal ion combination, and is markedly influenced by the solvent. The largest effect is displayed by 2-AcO-21C6, which reacts with EtOBaBr half a million times faster than with EtONMe$_4$. The conclusion was reached [6] that the huge rate enhancements observed in the ethanolysis reactions are a consequence of the fact that not only cation–anion electrostatic binding but also coordinative binding to the polyether chain in the metal-bound transition states are much more efficient in EtOH than in MeOH.

Not surprisingly, maximum rate enhancements of methanolysis caused by the addition of alkali metal ions from Na to Cs are much lower, from 2- to 20-fold. Given that monovalent cations bind significantly to the crown ether substrates, but negligibly so to the methoxide ion, the kinetics are consistent with a reaction scheme in which a free nucleophile reacts with a 1:1 complex of substrate and metal electrophile (Scheme 5.1, type c). The situation is more complex with the divalent metal ions, because association with methoxide is also significant. In terms of classical kinetics, there is no way to distinguish a mechanism in which a cation-paired nucleophile reacts with a free substrate (Scheme 5.1, type d) from one in which a cation-bound substrate reacts with a free methoxide (Scheme 5.1, type c). This mechanistic ambiguity can be overcome, at least conceptually, if one assumes that in all cases the reactions proceed through a ternary complex **V**, which is formed from reactants and metal ion with no obligatory order of combination. In the language of the enzyme kineticist, this is a random sequential mechanism [16]. Kinetic evidence was obtained that reactions in the EtONMe$_4$/EtOH base solvent system actually proceed through ternary complexes that have the same composition as the rate-limiting transition states, and decompose into products in a monomolecular step.

V

An example of a more structured aryl acetate is offered by the monoacetyl derivative **2** of *p-tert*-butylcalix[4]arene-crown-5 (**1**). The half-life for deacetylation in the presence of 1 mM MeONMe$_4$ in MeOH at 25 °C is 34 weeks, but drops to 8 s upon addition of 1 mM BaBr$_2$ [17]. Also, SrBr$_2$ accelerates the reaction, albeit to a somewhat smaller extent. Kinetic and UV spectroscopic data show that in the absence of metal ions MeO$^-$ reacts with the unionized form **2**, but in the presence of metal ions the reactive species is the metal complex of the ionized form **3**, which is present in significant amounts by virtue of the acidity enhancing effect of the metal ion complexed by the polyether chain.

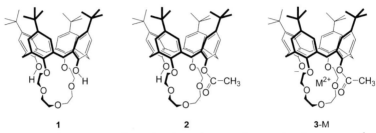

Rate enhancements brought about by metal ions, $k_{Sr}/k_o = 1.2 \times 10^6$ and $k_{Ba}/k_o = 2.1 \times 10^7$ translate into transition state stabilization of 8.3 and 10.0 kcal mol^{-1}, respectively. To the best of our knowledge, these are the most striking examples of electrophilic catalysis by Sr^{2+} and Ba^{2+} ions.

In conclusion, the data reported in this section illustrate the notion that transition state stabilization by alkaline-earth metal ions arises from cooperation of electrostatic binding to the negative charge being transferred from the incoming nucleophile to the carbonyl oxygen, and coordinative interactions with the oxygen donors. It appears, therefore, that the polyether chain serves to increase, even to a very significant extent, the available binding energy and, consequently, the portion of this binding energy that can be utilized in catalysis.

5.3
Nucleophilic Catalysts with Transacylase Activity

Host structures that, like many hydrolytic enzymes [16], contain an OH/SH group that acts as acyl-receiving/acyl-releasing unit have been reported by several authors [18]. Such host structures are acylated by suitably functionalized esters at enhanced rates thanks to the presence of tailored binding sites proximal to the nucleophilic group. However, since no provision was made to activate the cleavage of the acylated host, many of the investigated systems showed little or no turnover capability.

Scheme 5.2 outlines our design of nucleophilic transacylation catalysts based on crown-complexed alkaline-earth metal ions. By virtue of the acidity-enhancing effect of the complexed metal ion, dissociation of the proton-ionizable function XH should take place under moderately basic conditions. The metal ion assists acyl transfer from a reactant ester to the catalyst and its subsequent transfer from the acylated catalyst to an external nucleophile (solvent), thus restoring the active form of the

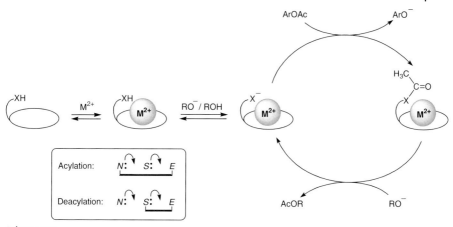

Scheme 5.2

catalyst and turning over. The catalytic cycle closely resembles the double-displacement or ping-pong mechanism of enzyme kinetics [16], with the difference that, unlike the natural enzymes, no binding site is present in the catalyst for substrate recognition.

5.3.1
Calixcrowns

A first example of nucleophilic catalyst with transacylase activity is given by an equimolar mixture of a Ba^{2+} salt (either bromide or perchlorate) and p-tert-butylcalix[4]arene-crown-5 (**1**), which catalyzes the methanolysis of aryl acetates in MeCN–MeOH (1:1, v/v) under slightly basic conditions (3:1 diisopropylethylamine-perchlorate salt buffer) at 25°C [19,20].

Calixcrown **1** remains in its unionized form in the given buffer solution, but upon addition of 1 molar equivalent of Ba^{2+} salt it is transformed into a mixture of the Ba^{2+} complexes of the singly and doubly ionized forms. There is evidence that the Ba^{2+} complex of the doubly ionized form **4-Ba** is the active form of the catalyst.

Figure 5.1 shows the spectrophotometrically determined liberation of p-nitrophenol (pNPOH) from pNPOAc in the absence (curve a) and presence (curve b) of metal catalyst. The catalyzed reaction exhibits biphasic kinetics, characterized by an initial "burst" of pNPOH release, followed by a slower linear phase.

Such behavior is consistent with the double-displacement mechanism of Scheme 5.2, in which fast acylation of the catalyst (cat) [Eq. (4)] is followed by slower deacylation of the acylated form catAc, [Eq. (5)].

$$cat + ArOAc \xrightarrow{k_1} catAc + ArOH \qquad (4)$$

$$catAc \xrightarrow[\text{AcOMe}]{MeO^-/MeOH,\ k_2} cat \qquad (5)$$

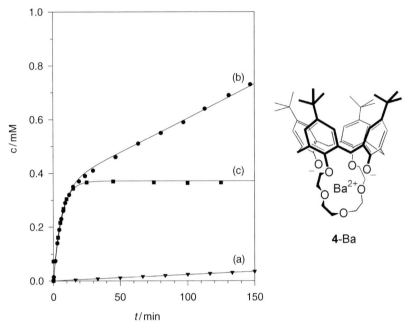

Figure 5.1 Methanolysis of 3 mM pNPOAc in buffered acetonitrile-methanol (9:1), 25 °C. Liberation of pNPOH (UV/Vis) in the presence of buffer alone (curve a) and of buffer plus 0.39 mM **1** and 0.39 mM Ba(ClO$_4$)$_2$ (curve b). Curve c is a plot of [catAc] (HPLC) obtained under identical experimental conditions.

The time-dependent concentration of pNPOH was analyzed on the basis of Eq. (6), which is the sum of an exponential (pre-steady state) phase, characterized by a first-order rate constant k, and a linear (steady state) phase obtained when the exponential term dies out ($kt > 5$). The equation reduces to the form of Eq. (7). The intercept of the linear portion defines the burst π of ArOH liberation.*

$$[\text{ArOH}] = \pi[1 - \exp(-kt)] + st \qquad (6)$$

$$[\text{ArOH}] = \pi + st \qquad (7)$$

As shown in Figure 5.1, there is an excellent agreement between data points and the solid line calculated from Eq. (6) and the best fit parameters $k = k_1 [\text{ArOAc}] = 0.15\,\text{min}^{-1}$ and $k_2 = s/[\text{cat}]_o = 7.0 \times 10^{-3}\,\text{min}^{-1}$. The value of the initial burst, $\pi = 0.35$ mM, corresponds to 90% of the initial catalyst concentration, $[\text{cat}]_o$.

A pure sample of **2** undergoes methanolysis under conditions identical to those of the catalytic experiment with a first-order rate constant of $7.7 \times 10^{-3}\,\text{min}^{-1}$, in good agreement with the value obtained from the two-parameter treatment of the kinetics

---

*A more extended form of Eq. (6) is given below, where $\tau = (k + k_2)^{-1}$ (see Ref. [16]): $[\text{ArOH}] = [\text{cat}]_o \tau k\{\tau k[1 - \exp(-t/\tau)] + k_2 t\}$. When the exponential term dies out, the equation reduces to the simple form $[\text{ArOH}] = [\text{cat}]_o (\tau k)^2 + [\text{cat}]_o \tau k k_2 t$, whose simplified expression is given by Eq. (7).

of pNPOH liberation. Furthermore, definite confirmation of the intermediacy of the monoacetylated form of the catalyst is obtained from HPLC analysis of acid-quenched samples of the reaction mixture (Figure 5.1, curve c). The time–concentration profile shows that **2** builds up until a steady concentration is reached, corresponding to 95% of $[cat]_o$. Kinetic analysis of the concentration data [20] gives $k = 0.16\,min^{-1}$ and $k_2 = 7.7\times10^{-3}\,min^{-1}$, in excellent agreement with the values obtained from independent experiments. Thus, the initial burst of pNPOH release in the 4-Ba catalyzed methanolysis of pNPOAc corresponds to a nearly complete accumulation of the acetylated intermediate, and closely resembles analogous phenomena occurring in the cleavage of p-nitrophenyl esters catalyzed by proteolytic enzymes such as chymotrypsin [21]. The linear phase is a consequence of an exact balance between formation and destruction of the acylated intermediate.

In the catalytically active complex **4-Ba** the negative poles and the polyether bridge act as working units that perform cooperatively in providing the driving force for the formation of the complex itself, whereas the metal ion serves as an electrophilic catalyst both in the acylation and deacylation steps. The crucial importance of the polyether bridge is demonstrated by the disappearance of any catalytic activity upon replacement by two methoxy groups.

The Sr^{2+} complex of **1** was also tested for catalytic activity and found to be slightly less effective than the Ba^{2+} complex. As found in many other instances, the Ca^{2+} complex could not be tested because of solubility problems.

5.3.1.1 Catalytic Efficiency vs. Ester Reactivity

The rates of acetylation of **4-Ba** (Figure 5.1) by a series of aryl acetates, including, in addition to pNPOAc and the parent phenyl acetate (POAc), the *meta*-nitro (mNPOAc) and *para*-chloro (pCPOAc) derivatives, is influenced by the nature of the substituents in such a way that there is a changeover from rate-determining deacetylation to rate-determining acetylation on going from the most reactive pNPOAc to the least reactive POAc (Table 5.3). A major reason for this is the high sensitivity exhibited by the acetylation step to the structure of the ester. By way of example the pNPOAc/POAc ratio is 500 for the acetylation step, but only 24 for the background reaction. Such a high sensitivity provides a strong indication that the slow step in the acetylation of the catalyst is the breakdown of the tetrahedral intermediate [20].

Table 5.3 Methanolysis of aryl acetates in the presence of 4-Ba. Rate acceleration relative to background reaction.[a]

Substrate	Pre-steady state (initial rate)	Steady-state	Slow step
pNPOAc	250	12	Deacylation
mNPOAc	80	19	⇓
pCPOAc	52	41	⇓
POAc	12	12	Acylation

[a] [Substrate] = 3 mM, [cat] = 0.4 mM.

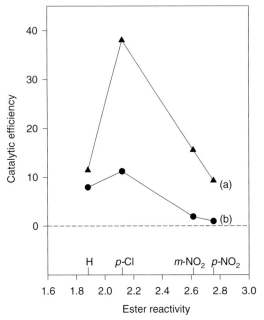

Figure 5.2 Catalytic efficiency of **4**-Ba in the methanolysis of aryl acetates calculated for a substrate-to-catalyst ratio of 10:1 (curve a) and 100:1 (curve b) versus ester reactivity (as measured by the log k_{bg} values).

The catalytic efficiency, as conveniently measured by the ratio of the catalyzed process under steady state conditions to the rate of background methanolysis (Figure 5.2), is a function of the substrate-to-catalyst ratio and reaches maximum values in the reaction of pCPOAc.

It is apparent that an increase in the substrate-to-catalyst ratio dramatically decreases the catalytic efficiency for the pNPOAc reactions, but affects to a much lower extent the POAc reaction. This is easily understood with reference to Table 5.3. Since in the reaction of pNPOAc the rate-determining step is mainly deacetylation, an increase in ester concentration causes a proportional increase in the rate of background methanolysis, but hardly affects the rate of deacetylation, with the result that catalytic efficiency varies inversely to ester concentration. Conversely, the reaction of POAc approaches a situation in which acetylation of the catalyst is rate determining, which implies that both acetylation and background reactions increase on increasing ester concentration.

Extrapolation of the above considerations to acetate esters less reactive than POAc, i.e., with leaving groups more basic than PO$^-$, leads to the prediction of lower catalytic efficiencies because the rate of acetylation of the catalyst is expected to decrease more rapidly than the rate of background methanolysis. In contrast, with esters of phenols more acidic than pNPOH, the catalytic reaction is predicted to be obscured by increased rates of background methanolysis. Thus, for one reason or another, the scope of **4-Ba** as catalyst of ester methanolysis is restricted to acetate esters whose reactivity lies in the range approximately defined by the POAc–pNPOAc pair.

5.3.1.2 Trifunctional Catalysis

Calixcrown **5**, featuring two diethylaminomethyl side-arms at the polyether bridge, testifies an attempt at a higher order multifunctional catalysis of ester cleavage, namely, from nucleophilic-electrophilic to nucleophilic-electrophilic-general acid catalysis [20].

As shown in the preceding section, acetylation of the catalyst becomes critically slow for esters of the less acidic phenols, most likely because decomposition of the tetrahedral intermediate is rate determining. The idea underlying the choice of **5**, whose C_2 symmetry avoids any complications arising from non-equivalence of the two faces, is that the monoprotonated form of the barium salt **6-Ba** could possibly exhibit enhanced acetylation rates due to general acid assistance to the departure of the aryl oxide leaving group. The results of the kinetic experiments are disappointing, in that the rate of acetylation of **6-Ba** by pNPOAc is 1/9th of that of **4-Ba**, and the corresponding figure with POAc is $1/2$. The results are explained by assuming that acetylation is hindered by the sterically demanding diethylaminomethyl side-arm. However, the fact that the two esters respond quite differently to the presence of the side-arm is taken as a circumstantial evidence that the protonated nitrogen actually assists departure of the worse leaving group PO^-, but it assist much less effectively or even to a negligible extent the departure of the better leaving group $pNPO^-$. In other words, the likely advantage of the additional catalytic group is overshadowed by the steric hindrance of the side-arm bearing the catalytic group itself.

5   **6H$^\pm$-Ba**

The lesson to be learned by the above experiments is that designed multifunctional catalysts may lead to unsuccessful results, and awareness of imperfections and drawbacks in the design comes only *a posteriori*. Progress towards a higher order multifunctional catalysis requires a most careful design of intra- and intermolecular interactions for optimal positioning of catalytic units in a molecular framework.

5.3.1.3 *p-tert*-Butylcalix[5]arene Derivatives

With the idea that crown ethers based on the *p-tert*-butylcalix[5]arene platform could provide an interesting extension of our catalytic studies, several calix[5]-crown-ethers were investigated as potential catalysts of ester methanolysis in the presence of a Ba^{2+} salt [22]. Of the various structures investigated, the calixcrown-5 derivative **7** gave the best results. The spectrophotometrically determined liberation of pNPOH from pNPOAc under typical conditions of our catalytic experiments shows biphasic kinetics, characterized by an initial burst of pNPOH release, followed by a slower steady-state phase. The rate of production of pNPOH in the exponential phase is 0.4 as that observed with 4-Ba under identical conditions. Combined HPLC and HPLC-ESMS analyses of the reaction progress show the quantitative conversion of 7 into a single monoacetylated derivative, which could be isolated from the reaction mixture and was shown to be the symmetrical product derived from acetylation of the isolated hydroxyl of ring 2. Analogous results were obtained with the calixcrown-6 homologue of 7.

Analysis of the kinetic data shows that the barium salt of 7, as well as the analogous salts of its higher homologues, perform much less efficiently than 4-Ba. The Ba complex of 7 turns over with a very low efficiency, caused by the extreme slowness of the deacylation step. Only a minor fraction of the liberated pNPOH in the steady-state phase is due to the expected double displacement mechanism. A larger fraction is most likely ascribable to the metal ion not sequestered by 7, and thereby available in solution for electrophilic assistance to direct methanolysis of the ester reactant.

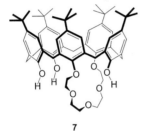

7

5.3.2
Thiol-Pendant Crown Ethers

Additional examples of type *d* (Scheme 5.1) bifunctional reactants are provided by the alkaline-earth metal ion complexes of lariat ethers 8–10, bearing a sulfhydryl side arm, instead of the phenolic hydroxyl of a calixcrown [23,24]. Here the acyl-receiving and acyl-releasing unit, like in papain and ficin, is a sulfhydryl group.

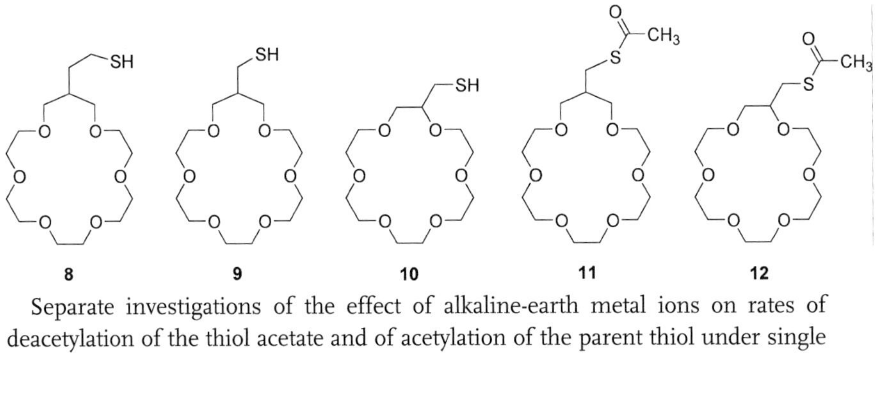

8   9   10   11   12

Separate investigations of the effect of alkaline-earth metal ions on rates of deacetylation of the thiol acetate and of acetylation of the parent thiol under single

5.3 Nucleophilic Catalysts with Transacylase Activity

Table 5.4 Methanolysis of **12** in MeCN–MeOH (9 : 1) at 25 °C.

Metal salt[a]	k_{rel}[b]
None	1
Ba(ClO$_4$)$_2$[c]	≥2700
SrBr$_2$	≥5400
CaBr$_2$	≥76 000

[a] Initial substrate and metal salt concentration 2.70 mM; 3 : 1 diisopropylethylamine-perchlorate salt buffer.
[b] In the absence of metal salt $k_{bg} \leq 7 \times 10^{-7}$ min^{-1} (no reaction after 50 days).
[c] $k_{rel} \geq 1000$ for the solvolysis of **11** under identical conditions.

Table 5.5 Effect of additives on the rate of liberation of pNPOH from 0.050 mM pNPOAc.[a]

Additives[b]	k_{obs} (min^{-1})	k_{rel}	k (M^{-1} min^{-1})
None	1.85 × 10^{-4}	1.0	
10	8.64 × 10^{-4}	4.7	0.32
Ba(ClO$_4$)$_2$	4.06 × 10^{-3}	22	
SrBr$_2$	1.10 × 10^{-2}	60	
CaBr$_2$	0.160	865	
10 + Ba(ClO$_4$)$_2$	0.116	627	43[c]
10 + SrBr$_2$	0.201	1 090	74
10 + CaBr$_2$	2.57	13 900	952

[a] Reaction conditions as in the experiments in Table 5.4.
[b] Concentration of all additives is 2.70 mM.
[c] k value for the Ba^{2+} complex of **9** under identical condition is 4.7 M^{-1} min^{-1}.

turnover conditions reveal significant rate enhancements in both reactions (Tables 5.4 and 5.5), which are remarkably high in the presence of Ca^{2+}. Notably, investigation of the rate enhancing effects of Ca^{2+} is not frustrated in these cases by solubility problems.

The control experiments reported in Table 5.5 show that (a) thiolysis of pNPOAc in the absence of metal ions is only 5 × faster than the very slow background reaction ($t_{1/2} \approx 3$ days) and (b) remarkable accelerations are brought about by the metal salts alone, which is ascribed to the formation of reactive metal bound methoxide species in equilibrium with the tiny amount of free methoxide in the buffered solution [Eq. (1a)]. Whereas these large accelerations are interesting per se, here it suffices to emphasize that mixtures of **10** and M^{2+} are much more effective than either thiol or metal ion alone. The reactivity order Ca ≫ Sr > Ba found in the acylation step closely parallels that found in deacylation.

The superiority of the 18-crown-6 scaffold in **10** and **12** to the 19-crown-6 scaffold in **9** and **11** is apparent from a comparison of acylation and deacylation rates in the presence of Ba^{2+} (footnote c in Tables 5.4 and 5.5).

The data reported in Tables 5.4 and 5.5 show that the alkaline-earth metal ion complexes are suitable to catalyze with turnover the methanolysis of pNPOAc via the double displacement mechanism of Scheme 5.2. Time–concentration profiles for the

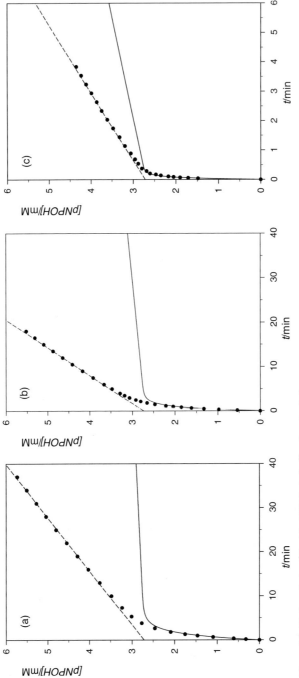

Figure 5.3 Methanolysis of 3 mM pNPOAc in buffered acetonitrile–methanol (9:1), 25 °C. Liberation of pNPOH (UV/Vis) in the presence of 2.7 mM **10** and (a) 2.7 mM Ba(ClO$_4$)$_2$, (b) 2.7 mM SrBr$_2$, (c) 2.7 mM CaBr$_2$.

liberation of pNPOH, calculated from Eq. (6), and the pertinent kinetic parameters from Tables 5.4 and 5.5 are compared with experimental data in Figure 5.3. There is a good adherence of data points to the calculated profiles in the exponential phases, and in all cases the initial burst closely corresponds to the catalyst concentration, in what amounts to ideal active-site titration experiments [25]. Consistently, HPLC analyses of reaction samples taken at times corresponding to the extinction of the exponential phase confirm a complete accumulation of **12** in all cases.

However, in the steady-state portion of the profiles, rates of liberation of pNPOH are much higher than those calculated on the basis of the directly determined deacylation rates. The discrepancies are clearly too large to be ascribed to experimental errors. Besides the thiol-mediated methanolysis proceeding via the expected acylation–deacylation catalytic cycle (Figure 5.4, mechanism A), the coexistence of a bypass mechanism is suggested (Figure 5.4, mechanism B) in which no covalently bound intermediate is involved because the crown-complexed metal ion directly delivers the methoxide nucleophile to the ester carbonyl. Differences between experimental and calculated rates of production of pNPOH at steady-state are taken as a measure of contributions of mechanism B to the overall catalytic process. Dissection of the turnover frequencies into separate contributions from the competing pathways (Table 5.6) shows that the bypass mechanism B accounts for 2/3 of the overall rate in the case of the reaction catalyzed by the Ca^{2+} complex of **10**, and for much larger fractions with the corresponding Ba^{2+} and Sr^{2+} complexes.

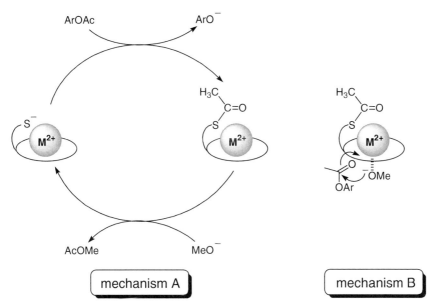

Figure 5.4 Competing catalytic mechanisms of methanolysis of pNPOAc. Mechanism A: thiol-mediated methanolysis via an acylation–deacylation cycle; mechanism B: direct delivery of complex-bound methoxide ion.

Table 5.6 Turnover frequencies (h^{-1}) in the methanolysis of pNPOAc catalyzed by alkaline-earth metal complexes of **10**.[a,b]

	Ba^{2+}	Sr^{2+}	Ca^{2+}
None	1.8	3.5	9.6
10	0.11	0.22	3.2
10 + $CaBr_2$	1.7	3.3	6.4

[a] Data refer to the catalytic experiments plotted in Figure 5.3.
[b] The corresponding data (h^{-1}) for the reaction catalyzed by the Ba^{2+} complex of **9** under identical conditions are: overall 0.16; mechanism A, 0.042; mechanism B, 0.12.

5.4
Bimetallic Catalysts

Many hydrolytic enzymes possess two or three metal ions in their active site. Common strategies in the mimicry of such enzymes are based on synthetic catalysts consisting of several ligated metal ions connected by suitable spacers [26].

Our design of bimetallic catalysts based on crown-complexed alkaline-earth metal ions, for use in reactions of ester and activated amides endowed with a distal carboxylate anchoring group, is based on the mechanistic hypothesis outlined in Scheme 5.3. Such hypothesis critically rests on the finding that in EtOH solution

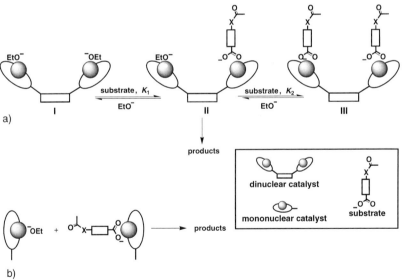

Scheme 5.3 (a) Mechanism of ester and anilide ethanolysis catalyzed by dinuclear M^{2+} complexes, showing productive (**II**) and non-productive (**I** and **III**) species and (b) the corresponding intermolecular model reaction based on monomolecular complexes.

AcO⁻ ion favorably competes with EtO⁻ ion for the crown-complexed Ba^{2+} [Eq. (8)], with an equilibrium constant $K = 69$ [27]. The corresponding figure for the Sr^{2+} complex is 35 [28].

$$[(18C6)BaOEt]^+ + AcO^- \rightleftharpoons [(18C6)BaOAc]^+ + EtO^- \quad (8)$$

Since the equilibrium constant for binding of EtO⁻ to the crown complexed Ba^{2+} and Sr^{2+} ions is $>10^4$ M^{-1} in both cases [9,28], the affinity of carboxylate anion for the ligated metal ions turns out to be large enough ($K \gg 10^5$ M^{-1}) to ensure a virtually complete anchorage to the metal in the dilute solutions of the catalytic experiments.

5.4.1
Azacrown Ligating Units

The rates of ethoxide induced ethanolysis of ester **14** and anilide **16** are enhanced by the bimetallic catalyst **17**-Ba₂ to a remarkably larger extent than those of the corresponding parent compounds **13** and **15** [27]. Furthermore, the bimetallic catalyst is far superior to the monometallic catalyst **18**-Ba in the reactions of **14** and **16**, but hardly so in the reactions of **13** and **15**. Clearly, bifunctional catalysis is observed when the dinuclear character of the metal catalyst is combined with the presence of a carboxylate anchoring group. Incidentally, we note that reactions of **14** and **16** are more sensitive to the presence of the mononuclear complex than the corresponding reactions of the parent compounds **13** and **15**, because binding to the barium ion transforms an electron-donating (rate-retarding) carboxylate into an electron-withdrawing (rate-enhancing) carboxylate-metal ion pair.

In the presence of a large excess of EtO⁻ ion, the bimetallic catalyst is fully saturated with EtO⁻ as shown by structure **I** in Scheme 5.3. Incremental additions of a carboxylate substrate would cause the gradual conversion of **I** into the 1:1 productive complex **II**, but further additions would yield the unproductive complex **III**. As expected from this mechanism a bell-shaped profile is observed in a plot of initial rate versus substrate concentration related to the catalyzed ethanolysis of **16** (Figure 5.5). The fairly good quality of the fit supports the validity of Scheme 5.3. Further confirmation comes from the finding that benzoate anions behave as competitive inhibitors of the reaction. Since the reaction product of the ethanolysis of **16** is also a benzoate anion, product inhibition is expected. Indeed, only four to five turnovers are seen in the ethanolysis of **16** before product inhibition shuts down the reaction. The first two turnovers are shown graphically in Figure 5.6.

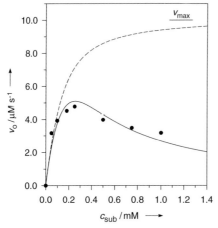

Figure 5.5 Plot of initial rate versus substrate concentration c_{sub} for the catalyzed ethanolysis of **16**. Reaction conditions: 0.10 mM **17**, 0.20 mM Ba(SCN)$_2$, 5.0 mM Me$_4$NOEt. Calculation of the solid line is based on Scheme 5.3 with $v_{max} = 1.0 \times 10^5$ M s^{-1}, $K_1 = 65$, and $K_2 = K_1/4$. The dashed line is a Michaelis–Menten plot calculated without the K_2 equilibrium step shown in Scheme 5.3.

In conclusion, the bis-barium complex of **17** catalyzes the ethanolysis of anilide and ester substrates endowed with a distal carboxylate anchoring group (Table 5.7). The catalyst shows recognition of the substrate, induces fairly high reaction rates with catalytic turnover, and is subjected to competitive inhibition by carboxylate anions, as

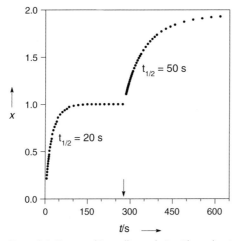

Figure 5.6 Turnover bimetallic catalysis with product inhibition in the ethanolysis of **16** under the conditions outlined in Figure 5.5. One molar equivalent of **16** is added at time zero. A second molar equivalent of **16** is added at the time indicated by the arrow. Further portions of **16** (1 mol equiv) solvolyze with increasingly lower rates (not shown in the plot). x = number of turnovers.

Table 5.7 Catalysis of ester and anilide ethanolysis by Ba^{2+}-complexes of azacrown ligands **17** and **18**;[a] k_{rel} values.[b]

Catalyst				
	13	14	15	16
18-Ba	41	61	4.8	23
17-Ba$_2$	35	1250	7.1	865

[a]Reaction conditions: 0.050 mM substrate, 1.00 mM Me$_4$NOEt, 2.00 mM Ba(SCN)$_2$, 2.00 mM **18** or 1.00 mM **17**, EtOH, 25 °C.
[b]Rate enhancements with respect to reactions carried out in the presence of 1.00 mM EtO$^-$ alone.

well as by the reaction product. A novel and distinctive feature of the catalyst is that substrate recognition takes place by means of a distal group, rather than the group that undergoes reaction. Given the extreme structural simplicity of the ditopic ligand **17**, and the fact that the active form of the catalyst is spontaneously self-assembled when the components are mixed, **17**-Ba$_2$ represents an extremely parsimonious realization of an artificial amidase/esterase. The function of the binding step is manifold. It selects the substrate, provides a (moderate) electronic activation to the amide/ester group undergoing reaction, and converts an otherwise intermolecular reaction into an intramolecular (intracomplex) one.

The dinuclear Sr^{2+} complex of the ditopic ligand **17** increases the rate of basic ethanolysis of the malonate derivative **19** by a surprising 5700-fold, but increases by only 9.5-fold the rate of cleavage of **14** [28]. It is remarkable that such a huge rate-enhancement occurs under extremely dilute conditions, namely 15 µM **19** and 30 µM **17**-Sr$_2$. A slightly lower rate enhancement is observed in the presence of **17**-Ba$_2$. It seems likely that under the dilute conditions of the catalytic experiments several crown-complexed metal species occur simultaneously (Scheme 5.4). Given the plethora of species involved in such a complicated system of multiple equilibria, quantitative kinetic treatment is out of reach. Nevertheless, a comparison with the reactivity of model compounds, particularly that of the malonate derivative **20**, provides insight into the composition of the reactive intermediate (Table 5.8).

19   **20**

The low rate-enhancement brought about by the dinuclear complex in the reaction of **14** suggests a very modest formation, in the very dilute solution, of the productive intermediate **17** [SrOEt] [SrO$_2$CR], a supramolecular complex composed of one molecule of ditopic ligand **17**, two Sr^{2+} ions, one EtO$^-$ ion, and one substrate molecule. In contrast, the absence of any difference between mononuclear and dinuclear catalyst in the cleavage of **20** demonstrates that only one metal

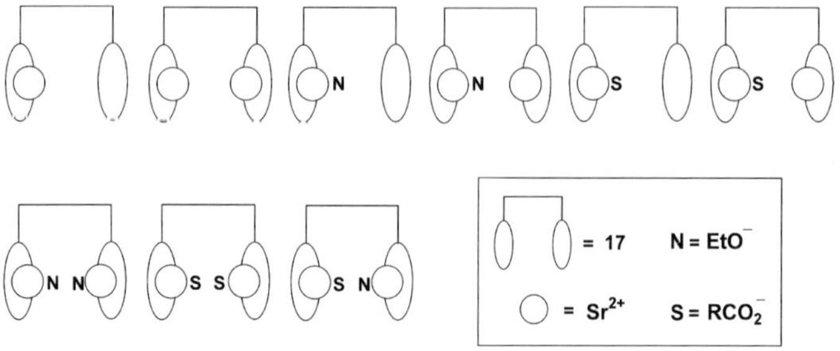

Scheme 5.4 Schematics of the crown complexed metal species occurring in solutions of ligand **17**, strontium ion, ethoxide nucleophile, and carboxylate substrate.

Table 5.8 Catalysis by mononuclear **18**-Sr and dinuclear **17**-Sr$_2$ metal complexes in the basic ethanolysis of esters;[a] k_{rel} values.[b]

Ester	18-Sr, 60 µM	17-Sr$_2$, 30 µM	k_{di}/k_{mono}
13	1.2	1.2	1.0
14	3.2	9.5	3.0
20	850	910	1.1
19	600	5700	9.5

[a]Reaction conditions: 15 µM ester, 1.00 mM Me$_4$NOEt, EtOH, 25 °C.
[b]Rate enhancements with respect to reactions carried out in the presence of EtO$^-$ alone.

ion is required in the catalyzed ethanolysis. It seems likely that a chelate interaction of the carboxylate-paired metal ion with the neighboring carbonyl enhances the binding, and that ethanolysis takes place via ethoxide addition to the metal activated ester carbonyl as shown in **VI**. Clearly, the two metal ions in the dinuclear complex act as independent units. The malonate moiety of the dicarboxylate ester **19** should also be involved in a chelate interaction of the same kind. However, in this case the catalytic efficiency of the dinuclear complex is much higher than that of the mononuclear complex, which clearly indicates that in the productive complex **VII** one metal ion is chelated to the malonate moiety, while the other binds to the distal carboxylate.

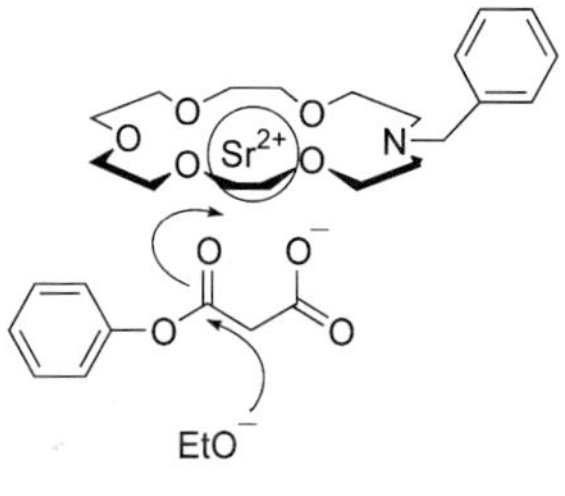

VI 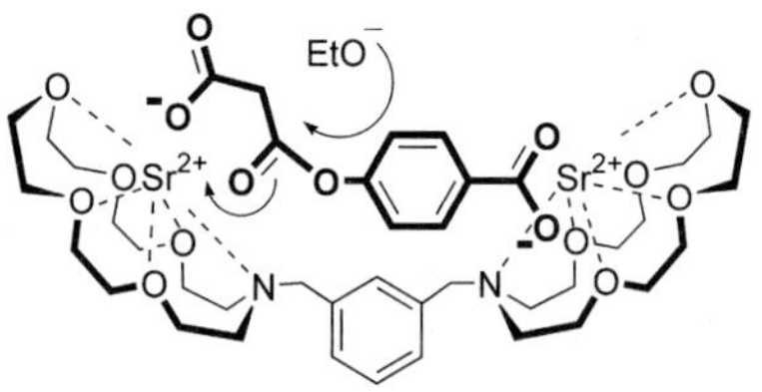   VII

5.4.1.1 Azacrown Decorated Calixarenes

The catalytic mechanism depicted in Scheme 5.3 implies that the target substrate and catalyst must form a well-matched pair in terms of size and geometrical features and, consequently, that catalytic efficiency should critically depend on the choice of the scaffold on which the metal ion ligating units are implanted.

The use of the upper rim of calix[4]arenes, blocked in the *cone* conformation, as convenient platforms for the introduction of metal ion binding sites is well documented [26e,29]. Table 5.9 shows the catalytic performances of regioisomeric vicinal **21**-Ba$_2$ and distal **22**-Ba$_2$, and of the *m*-xylylene derivative **17**-Ba$_2$ in the basic ethanolysis of esters **14** and **23–25** [30].

The dinuclear catalyst **21**-Ba$_2$, in which the azacrown ether units are linked to vicinal positions of the calix[4]arene scaffold, is not only superior to its diagonal regioisomer **22**-Ba$_2$ in all cases, but is also superior to **17**-Ba$_2$ in the reactions of esters **14**, **23**, and **24**. Modest levels of cooperation between metal ions are seen in the catalyzed reactions of the "longest" substrate **25**, which indicates that dinuclear complexes cannot expand their intermetal distances to adapt to the long carboxylate-carbonyl distance in **25**.

There is no doubt that a close fit of ester size to intermetal distance is an important prerequisite for efficient catalysis. However, the superiority of **21**-Ba$_2$ to **22**-Ba$_2$ can hardly be ascribed to a more suitable intermetal distance in the former, which indicates that other factors, whose origin is still poorly understood, may come into play.

5.4.2
Stilbenobis(18-Crown-6) Ligands

A deeper insight into the question of size selectivity in ester cleavage by dinuclear complexes is obtained from an investigation of the catalytic properties of the Ba^{2+} complexes of isomeric stilbenobis(18-crown-6) ditopic ligands **26** and **27** [30,31]. The two catalysts have quite different shapes as a consequence of the configurational change in the spacer connecting the two crown ether units. Although the geometries are not rigidly defined, because of the flexibility of the crown ether moieties and of the possibility of restricted rotations around the bonds connecting the aromatic rings to the double-bond carbons, there is little doubt that the trans isomer has a flat, more extended shape, with a broader gap between the metals.

Table 5.9 Basic ethanolysis of esters **14** and **23–25** catalyzed by the dinuclear Ba^{2+} complexes of ligands **21**, **22**, and **17**.[a]

Ester	Ligand	k_{rel}[b]	k_{di}/k_{mono}[c]
23	21	35 100	120
	22	1 340	4.5
	17	2 760	9.2
14	21	22 000	1100
	22	420	21
	17	1 220	61
24	21	20 000	385
	22	570	11
	17	4 240	82
25	21	170	14
	22	24	2
	17	300	25

[a] Runs carried out in EtOH, 25 °C, 0.025 mM substrate, 1.00 mM EtONMe$_4$, 0.10 mM ditopic ligand, 0.20 mM Ba(SCN)$_2$.
[b] Relative to background ethanolysis in the presence of 1.00 mM Me$_4$NOEt alone.
[c] k_{mono} refers to reactions catalyzed by 0.20 mM Ba(SCN)$_2$ and 0.20 mM monotopic ligand **18**.

As shown in Table 5.10, the highest catalytic efficiency is seen in the reaction of the longest ester **25** catalyzed by **26**-Ba$_2$. As suggested by molecular models (Figure 5.7), there appears to be a good fit of *trans*-stilbene catalyst **26**-Ba$_2$ to the dianionic tetrahedral intermediate involved in the ethanolysis of **25**. This nicely explains why **26**-Ba$_2$ is very effective in the cleavage of **25**, but much less so in the cleavage of the shorter esters **14** and **24**. In contrast, the performance of the *cis*-stilbene catalyst **27**-Ba$_2$ does not appear to be affected significantly by ester size, pointing to a remarkable adaptability of the intermetal distance to substrates of varying size.

26   **27**

The net result is that the *trans*-stilbene catalyst is much more effective than its cis isomer in the cleavage of the long substrate **25**, but the reverse holds in the reactions of the shorter esters **14** and **24**. Similar results are obtained in the ethanolysis of trifluoroacetanilide substrate **16** and of its difluoro analogue, which are cleaved by **27**-Ba$_2$ six times more rapidly than by **26**-Ba$_2$.

Table 5.10 Basic ethanolysis of esters **14**, **24**, and **25** catalyzed by the dinuclear Ba^{2+} complexes of ligands **26** and **27**.[a]

Ester	Ligand	k_{rel}	k_{di}/k_{mono}[b]
14	26	460	12
	27	4 260	107
24	26	860	25
	27	2 450	70
25	26	27 500	1 450
	27	2 835	150

[a]Reaction conditions as in Table 5.9.
[b]k_{mono} refers to reactions catalyzed by 0.20 mM Ba(SCN)$_2$ and 0.20 mM benzo-18-crown-6.

Figure 5.7 Computer drawn molecular model of the complex between catalyst **26**-Ba$_2$ and the tetrahedral intermediate involved in the addition of EtO$^-$ to ester **25**.

5.4.3
A Phototunable Dinuclear Catalyst

The bisbarium complex of azobis(benzo-18-crown-6) ether **28** exhibits catalytic properties that can be reversibly activated–deactivated by light-induced changes in molecular geometry [32]. The azobenzene unit is a well-known photochromic [33], often used in the construction of molecular switches [34].

Trans-**28** displays photoisomerization behavior typical of an azobenzene (Figure 5.8). Irradiation of the bis-barium complex at 370 nm in EtOH/MeCN (65/35, v/v) leads after 40 s to a stationary state – quasi-*cis*-**28**-Ba$_2$ – in which the cis/trans ratio is as high as 95:5. In turn, irradiation of the quasi-cis mixture leads to a

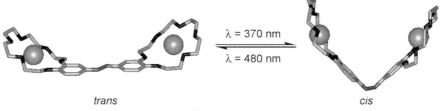

trans   cis

Figure 5.8 Computer-generated structures of interswitchable trans and cis forms of **28**-Ba$_2$.

new photostationary state – quasi-*trans*-**28**-Ba$_2$ – whose cis/trans composition is 19:81. The photostationary mixtures slowly revert in the dark to the pure trans isomer, with a half-life of 7.9 h at 25 °C, which prevents isolation of the cis isomer in a pure form by ordinary column chromatography.

The bis-barium complex of pure *trans*-**28** (0.10 mM) increases the rate of ethanolysis of **29** (65/35 EtOH/MeCN, 1 mM Me$_4$NOEt, 25 °C) by 230-fold. The corresponding figures for quasi-*trans*-**28**-Ba$_2$ and quasi-*cis*-**28**-Ba$_2$ are 420- and 1280-fold, respectively. Thus, the behavior of trans and cis isomeric catalysts based on azobenzene closely parallels the behavior of the corresponding stilbene derivatives **26** and **27**, in that in both cases the cis form is a superior catalyst in the cleavage of anilide **29**.

The photoregulation of the catalytic activity of **28**-Ba$_2$ in the ethanolysis of **29** is illustrated by an experiment (Figure 5.9) in which the catalyst is phototuned "HIGH"/"LOW" several times during the reaction by virtue of the complete interconvertibility of stationary states.

5.4.4
Effective Molarity and Catalytic Efficiency

Since the dinuclear catalysts transform the intermolecular reaction of ethoxide with substrate into an intramolecular reaction within a supramolecular complex (Scheme 5.3), the effective molarity (*EM*) parameter, defined as k_{intra}/k_{inter}, strictly applies to the catalytic process at hand and, more in general, to processes in which molecular receptors promote the reaction of two simultaneously complexed reactants [35].

The measurement of reliable *EM* values requires the composition and structure of the productive (Michaelis) complex to be known, and its kinetics of decomposition into products accurately measured. The quantity k_{inter} must be measured under identical conditions on a carefully chosen intermolecular reaction (Scheme 5.3). Unlike the rate accelerations measured under a given set of experimental conditions, usually taken as a measure of catalytic efficiency, the *EM* is independent of concentrations of reactants and catalysts. More importantly, the *EM* is also independent of the chemical activation possibly provided by catalytic groups, as well as of the chemical nature of reacting groups, because k_{inter} in the denominator corrects the intramolecular reactivity for the inherent reactivity of reacting groups. As such, the *EM* sets on a common scale reactivity data for different reactions, and provides an absolute measure of catalytic efficiency because it depends solely on the way in which

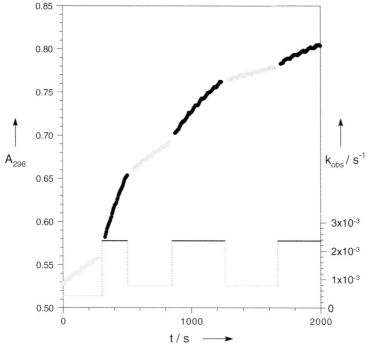

Figure 5.9 Basic ethanolysis of 0.025 mM **29** in EtOH/MeCN in the presence of 1.00 mM Me$_4$NOEt and 0.10 mM *trans*-**28**-Ba$_2$. Repeated photoconversions into quasi-cis and quasi-trans forms are obtained upon alternate irradiations at 370 and 480 nm for 40 s. Specific rates are reported on the right-hand ordinate axis.

the catalytic template assembles the two reactants in a position suitable for the reaction to occur.

As an illustration, Figure 5.10 presents a collection of available *EM* data related to two-substrate reactions catalyzed by bimetallic catalysts.[†] The *EM* values quoted in Figure 5.10 for the bis-barium complexes of bis(azacrown)calix[4]arene and *trans*-stilbenobis(18-crown-6) ligands are the best values in the respective series [30]. A glance to Figure 5.10 reveals *EM* values much lower than the high values, often amounting to several powers of ten, reported for intramolecular processes involving small and common-sized cyclic species [36]. The number of skeletal single bonds has an important influence on the efficiency of intramolecular processes, because a part of the torsional entropy is lost upon cyclization. The EM^S values also reported in Figure 5.10 are idealized values calculated on the basis of the number of essential

---

[†] All *EM* data related to catalytic systems reported by other authors have been calculated by us (see Ref. [35]) utilizing the kinetic data reported in the original papers. We take full responsibility for the quoted values.

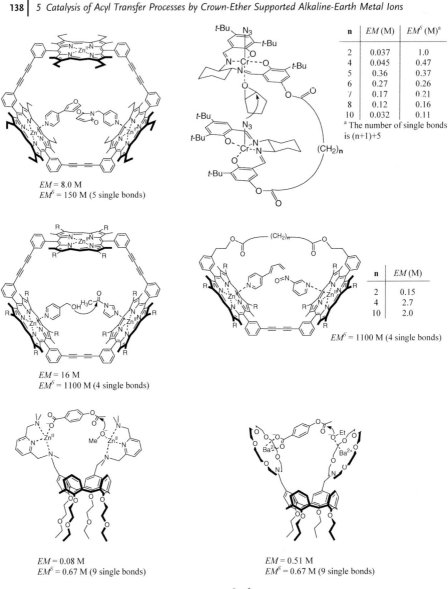

Figure 5.10 EM and EM^S data for two-substrate reactions catalyzed by bimetallic catalysts.

single bonds in the productive complex undergoing intramolecular reaction in the absence of adverse strain effects [35]. For any reaction the number of essential single bonds determines the value of the EM^S, which is viewed as the maximum EM to be obtained in that reaction.

The epoxide ring-opening catalyzed by Jacobsen bischromium catalysts [37] reveals a generally excellent agreement between predicted and experimental values, with the exception of the shortest chains. Relatively large EM values characterize the various reactions catalyzed by Sanders' zinc porphyrin hosts [38], but these values are in all cases much smaller than the predicted ones on the basis of the number of single bonds involved. Interestingly, there is an almost exact coincidence between experimental and calculated values for the best combinations of bis (barium) catalyst and ester substrate [30] but the dinuclear Zn^{2+} complex of calix[4] arene decorated with bis(dimethylaminomethyl)pyridine ligands [39] cleaves ester **14** with an efficiency lower than calculated by an order of magnitude.

Although too much emphasis cannot be put on exact figures, in view of the approximate nature of the treatment on the basis of which the EM^S values are calculated, the close adherence of experimental to predicted values provides an indication of the existence of a virtually ideal match between bimetallic catalyst and transition state. Lower-than-predicted EM values are strongly suggestive of a more or less pronounced mismatch between catalyst and transition state.

5.5
Concluding Remarks

Alkaline-earth metal ions greatly influence the reactions of esters (and anilides) with anionic nucleophiles. Their action ranges from simple promotion to turnover catalysis, either without or with substrate recognition. Dinuclear complexes of structurally simple ditopic crown-ether ligands perform as turnover catalysts with substrate recognition. Cation–anion electrostatic attractions between the doubly charged metal ions and their counter-anions, and coordinative interactions with the crown-ether units provide the driving forces for the self-assembly of the productive (Michaelis) complex. This is composed of one molecule of ditopic ligand, two metal ions, one ethoxide ion, and one substrate molecule. One of the metal ions serves as an anchoring group for a carboxylate-functionalized substrate, whereas the other delivers the ethoxide ion to the ester carbonyl, while providing electrophilic assistance. In the reactions of anilides, a metal-bound EtOH molecule provides the C—N bond undergoing rupture in the slow step with general acid assistance. A good fit of substrate size to intermetal distance is an important prerequisite for effective catalysis. In a sense, therefore, the dinuclear catalysts mimic just those features that are essential for enzyme-like catalytic activity, in that they recognize – with a certain degree of selectivity – their substrates and promote their transformation into products.

References

1 Cacciapaglia, R. and Mandolini, L. (1993) Catalysis by metal ions in reactions of crown ether substrates. *Chem. Soc. Rev.*, **22**, 221.

2 Cacciapaglia, R. and Mandolini, L. (1996) "Alkaline-Earth Metal Ion Catalysis of Ester Cleavage" in *Molecular Design and Bioorganic Catalysis, NATO ASI Series,* (eds C.S. Wilcox and A.D. Hamilton) Kluwer Academic Publishers, The Netherlands, pp. 71–86.

3 Cacciapaglia, R., Lucente, S., Mandolini, L., van Doorn, A.R., Reinhoudt, D.N. and Verboom, W. (1989) Catalysis by alkali and alkline-earth metal ions in nucleophilic attack of methoxide ion on crown ethers bearing an intra-annular acetoxy group. *Tetrahedron*, **45**, 5293.

4 Ercolani, G. and Mandolini, L. (1990) Alkali and alkaline-earth metal ion catalysis in the reaction of aryl acetates with methoxide ion. Effect of a poly (oxyethylene) side arm. *J. Am. Chem. Soc.*, **112**, 423.

5 Kraft, D., Cacciapaglia, R., Böhmer, V., El-Fadl, A.A., Harkema, S., Mandolini, L., Reinhoudt, D.N., Verboom, W. and Vogt, W. (1992) New crown ether-like macrocycles containing a nitrophenol unit. Synthesis and metal ion effects on the reactivity of their acetates in transcylation reactions. *J. Org. Chem.*, **57**, 826.

6 Cacciapaglia, R., Mandolini, L., Reinhoudt, D.N. and Verboom, W. (1992) Alkaline-earh metal ion catalysis of alcoholysis of crown-ether aryl acetates. Effects of the base-solvent system. *J. Phys. Org. Chem.*, **5**, 663.

7 Gordon, J.E. (1975) *The Organic Chemistry of Electrolyte Solutions*, Wiley, New York, pp. 466–469.

8 Cacciapaglia, R., van Doorn, A.R., Mandolini, L., Reinhoudt, D.N. and Verboom, W. (1992) Differential metal ion stabilization of reactants and transition states in the transacylation of crown ether aryl acetates. *J. Am. Chem. Soc.*, **114**, 2611.

9 Cacciapaglia, R., Mandolini, L., and Van Axel Castelli, V. (1997) Effect of cation complexing agents on the Ba(II)-assisted basic ethanolysis of phenyl acetate: from cation deactivation to cation activation. *J. Org. Chem.*, **62**, 3089.

10 Lefour, J.-M. and Loupy, A. (1978) The effect of cations on nucleophilic additions to carbonyl compounds: carbonyl complexation control versus ionic association control. Application to the regioselectivity of addition to α-enones. *Tetrahedron*, **34**, 2597.

11 Suh, J. and Mun, B.S. (1989) Crown ethers as a mechanistic probe. 1. Inhibitory effects of crown ethers on the reactivity of anionic nucleophiles toward diphenyl p-nitrophenyl phosphate. *J. Org. Chem.*, **54**, 2009.

12 Suh, J. and Heo, J.S. (1990) Crown ethers as mechanistic prob. 2. The mechanism of acetyl transfer between phenolates. *J. Org. Chem.*, **55**, 5531.

13 Cacciapaglia, R., Di Stefano, S., Kelderman, E., Mandolini, L. and Spadola, F. (1998) Catalysis of anilide ethanolysis by barium- and strontium-eathoxide pairs and their complexes with 18-crown-6. *J. Org. Chem.*, **63**, 6476.

14 Schowen, R.L., Hopper, C.R. and Bazikian, C.M. (1972) Amide hydrolysis. V. Substituent effects and solvent isotope effects in the basic methanolysis of amides. *J. Am. Chem. Soc.*, **94**, 3095.

15 Swain, C.G. (1950) Concerted displacement reactions. V. The mechanism of acid-base catalysis in water solution. *J. Am. Chem. Soc.*, **72**, 4578.

16 Fersht, A. (1985) *Enzyme Structure and Mechanism*, W.H. Freeman, New York.

17 Cacciapaglia, R., Casnati, A., Mandolini, L. and Ungaro, R. (1992) Remarkable metal ion catalysis of methanolysis of p-tert-butylcalix[4]arene-crown-5 monoacetate. *Chem. Commun.*, 1291.

18 (a) Breslow, R., Trainor, G. and Ueno, A. (1983) Optimization of metallocene

substrates for β-cyclodextrin reactions. *J. Am. Chem. Soc.*, **105**, 2739.
(b) Lehn, J.-M. and Sirlin, C. (1987) Supramolecular catalysis: cleavage of reactive amino acid esters bound by a macrocyclic receptor bearing cysteinyl residues. *New J. Chem.*, **11**, 693. (c) Suh, J. (1990) Model studies of carboxypeptidase A. *Bioorg. Chem.*, **18**, 345. (d) Diederich, F. (1991) *Cyclophanes*, The Royal Society of Chemistry, Cambridge ch. 8. (e) Murakami, Y., Kikuchi, J. and Hisaeda, Y. (1991) Catalytic applications of cyclophanes, in *Inclusion Compounds* (eds J.L. Atwood, J.E.D. Davies and D.D. Mac Nicol), Oxford University Press, New York, vol. 4. (f) Tecilla, P., Jubian, V. and Hamilton, A.D. (1995) Synthetic hydrogen bonding receptors as models of transacylase enzymes. *Tetrahedron*, **51**, 435. (g) Cram, D.J. (1995) *Container Molecules and Their Guests*, The Royal Society of Chemistry, Cambridge.

19 Cacciapaglia, R., Casnati, A., Mandolini, L. and Ungaro, R. (1992) The barium(II) complex of p-tert-butylcalix[4]arene-crown-5: a novel nuclephilic catalyst with transacylase activity. *J. Am. Chem. Soc.*, **114**, 10956.

20 Baldini, L., Bracchini, C., Cacciapaglia, R., Casnati, A., Mandolini, L. and Ungaro, R. (2000) Catalysis of acyl group transfer by a double-disiplacement mechanism: the cleavage of aryl esters catalyzed by calixcrown-Ba^{2+} complexes. *Chem.–Eur. J.*, **6**, 1322.

21 Hartley, B.S. and Kilby, B.H. (1954) Reaction of p-nitrophenyl esters with chymotrypsin and insulin. *Biochem. J.*, **56**, 288.

22 Cacciapaglia, R., Mandolini, L., Arnecke, R., Böhmer, V. and Vogt, W. (1998) Ba(II) complexes of calixcrowns derived from *p-tert*-butylcalix[5]arene as potential transa-cylation catalysts. Regio- and stereoselective monoacylation of the calixcrown. *J. Chem. Soc., Perkin Trans.*, **2**, 419.

23 Cacciapaglia, R., Mandolini, L. and Spadola, F. (1996) The Ba(II) complex of a crown ether bearing a sulphydryl side-arm as turnover catalyst of ester of cleavage. *Tetrahedron*, **52**, 8867.

24 Breccia, P., Cacciapaglia, R., Mandolini, L. and Scorsini, C. (1998) Alkaline-earth metal complexes of thiol pendant crown ethers as turnover catalysts of ester cleavage. *J. Chem. Soc., Perkin Trans.*, **2**, 1257.

25 Bender, M.L. (1971) *Mechanism of Homogeneous Catalysis from Protons to Proteins*, Wiley-Interscience, New York, ch. 12.

26 (a) Mancin, F. and Tecilla, P. (2007) Zinc (II) complexes as hydrlytic catalysts of phosphate diester cleavage: from model substrates to nucleic acids. *New J. Chem.*, **31**, 800. (b) Niittymäki, T. and Lönnberg, H. (2006) Artificial ribonucleases. *Org. Biomol. Chem.*, **4**, 15. (c) Mancin, F., Scrimin, P., Tecilla, P. and Tonellato, U. (2005) Artificial metallonucleases. *Chem. Commun.*, 2540. (d) Morrow, J.R., Iranzo, O. (2004) Synthetic metallonucleases for RNA cleavage. *Curr. Opin. Chem. Biol.*, **8**, 192. (e) Molenveld, P., Engbersen, J.F.J. and Reinhoudt, D.N. (2000) Dinuclear metallophosphodiesterase models: application of calix[4]arenes as molecular scaffold. *Chem. Soc. Rev.*, **29**, 75. (f) Kimura, E. (2000) Dimetallic hydrolases and their models. *Curr. Opin. Chem. Biol.*, **4**, 207. (g) Williams, N.H., Takasaki, B. and Chin, J. (1999) Structure and nuclease activity of simple dinuclear metal complexes: quantitative dissection of the role of metal ions. *Acc. Chem. Res.*, **32**, 485. (h) Hegg, E.L. and Burstyn, J.N. (1998) Toward the development of metal-based synthetic nucleases and peptidases: a rationale and progress report in applying the principles of coordination chemistry. *Coord. Chem. Rev.*, **173**, 133.

27 Cacciapaglia, R., Di Stefano, S., Kelderman, E. and Mandolini, L. (1999) Supramolecular catalysis of ester and amide cleavage by a dinuclear barium(II) complex. *Angew. Chem., Int. Ed.*, **38**, 348.

28 Cacciapaglia, R., Di Stefano, S. and Mandolini, L. (2001) A dinuclear

strontium(II) complex as substrate-selective catalyst of ester cleavage. *J. Org. Chem.*, **66**, 5926.

29 (a) Gutsche, C.D. (1998) *Calixarenes Revisited*, The Royal Society of Chemistry, Cambridge. (b) L. Mandolini and Ungaro, R. (eds) (2000) *Calixarenes in Action*, Imperial College Press, London. (c) Z. Asfari, V. Böhmer, J. Harrowfield and Vicens, J. (eds) (2001) *Calixarenes*, Kluwer, Dordrecht.

30 Cacciapaglia, R., Casnati, A., Di Stefano, S., Mandolini, L., Paolemili, D., Reinhoudt, D.N., Sartori, A. and Ungaro, R. (2004) Dinuclear barium(II) complexes based on a calix[4]arene scaffold as catalysts of acyl transfer. *Chem.–Eur. J.*, **10**, 4436.

31 Cacciapaglia, R., Di Stefano, S. and Mandolini, L. (2002) Size selective catalysis of ester and anilide cleavage by the dinuclear barium(II) complexes of *cis*- and *trans*-stilbeno-bis-18-crown-6. *J. Org. Chem.*, **67**, 521.

32 Cacciapaglia, R., Di Stefano, S. and Mandolini, L. (2003) The bis-barium complex of a butterfly crown ether as a phototunable supramolecular catalyst. *J. Am. Chem. Soc.*, **125**, 2224.

33 Balzani, V., Moggi, L., Scandola, F. (1987) "Towards a supramolecular photochemistry: assembly of molecular components to obtain photochemical molecular devices" in *Supramolecular Photochemistry*, NATO Advanced Science Institutes Series: Series C: Mathematical and Physical Sciences (ed. V. Balzani), D. Reidel Publishing Company, Dordrecht, vol. 214, pp. 1–28.

34 B. Feringa (ed.) (2001) *Molecular Switches*, Wiley-VCH, Weinheim.

35 Cacciapaglia, R., Di Stefano, S. and Mandolini, L. (2004) Effective molarities in supramolecular catalysis of two-substrate reactions. *Acc. Chem. Res.*, **37**, 113.

36 Kirby, A.J. (1996) Enzyme mechanisms, models, and mimics. *Angew. Chem., Int. Ed.*, **35**, 707.

37 Jacobsen, E.N. (2000) Asymmetric catalysis of epoxide ring-opening reactions. *Acc. Chem. Res.*, **33**, 421.

38 (a) Walter, C.J., Anderson, H.L. and Sanders, J.K.M. (1993) *Exo*-selective acceleration of an intermolecular Diels-Alder reaction by a trimeric porphyrin host. *Chem. Commun.*, 458. (b) Mackay, L.G., Wylie, R.S. and Sanders, J.K.M. (1994) Catalytic acyl transfer by a cyclic porphyrin trimer: efficient turnover without product inhibition. *J. Am. Chem. Soc.*, **116**, 3141. (c) Nakash, M., Clyde-Watson, Z., Feeder, N., Davies, J.E., Teat, S.J. and Sanders, J.K.M. (2000) Product-induced distortion of a metalloporphyrin host: implications for acceleration of Diels-Alder reactions. *J. Am. Chem. Soc.*, **122**, 5286. (d) Nakash, M. and Sanders, J.K.M. (2000) Structure-activity realtionships in the acceleration of a hetero Diels-Alder reaction by metalloporphyrin hosts. *J. Org. Chem.*, **65**, 7266

39 Cacciapaglia, R., Casnati, A., Mandolini, L., Reinhoudt, D.N., Salvio, R., Sartori, A. and Ungaro, R. (2005) Calix[4]arene-based Zn^{2+} complexes as shape- and size-selective catalysts of ester cleavage. *J. Org. Chem.*, **70**, 5398.

6
Bio-Inspired Supramolecular Catalysis

Johannes A.A.W. Elemans, Jeroen J.L.M. Cornelissen, Martinus C. Feiters, Alan E. Rowan, and Roeland J.M. Nolte

6.1
Introduction

Since the first ideas about catalysis were postulated by Berzelius in the mid-1800s the design and synthesis of new catalysts and the study of catalytic processes have been major themes of research in chemistry. Starting with simple acid–base catalysis, the field has broadened enormously and a wide variety of increasingly complex catalytic systems has been developed over the years, ranging from homogenous catalysts in which all components including the catalyst itself exist in a single phase to heterogeneous catalysts that are present in a different phase and therefore can be separated from the starting materials and the products of the reaction. In the last decade chemists have become more and more inspired by nature's catalysts, i.e., the enzymes that are involved in the regulation of the myriads of chemical processes that are essential to life. While nature has been able to use billions of years to develop the highly efficient and selective enzyme systems, the construction of bio-inspired artificial catalytic systems is a research area that has only been explored recently. This is because the construction of biomimetic catalysts by means of conventional covalent synthesis is a complex and time-consuming process, since the natural catalytic systems usually contain a great variety of components, the actions of which are coupled both in space and time. With the advent of supramolecular chemistry in the early 1970s it became possible to construct large, complex molecular systems by assembling molecular components with the help of noncovalent interactions, and since that time a wealth of information on how to construct biomimetic catalytic systems, some of them already displaying efficiencies and selectivities approaching those of their natural counterparts, has been obtained [1–7].

This chapter gives an overview of bio-inspired supramolecular catalytic systems that have been developed in our group since the early 1990s. Initially, attention was focused on the construction of relatively simple host–guest systems that were able to mimic certain aspects of enzymatic catalysis, e.g., substrate binding in a cavity and

Supramolecular Catalysis. Edited by Piet W. N. M. van Leeuwen
Copyright © 2008 WILEY-VCH Verlag GmbH & Co. KGaA, Weinheim
ISBN: 978-3-527-32191-9

conversion at a nearby catalytic center. To this end new cavity containing molecules based on diphenylglycoluril were developed possessing metal complexes, with the aim of achieving substrate selective conversions. Section 6.2 describes these efforts and shows that this research is still in a focal point of attention with the development of the processive enzyme mimics in which a cavity-containing catalyst threads onto a polymer chain and modifies it while gliding along it. Parallel to the enzyme mimics based on cage molecules, a second line of research was initiated aimed at mimicking the complex enzyme system cytochrome P450, a naturally occurring oxidation catalyst that uses molecular oxygen as oxidant at ambient temperature. Section 6.3 describes several cytochrome P450 model systems, varying in complexity, that have been developed by our group, which are based on artificial membrane systems in which the various components required for the catalytic reaction are captured. Very recently this research has gained further momentum as a result of new scanning probe techniques that allow the catalytic oxidation reaction to be studied at the level of single molecules, with atomic resolution.

In the last Section 6.4 new supramolecular approaches to construct synthetic biohybrid catalysts are described. So-called "giant amphiphiles" composed of a (hydrophilic) enzyme headgroup and a synthetic apolar tail have been prepared. These biohybrid amphiphilic compounds self-assemble in water to yield enzyme fibers and enzyme reaction vessels, which have been studied with respect to their catalytic properties. As part of this project, catalytic studies on single enzyme molecules have also been carried out, providing information on how enzymes really work. These latter studies have the potential to allow us to investigate in precise detail how slight modifications of the enzyme, e.g., by attaching a polymer tail, or a specific mutation, actually influence the catalytic activity.

6.2
Host–Guest Catalysis

For many years our group has been using host molecules to construct substrate-selective catalysts. Taking natural enzymes as a blueprint, various catalytically active metal centers have been linked to receptor cavities in which reactive substrates can be bound via noncovalent interactions. These supramolecular systems were synthesized to mimic processes typically encountered in enzymatic systems, such as selective binding of substrates, their catalytic conversion, and the release of the products. The basic host molecule used by our group, often referred to as a "molecular clip", consists of a diphenylglycoluril framework to which two aromatic side-walls are connected via four methylene linkers (Figure 6.1a, compound 1) [8]. ^{1}H NMR studies have demonstrated that in organic solvents these hosts are excellent receptors for neutral aromatic guest molecules, such as (di-)hydroxybenzenes (Figure 6.1b). The binding strength toward these types of guests can reach values of $K_a > 10^5 \, M^{-1}$, depending on the binding functions in the receptor molecule and the substitution pattern in the guest [9]. The binding is a result of three cooperative effects, viz. hydrogen bonding, π–π stacking

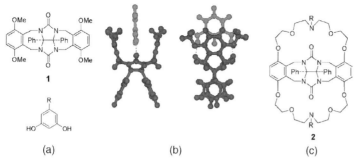

Figure 6.1 (a) Structure of molecular clip **1** and a dihydroxybenzene guest molecule. (b) Molecular models showing the binding of a resorcinol molecule in clip **1** (front and side view). (c) Structure of molecular basket **2**.

interactions, and a so-called "cavity effect", which can be simply described as the entropically favorable filling of the empty cavity by the guest [10]. The cavity sidewalls of the clips can be further functionalized with crown ether moieties to give host molecules such as **2** (Figure 6.1c), which have been named "molecular baskets" because of their basket- or bowl-like shape [11]. In addition to alkali metal ions and diammonium salts, these basket molecules are also excellent receptors for charged aromatic compounds such as viologens (N,N'-disubstituted 4,4'-dipyridinium compounds) and polymeric derivatives thereof [12].

6.2.1
Rhodium-based Receptors

Rhodium complexes are well-known hydrogenation, hydroformylation, and isomerization catalysts and to combine them with a substrate binding cavity a tetrakis(triphenylphosphite)rhodium(I) hydride complex was attached to a molecular basket (**3**, Figure 6.2) [13,14]. A first series of experiments investigated whether 4-allylcatechol, a guest molecule that is known to bind in the cavity of **3**, was isomerized more efficiently than allylbenzene, which is a non-binding guest. As expected, in chloroform solution **3** converted 66% of an equimolar amount of 4-allylcatechol within 2 h, whereas in a separate experiment only 12% of allylbenzene was converted during the same period. When, instead of **3**, $HRh(CO)[P(OPh_3)]_3$, a catalyst without a substrate binding cavity, was used no significant difference in reactivity between the two substrates was observed.

In addition to isomerization, the hydrogenation of styrene derivatives **5–7** was also investigated, using hydride complex **4** as the catalyst (Figure 6.3). Compared to reference catalyst $HRh[P(OPh_3)]_4$, complex **4** accelerated the conversion of substrates **6** and **7**, which bind in the receptor cavity, whereas the conversion of the nonbinding substrate **5** was delayed. For the supramolecular catalyst and the model compound, the ratio of the initial rates $\{v_{init}(4)/v_{init}(HRh[P(OPh)_3]_4)\}$ for the conversion of substrates **5–7** amounted to 0.1, 2.0, and 4.7, respectively, highlighting the defined effect of substrate binding on the rate of catalysis.

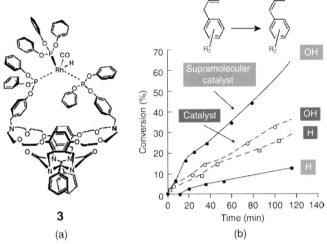

Figure 6.2 (a) Structure of rhodium host **3**. (b) Catalytic isomerization of allylbenzene derivatives carried out by supramolecular catalyst **3** (solid lines) or by reference catalyst HRh(CO)[P(OPh$_3$)]$_3$ (dashed lines), depending on the R-group of the substrate (◊ and ♦: 4-allylcatechol; ■ and □: allylbenzene).

To obtain more detailed information about the kinetics of the reactions using the supramolecular catalyst, the hydrogenation of **7** was monitored at different substrate concentrations. A double reciprocal plot of the initial substrate concentration versus the initial rate gave a straight line, which indicates Michaelis–Menten-like kinetics, reminiscent of enzyme-like behavior. To further demonstrate such behavior, competition experiments were carried out in which equimolar amounts of substrates **5–7** were added to host **4** (Figure 6.3c). Whereas the hydrogenation of resorcinol derivative **7** (the substrate that binds strongest in **4**) started immediately, the hydrogenation of catechol derivative **6** only began after an induction period of 3 min, while the non-binding substrate **5** only started to react after 22 min. These observations clearly demonstrate that **4** behaves as an enzyme mimetic substrate-selective catalyst.

Remarkably, the hydrogenation of **6** by **4** appeared to be accelerated in the presence of added resorcinol (Figure 6.3d), while the reference catalyst did not exhibit this behavior. This feature was ascribed to a cooperative binding of resorcinol and **6** in the cavity of **4**. As a result of this cooperative binding the catechol moiety of **6** can preserve its intramolecular hydrogen bond and more favorably bridge the distance between the carbonyl groups of the receptor molecule which act as hydrogen bond acceptor sites (Figure 6.3e).

6.2.2
Copper-based Receptors

In an approach similar to that used for the supramolecular hydrogenation catalysts, supramolecular oxidation catalysts were also designed by connecting pyrazole-based

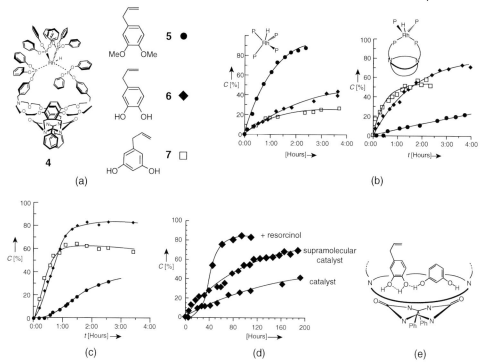

Figure 6.3 (a) Structure of rhodium host **4** and allylbenzene derivatives **5–7**. (b) Hydrogenation of **5–7** by **4** and a reference catalyst, highlighting the substrate selectivity of the metallo-host. (c) Competition experiment in which equimolar amounts of **5–7** are hydrogenated by **4**. (d) Effect of the addition of resorcinol during the hydrogenation of **6**. (e) Schematic representation of the cooperative binding of resorcinol and **6** in the cavity of catalyst **4**.

ligands to molecular baskets. Upon the addition of Cu^{II} salts, metallohosts of type **8** were formed that were still able to bind dihydroxybenzene guests (Figure 6.4a) [15]. Since the Cu^{II} ions in these types of complexes can be reduced to Cu^{I} by the addition of methanol, which is concomitantly oxidized to formaldehyde, it was investigated whether alcohols that bind in the cavity of **8** would react more efficiently than alcohols that do not bind. Upon oxidizing a series of substituted benzyl alcohols, 3-hydroxybenzyl alcohol and 3,5-dihydroxybenzyl alcohol, which are guests that bind in the cavity, were oxidized several orders of magnitude faster than the other alcohols (Figure 6.4b). With the latter guest, a rate enhancement greater than 50 000 was observed. These observations clearly illustrate the effect of substrate selectivity and the ideal preorganization of the alcohol function with respect to the copper centers by the cavity of the catalyst.

The reaction with **8** only occurred in stoichiometric ratios, hence we searched for a way to use copper-based molecular clips as true oxygenation catalysts. To this end the host molecule was altered by changing the pyrazole ligands for pyridine to obtain a model system that could mimic dicopper proteins, which can bind molecular oxygen between the copper centers in a bridging fashion [16]. After the binding of two Cu^{I}

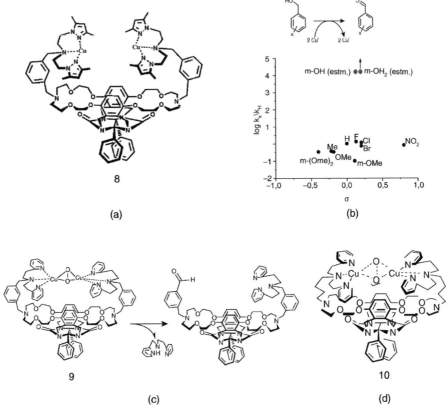

Figure 6.4 (a) Structure of bis-Cu^{II} host **8**. (b) Oxidation of benzyl alcohols by **8**. The relative rate of conversion is plotted versus the Hammett constant of the substituent on the alcohol. Note the dramatic rate acceleration when the substrate contains hydroxy substituents. (c) Structure of bis-Cu^{I} host **9**, binding an O$_2$ molecule in a bent μ-η^2:η^2 binding mode, and the subsequent autodestruction reaction in which one of the ligand sets is split off. (d) Structure of complex **10** containing aliphatic spacers between the receptor and the ligand moieties.

ions, host **9** formed metastable O$_2$ adducts in CH$_2$Cl$_2$ at −85 °C; based on spectroscopic measurements it was proposed that the oxygen molecule was bound between the copper centers in a bent μ-η^2:η^2 binding mode (Figure 6.4c). As with the pyrazole-derived clip molecule, guests can still be complexed in the cavity of **9**. However, the dicopper complex did not work as a catalyst for the oxidation of bound guests, since oxidative splitting of the attached pyridine ligands occurred first. Molecular modeling calculations revealed that in the bridged dioxygen complex the benzylic protons linking the cavity and the pyridine functions were positioned next to the bound oxygen, and, consequently, the catalyst itself was oxidized in preference to the guest (Figure 6.4c).

To circumvent this problem, new catalysts were developed containing simple alkyl spacers instead of benzylic linkers between the cavity and the ligands [17]. The ligand system of oxygenated metallohost **10** (Figure 6.4d) indeed appeared to be stable, although above −80 °C the dioxygen complex degraded rapidly. The catalytic activity of **10** was investigated at −80 °C but was unfortunately inactive either in the hydroxylation of (di)hydroxybenzene-derived guest molecules or the epoxidation of simple alkenes like styrene and limonene. With the phenolic substrates, however, a polymerization reaction occurred. In a control experiment using the hindered phenol derivative 2,4-di-*tert*-butylphenol, complex **10** was catalytically active in a radical C–C coupling reaction, giving a dimeric product, and evidence was provided that this dimerization occurred via a radical oxidation process [17].

6.2.3
Porphyrin-based Receptors

As a next approach to cavity-functionalized catalysts a porphyrin "roof" was fixed, via four spacers, on top of the molecular clip. The resulting molecule (**11**) has a rigid cavity with a diameter of ca. 9 Å and is perfectly suited to complex small aromatic guests (Figure 6.5a) [18]. It appeared to be an excellent host for viologen derivatives, which are strongly complexed in its cavity by a combination of electrostatic and π–π stacking interactions ($K_a = 10^5–10^6 \, M^{-1}$ in acetonitrile/chloroform mixtures) [19]. Insertion of a zinc ion in the porphyrin gave **Zn11**, a host that can strongly complex pyridine-based axial ligands inside its cavity. As a result of stabilizing cavity effects, pyridine was bound much more strongly to the porphyrin in **Zn11** ($K_a = 1.1 \times 10^5 \, M^{-1}$ in chloroform) than to a porphyrin without a cavity ($K_a \approx 1000 \, M^{-1}$) [19]. This strong binding of axial ligands was beneficially used in the epoxidation of alkenes with porphyrin clip **Mn11** as a catalyst, since only one equivalent of added pyridine sufficed to give a fully (>99+%) coordinated **Mn11**-pyridine complex. The resulting catalyst appeared to be highly active in the epoxidation of alkenes in a biphasic system

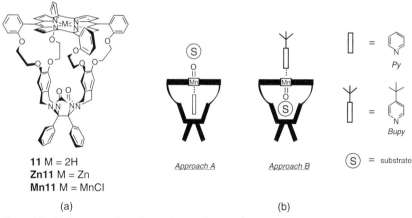

Figure 6.5 (a) Structure of porphyrin clips **11**, **Zn11** and **Mn11**. (b) Two approaches in which **Mn11** is used as an epoxidation catalyst in combination with pyridine or 4-*tert*-butylpyridine as the axial ligand.

using sodium hypochlorite as the oxygen donor (Figure 6.5b, Approach A) [20,21]. When *cis*-stilbene was used as the substrate, *cis*- and *trans*-stilbene oxide were obtained with a high selectivity for the cis-isomer (>95%). This effect can be rationalized by the strong coordination of the axial ligand, which pulls the metal center into the plane of the porphyrin, thereby hindering the rotation of the C–C bond of the *cis*-stilbene substrate coordinated to the other face of the porphyrin in the transition state. The only disadvantage of this porphyrin clip catalyst appeared to be its moderate stability. In the absence of substrate, the catalyst rapidly decomposed as a result of the formation of μ-oxo manganese porphyrin dimeric species, which are inactive in catalysis and destroy the porphyrin. To prevent this autodestruction, **Mn11** was applied in an alternative approach in which 4-*tert*-butylpyridine was used as the axial ligand (Figure 6.5b, Approach B). Owing to its bulkiness, this ligand does not fit inside the cavity of the host and only coordinates to the outside, effectively shielding the catalyst from forming the inactive μ-oxo dimeric species. As a result, no catalyst decomposition due to the formation of the dimer can occur, providing an extremely stable catalyst that can convert multiple batches of added alkene. The fact that catalysis now occurs inside the cavity had its impact on the stereochemistry of the reaction. For example, *trans*-stilbene was converted twice as fast as *cis*-stilbene. This observation can be rationalized by the fact that the rather bulky *cis*-stilbene experiences severe steric hindrance by the confinement of the host cavity, which it has to enter completely to reach the catalytic metal center. Molecular modeling studies show that for the much flatter trans-isomer this steric hindrance is almost negligible.

Host **Mn11** was also used as a mimic of processive enzymes. In nature these enzymes play an essential role in DNA synthesis and degradation, examples being DNA polymerase and λ-exonuclease, which operate by threading the biopolymer through the hole in their toroidal structure. After threading, several rounds of catalysis (e.g., replication or degradation) take place before the enzyme dissociates from the biopolymer, resulting in a process that occurs with high fidelity. To mimic the action of such enzymes, **Mn11** was used in combination with a synthetic polymer, i.e., polybutadiene, with the aim of epoxidizing this polymer in a processive fashion (Figure 6.6) [22]. 4-*tert*-Butylpyridine was used as the axial ligand to ensure that catalysis would take place inside the cavity. In the presence of the oxygen donors iodosylbenzene or sodium hypochlorite, polybutadiene with an average molecular weight of 300 000 (98% cis) was completely converted into the polyepoxide within 1 h. This suggests that **Mn11** is a very efficient catalyst that operates in a pseudo-rotaxane-like topology, in which it slides over the polymer chain while performing its catalytic action. Several control experiments were undertaken to prove this catalyst–polymer topology. The addition of N,N'-dimethylviologen, a guest that strongly binds inside the cavity of **Mn11**, to the reaction mixture appeared to dramatically lower the rate of epoxidation. A second indication that the reaction occurs inside the cavity came from the cis/trans ratio of the epoxide product. Whereas a reference manganese porphyrin catalyst yielded 78% cis- and 22% trans-epoxide, the cis/trans ratio of the product using **Mn11** as the catalyst was nearly reversed, i.e., 80% trans and 20% cis. This ratio suggests that the sterically demanding cavity has a strong effect on the catalytic reaction, favoring the sterically less-demanding transition state from which the trans-product results. An

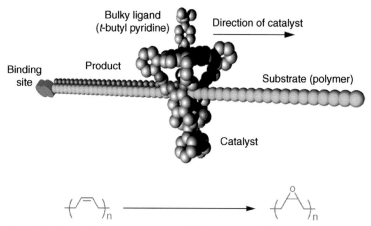

Figure 6.6 Processive enzyme mimic in which host **Mn11** slides over a polybutadiene chain and at the same time oxidizes its double bonds.

alternative for the threading mechanism would be a reaction at the outside of the cavity of **Mn11**, which is, however, rather unlikely given the blocking of the manganese center in the presence of the excess of the bulky axial ligand (molecular modeling studies indicate that an alternative possibility could be a looping of the polymer chain within the cavity without the occurrence of threading, see Figure 6.7a).

To get more insight into the mechanism of the reaction, and to exclude the occurrence of looping, detailed studies were carried out on the interactions between polymeric substrates and the free base porphyrin host **11** [23]. A series of three well-defined polytetrahydrofuran polymers with a different degree of polymerization was synthesized, containing a viologen trap at one side of the polymer chain as well as a 3,5-di(*tert*-butyl)phenyl blocking group, and an open end at the other side (Figure 6.7b). Hosts of type **11** are unable to slip over the blocking group and therefore they must traverse the whole polymer chain before they can reach the viologen trap, which upon binding in the cavity of **11** quenches the fluorescence originating from the porphyrin and thus serves as a probe for threading. By monitoring the fluorescence intensity versus time upon the addition of **11** to one of the polymers (Figure 6.7c), the kinetics of the threading process could be obtained. These followed a second-order process, and, as expected, the rate of threading became slower when the polymer length increased (Figure 6.7d). The reverse process of dethreading, which was achieved by rapidly diluting the threaded complexes and monitoring the increase in fluorescence, appeared to be a first-order process. From the threading and de-threading experiments it became clear that a length-dependent barrier, which increases with 61 J nm^{-1}, has to be surmounted by **11** to reach the viologen trap. The threading mechanism is thought to be similar to the one proposed by Muthukumar for the translocation of a DNA chain through the opening in a virus particle [24]. After finding the hole, host **11** has to thread the polymer over a certain critical length before the process can continue. According to this mechanism,

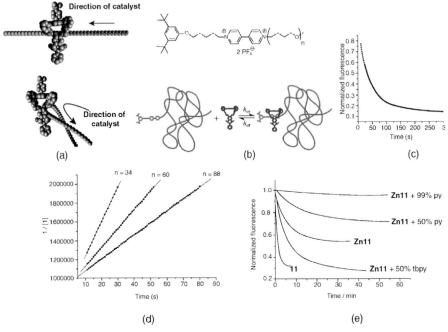

Figure 6.7 (a) Difference between threading (top) and looping (bottom) mechanism in the translocation of catalyst **Mn11** along the polymer chain. (b) Structure of the polytetrahydrofuran polymers with trap and blocking group (top) and a schematic representation of the threading of host **11** over these polymers. (c) Typical changes in fluorescence observed when host **11** is threading onto one of the polymers. (d) Second-order rate plots of the threading of **11** on polymers with different degrees of polymerization. (e) Normalized fluorescence intensity of **11** and of **Zn11** in the absence and presence of different axial ligands as a function of time upon threading onto a polymer.

the barrier that must be overcome is entropic in origin and should depend on the length of the polymer chain. Threading experiments at different temperatures gave positive values for $\Delta H^{\ddagger}_{on}$ and strongly negative values for $\Delta S^{\ddagger}_{on}$, while the absolute value of $\Delta S^{\ddagger}_{on}$ increased with polymer length. These results are in line with the nucleation mechanism of Muthukumar and the presence of an entropic barrier, related to the stretching of the end of the (bio)macromolecular chain, that has to be overcome.

To investigate the role of the presence of a metal in the porphyrin on the threading behavior, experiments were also carried out with **Zn11** (experiments with **Mn11** were not possible due to the lack of fluorescence of this catalytic host) [25]. This host turned out to thread much more slowly onto the polytetrahydrofuran polymers, which was attributed to a blocking of the cavity by a solvent molecule that is initially coordinated to the zinc ion in the cavity of the host. This assumption was confirmed by an experiment in which the cavity was intentionally blocked by a strongly binding pyridine ligand, which resulted in a dramatic decrease in the rate of threading. In contrast, when 4-*tert*-butylpyridine was added to **Zn11**, a ligand that exclusively binds to the outside of the cavity and thereby expels any bound solvent molecule from the inside, a large increase in threading rate was observed.

6.3
Cytochrome P450 Mimics

A very challenging topic in the field of biomimetic chemistry is the construction of a supramolecular cytochrome P450 mimic that uses *molecular oxygen* as an oxidant. The natural mono-oxygenase, which serves as a catalyst for the (ep)oxidation of xenobiotics, contains an Fe^{III} protoporphyrin to which a cysteinate axial ligand is coordinated [26]. After a stepwise two-electron reduction, this complex is able to bind O_2 and to cleave it reductively, leaving one oxygen atom at the porphyrin to produce a reactive high-valent iron-oxo intermediate, and the other one to combine with two protons to produce a molecule of water. The relative instability of iron protoporphyrins means that artificial systems generally make use of the much more stable manganese porphyrins and single oxygen donors as oxidants – the latter because it is very difficult to control the two subsequent reduction steps that are required to activate molecular oxygen.

6.3.1
Membrane-based Catalysts

The first model system investigated by our group involved a manganese porphyrin with a coordinated N-methylimidazole as axial ligand, which was incorporated in the bilayer of vesicles formed by a polymerizable analogue of dihexadecyldimethylammonium bromide (Figure 6.8) [27]. In the same vesicles, colloidal platinum as an electron donor and methylene blue as an electron carrier were incorporated. Using this system, all the features of the natural system were closely mimicked. Under an atmosphere of H_2 and O_2 (1:1) the artificial system catalyzed the oxidation of both

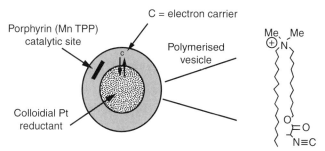

Figure 6.8 Schematic representation of a cytochrome P450 mimic in which catalytic manganese porphyrins are captured in the bilayer of polymerized vesicles. Colloidal platinum encapsulated in the vesicles in combination with molecular hydrogen serves as a reductant.

water-soluble as water-insoluble alkenes with molecular oxygen, with turnover numbers of 8 and 1.3 mol product per mol of porphyrin per hour, respectively.

The inefficiency of the platinum/hydrogen reduction system and the dangers involved with the combination of molecular oxygen and molecular hydrogen led to a search for alternatives for the reduction of the manganese porphyrin. It was, for example, found that a rhodium complex in combination with formate ions could be used as a reductant and, at the same time, as a phase-transfer catalyst in a biphasic system, with the formate ions dissolved in the aqueous layer and the manganese porphyrin and the alkene substrate in the organic layer [28].

In a somewhat more sophisticated approach, a so-called "picket fence" manganese porphyrin (**12**) was combined with the amphiphilic rhodium complex **13** in both dihexadecyl phosphate (DHP) and dioctadecylammonium chloride (DODAC) vesicles (Figure 6.9a–c) [29]. The reduction of **12** by formate under argon in the presence of N-methylimidazole as an axial ligand appeared to be faster in DODAC than in DHP vesicles, because of the accumulation of negatively charged formate ions on the positively charged surface of the former vesicles. Nevertheless, the DHP vesicles appeared to be a more efficient support, and the epoxidation of styrene in the presence of O_2 with **13**-formate as reducing agent occurred with a turnover of 60 mol per mol porphyrin per hour at 70 °C. The reason for this higher efficiency compared to the DODAC-based system was proposed to be the rate limiting availability of protons (*vide infra*).

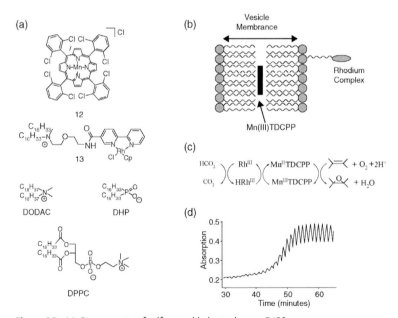

Figure 6.9 (a) Components of self-assembled cytochrome P450 mimics. (b) Schematic representation of such a mimic. (c) Catalytic cycle. (d) Change in absorbance at 435 nm (Mn^{II}) versus time under the conditions at which the oscillating reduction of the Mn^{III} porphyrin takes place.

Several additional studies were carried out to obtain information about the precise behavior of the various components in the model system. The interplay between the manganese porphyrin and the rhodium cofactor was found to be crucial for an efficient catalytic performance of the whole assembly and, hence, their properties were studied in detail at different pH values in vesicle bilayers composed of various types of amphiphiles, viz. cationic (DODAC), anionic (DHP), and zwitterionic (DPPC) [30]. At pH values where the reduced rhodium species is expected to be present as Rh^I only, the rate of the reduction of **13** by formate increased in the series DPPC < DHP < DODAC, which is in line with an expected higher concentration of formate ions at the surface of the cationic vesicles. The reduction rates of **12** incorporated in the vesicle bilayers catalyzed by **13**-formate increased in the same order, because formation of the Rh-formate complex is the rate-determining step in this reduction. When the rates of epoxidation of styrene were studied at pH 7, however, the relative rates were found to be reversed: DODAC ≪ DPPC < DHP. Apparently, for epoxidation to occur, an efficient supply of protons to the vesicle surface is essential, probably for the step in which the Mn^{II}-O_2 complex breaks down into the active epoxidizing Mn^V=O species and water. Using α-pinene as the substrate in the DHP-based system, a turnover number of 360 was observed, which is comparable to the turnover numbers observed for cytochrome P450 itself.

Under specific conditions, the self-assembled catalytic system displayed an intriguing oscillatory behavior involving a delicate interplay between the manganese porphyrin catalyst and the rhodium coreductor (Figure 6.9d) [31]. At 48 °C and a formate concentration of 0.25 M, no reduction of Mn^{III} porphyrin was observed at [**13**]/[**12**] ratios lower than 10. At higher ratios, reduction took place but at a ratio of exactly 10 oscillations in the concentration of the Mn^{II} species were detected. After an initial induction period of 30 min, the Mn^{II} species was generated and after a short while converted back to the Mn^{III} species by oxidation with O_2. After approximately 50 min, a stable oscillating process set in. This process was found to be exceptionally sensitive to slight changes in the temperature and the ratio between [**13**] and [**12**]. At a [**13**]/[**12**] ratio <10, the reoxidation of Mn^{II} is faster than the reduction, whereas at a [**13**]/[**12**] ratio >10 the situation is reversed. At a ratio of precisely 10 the manganese center shuttles between the 2+ and 3+ states. UV/Vis studies suggested that upon its formation the overall neutral Mn^{II} porphyrin moves toward the middle of the bilayer where the environment is less polar.

6.3.2
Single Molecule Studies on Epoxidation Catalysts

To obtain better information about the mechanisms by which manganese porphyrins act as catalysts in the aerobic epoxidation of alkenes, we have very recently applied a new technique to study this reaction, viz., scanning tunneling microscopy (STM) at a liquid–solid interface. Instead of using conventional spectroscopic techniques, which average the behavior of millions of molecules, STM allows the study at the level of a single molecule and monitors changes in real-time and -space. Moreover, in addition to topographic details STM also gives information about changes in the electronic density of states of the molecules. It was the aim to mount porphyrin catalysts with the π-surface parallel to the surface, thereby exposing their catalytic metal center to the STM tip. For this reason a manganese porphyrin was prepared that lacked the *meso*-phenyl rings to ensure good contact to the surface (**14**, Figure 6.10a) [32]. Experiments were carried out in a home-built STM setup, containing a liquid-cell covered with a bell-jar that allows

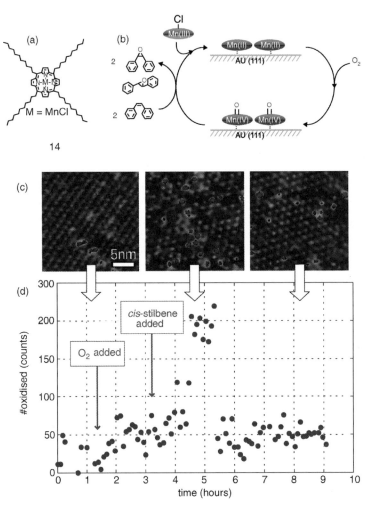

Figure 6.10 (a) Structure of manganese porphyrin catalyst **14**. (b) Catalytic cycle; Mn^{III}-porphyrins adsorb to a Au(111) surface and are reduced to Mn^{II}; in the next step one molecule of O_2 is homolytically cleaved, distributing the two oxygen atoms over two Mn-porphyrins to generate two Mn^{IV}=O species; these are catalytically active in the epoxidation of added *cis*-stilbene to give as products *cis*- and *trans*-stilbene oxide. (c) Series of STM images of a monolayer of Mn-porphyrins at the interface of Au(111) and n-tetradecane during the catalytic cycle; left: Mn^{II} porphyrins under argon; middle: after the addition of O_2 to the system; right: after the addition of *cis*-stilbene to the liquid cell; in all cases $V_{bias} = -200$ mV, $I_{set} = 10$ pA.

control over the atmosphere. Under an argon atmosphere catalyst **14** formed extended monolayers at the interface of *n*-tetradecane and a Au(1 1 1) surface (Figure 6.10c). Using this potentially catalytic surface, the complete catalytic cycle (Figure 6.10b) in which *cis*-stilbene is epoxidized into its *cis*- and *trans*-epoxides was investigated *in situ* at the single catalyst level.

On the monolayer formed by **14**, occasionally a porphyrin molecule was observed with a much higher apparent height (bright spots in Figure 6.10c, left). Realizing that manganese porphyrins can react with molecular oxygen, a contamination of the system with this gas was suspected. To prove this assumption, the bell-jar was purposely filled with O_2. Nearly instantaneously numerous porphyrins in the monolayer were altered (displayed different tunneling characteristics), clearly indicating a reaction with the added oxygen. This was not expected, because before a reaction with O_2 can take place Mn^{III} porphyrins first need to be activated by reducing them to Mn^{II}. Spectral changes in the reflection UV/Vis measurements on the monolayer of **14** at the liquid–solid interface, however, indicated that such a reduction had occurred upon adsorption of the catalysts to the Au(1 1 1) surface. A very remarkable aspect of the catalytic oxidation was the observation that each molecule of O_2 appeared to oxidize two adjacent catalysts on the surface, which highlights the unique information that can be obtained by single molecule imaging by STM. Statistical measurements on thousands of molecules of **14** indeed confirmed that adjacent catalysts were oxidized preferentially. In each case two identical species were obtained, and hence a homolytic splitting of O_2 was proposed, resulting in the generation of two $Mn^{IV}=O$ particles. This assumption was again confirmed by spectral changes in the reflectance UV/Vis spectra of the monolayer. Because such species are highly reactive in epoxidation catalysis, an aliquot of *cis*-stilbene was carefully added to the subphase of the STM liquid-cell while scanning was continued. These *in situ* STM measurements showed that at the moment the alkene reached the liquid–solid interface the number of Mn^{IV}-oxo species abruptly decreased, accompanied by regeneration of the non-oxidized catalysts that appeared to be able to react with O_2 again. After four days a sample was taken from the liquid-cell and analyzed by gas chromatography. It appeared to contain increased amounts of *cis*- and *trans*-stilbene oxide, proving that the surface was indeed catalytically active.

6.4
Biohybrid Catalytic Systems

6.4.1
Bioamphiphiles

In nature many enzymes are embedded in membranes, which not only serve as a scaffold but also regulate the transport of substrates and products, and control the concentrations of protons and other ions. Instead of embedding molecular catalysts into artificial membranes, as was done for the cytochrome P450 mimic, it is also possible to make amphiphiles that constitute the membranes catalytic themselves, e.g., by using a catalyst molecule as the polar head-group of the amphiphile. We have been elaborating on this approach by developing a new type of macromolecular surfactants, so-called "giant amphiphiles", in which a protein or an enzyme acts as the polar head group and a synthetic hydrophobic polymer as the apolar chain. The concept is particularly interesting when an enzyme is used as the head group, because in that case, upon assembly, superstructures with catalytic properties will be obtained. Three approaches were followed to couple a polymeric chain to an enzyme head-group: (a) covalent attachment, (b) attachment via cofactor reconstitution, and (c) attachment via biotin–streptavidin interactions.

In the first approach, a maleimide-functionalized polystyrene of 40 repeat units (PDI = 1.04) was covalently coupled to the sulfur atom of a reduced disulfide bridge on the surface of the lipase B from *Candida antarctica* (CALB) (Figure 6.11a) [33]. TEM studies of the reaction mixture in water/THF revealed well-defined enzyme fibers with a length of several micrometers. The fibers consisted of bundles of rods, of which the smallest had diameters between 25 and 30 nm. These diameters were in

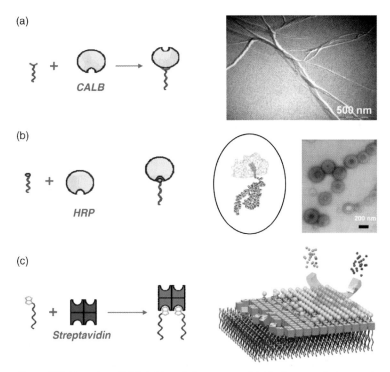

Figure 6.11 Examples of biohybrid catalytic systems. (a) Covalent coupling of a polystyrene tail to the enzyme CALB lipase; the resulting biohybrid forms micellar fibers. (b) Cofactor reconstitution: A polystyrene tail is connected to the horseradish peroxidase enzyme via the cofactor ferri-protoporphyrin IX.; the resulting amphiphile forms vesicular structures. (c) Noncovalent coupling of two biotinylated polystyrene tails to the protein streptavidin, to which in a second assembly step the enzyme horseradish peroxidase is coupled. Reproduced with kind permission [34,36].

good agreement with the predicted diameter of micellar architectures formed by the giant amphiphiles. To test the catalytic activity of the supramolecular aggregates, the hydrolysis of the pro-fluorescent substrate 6,8-difluoro-4-methylumbelliferyl octanoate (DiFMU-octanoate) was investigated. The CALB-containing fibers were found to be still catalytically active, but exhibited only ∼5% of the activity of the native enzyme. This reduced activity was attributed to destabilization of the active conformation of the enzyme due to its coupling to the polystyrene chain and to limited accessibility of the enzyme molecules in the fibers to substrate molecules.

In the second approach, the cofactor of horseradish peroxidase (HRP), ferriprotoporphyrin IX, was connected via one of its carboxylic acid groups to a single polystyrene chain (90 repeat units, PDI = 1.05) via a hydrophilic bis(aminoethoxy) ethane spacer (Figure 6.11b) [34]. In the next step, the modified cofactor was reconstituted with the apo-enzyme by mixing both components in THF/water. TEM studies showed that in aqueous solution the resulting giant amphiphiles self-assembled into vesicle-like structures with a diameter of 80–400 nm. The aggregates appeared to still exhibit some catalytic activity, albeit much less than that displayed by the native enzyme under the same conditions. Current work involves varying the hydrophobicity and length of the polymer chain to tune the catalytic activity [35].

The third approach made use of the streptavidin–biotin couple to connect enzymes to hydrophobic polymer tails (Figure 6.11c) [36]. In the first step of the self-assembly, two biotin-functionalized polystyrene chains (90 repeat units, PDI = 1.03) were complexed to one face of a streptavidin protein at the water–air interface by using the Langmuir–Blodgett technique. In this way, the opposite face of the protein remained available for the connection of other functional molecules. To the remaining two binding sites of the streptavidin, a doubly biotinylated HRP was connected by adding this modified enzyme to the subphase underneath a compressed monolayer of the polymer–protein hybrid. The resulting polymer–enzyme hybrid retained catalytic activity, which was assessed by using the standard ABTS/H_2O_2 assay.

6.4.2
Single Enzyme Catalysis

Although the bioamphiphile approach can lead to interesting new directions with regard to combining supramolecular self-assembly and catalysis, the above initial attempts made it clear that the study of the catalytic activity of the biohybrid catalysts is not straightforward because of the self-assembly process that takes place. For this reason, a better understanding was sought of how enzymes and the polymer-modified enzymes work at the single molecule level [37]. As a first study, the behavior of CALB was investigated with the help of confocal fluorescence microscopy (CFM). This enzyme was immobilized at a very low concentration at a hydrophobic glass surface, and a hydrolysis reaction was carried out in which the profluorescent substrate **15** was converted into the fluorescent product **16** (Figure 6.12a) [38]. When studied with the CFM, the start and progress of this reaction could be easily visualized, because points of fluorescence appeared that remained present over

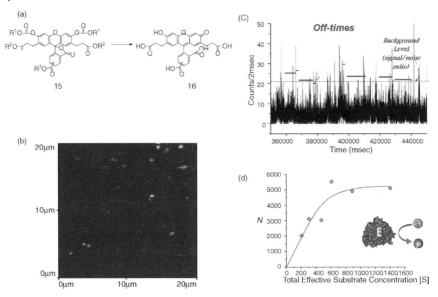

Figure 6.12 Studies on the catalytic behavior of single CALB enzymes. (a) Hydrolysis reaction catalyzed by the enzyme in which the fluorescent compound **16** is generated. (b) CFM image showing bright spots corresponding to single enzymes in action on a glass surface. (c) Fluorescence emission as a function of time for a single CALB enzyme during catalysis, highlighting the "off and on"-times. (d) Plot of the number of single enzyme turnover cycles (N) as a function of the substrate concentration ([S]); the curve shows a saturation profile, as expected for Michaelis–Menten behavior. Reproduced with kind permission [38].

time (Figure 6.12b). The observed fluorescence spots appeared to display blinking, a behavior attributed to enzyme turnover cycles, which is a direct measure of single enzyme catalysis. Since the formed product molecules diffused away rapidly compared to the formation of the fluorescent product in the active site of the enzyme, the reaction events of the single CALB enzymes could be measured over long periods of time, even up to 6 hours, allowing a detailed study of the kinetics of the CALB enzyme (Figure 6.12c) [39]. By setting a cut-off value for the background noise the fluorescence events originating from the catalytic turnovers could be separated and followed as a function of time. From the statistics of the waiting time distributions ("off"-states of the enzyme) we observed that single CALB enzymes exhibit a breathing motion in which they move between various conformational states, of which only a few display catalytic activity, in the sense that short bursts of top activity (~30 ms) are interchanged by long periods of inactivity (970 ms). This observation clearly highlights the unique information that can be obtained when single molecules instead of ensembles are studied. In further studies the catalytic activity of CALB was measured as a function of the substrate concentration and we were able to demonstrate for the first time that the kinetics of single CALB enzyme molecules exhibit Michaelis–Menten-like behavior (Figure 6.12d). The average k_{cat} and K_m values calculated from the saturation curve were determined to be $4\,\text{s}^{-1}$ and 2.5×10^{-7} M, respectively. Interestingly,

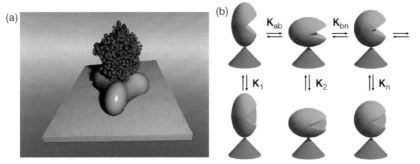

Figure 6.13 (a) Schematic representation of a enzyme–protein heterodimer in which the BSA protein serves as a "foot" for the TLL enzyme. (b) The fluctuating enzyme model, in which an enzyme adopts various interconverting conformations, each of which has a different catalytic activity ((b) Reprinted with kind permission of RSC).

the k_{cat} measured in solution differs by nearly two orders of magnitude from the top activity monitored for a single enzyme molecule ($k_{cat} = 125$ s^{-1}). The enhanced activity is tentatively attributed to surface-induced changes.

To further control the adsorption of the enzymes to the surface, with the aim of eliminating a multitude of non-active conformations, an enzyme–protein heterodimer was constructed by means of "click" chemistry, consisting of *Thermomyces lanuginoas* lipase (TLL) and bovine serum albumin (BSA) (Figure 6.13a) [40]. In this concept the BSA acts as a so-called "protein foot" that lifts the TLL enzyme from the surface, and as a result nearly all the immobilized enzymes exhibited activity in the hydrolysis of a profluorescent dye, in contrast to the enzyme molecules that were just adsorbed to the glass surface (*vide infra*). In addition, memory effects could now be observed in which the lipase enzyme carries out many catalytic conversions in the same geometry, before it changes to another geometry and becomes less active or inactive for a while. This behavior is in agreement with a "fluctuating enzyme" model in which the enzyme slowly moves through several conformations, of which only a few exhibit activity (Figure 6.13b).

6.5
Outlook

Heterogeneous and homogeneous catalysis nowadays are well-established fields of research and the same can be said for biocatalysis. The field of supramolecular catalysis is still in its infancy, but encouraging results have been obtained already, as shown in this chapter describing our own contributions and in reviews reporting on work published by other groups. It can be predicted that weak, noncovalent interactions will increasingly be used as a design tool to develop more active and more selective catalytic systems. This will require a better understanding of how these interactions can be controlled and manipulated under the various conditions of a chemical transformation. In this respect Nature will continue to be a good guide and a source of great inspiration for blueprints of future catalytic systems. Further help will come from new technologies, such as scanning probe techniques and laser spectroscopic methods that have been developed recently, which will allow a more precise understanding of catalytic mechanisms. They will turn out to be extremely useful in unraveling details of the structures of catalysts and in providing deeper insight into how catalytic reactions proceed at the level of individual molecules, both in solution and, an even greater challenge, at interfaces. Another line of research that can be foreseen as becoming more important in the field of bio-inspired or biomimetic supramolecular catalysis is the use of proteins and assemblies of proteins, perhaps even viruses, as structural building blocks to construct catalytic devices. The first steps in this direction have already been made, as outlined in this chapter.

References

1 Feiters, M.C. (1996) In: *Comprehensive Supramolecular Chemistry* (eds J.L. Atwood J.E. D.Davies, D.D. Mac Nicol, F. Vögtle, D.N. Rein houdt and J.-M. Lehn), Elsevier Science Ltd., Pergamon, Elmsford. vol. 10, 267–360.

2 Vriezema, D.M., Comellas Aragonès, M., Elemans, J.A.A.W., Cornelissen, J.J.L.M., Rowan, A.E. and Nolte, R.J.M. (2005) Self-assembled nanoreactors. *Chem. Rev.*, **105**, 1445–1490.

3 Rebek, J. Jr (2005) Simultaneous encapsulation: Molecules held at close range. *Angew. Chem., Int. Ed.*, **44**, 2068–2078.

4 Kirby, A.J. (1996) Enzyme mechanisms, models, and mimics. *Angew. Chem., Int. Ed. Engl.*, **35**, 706–724.

5 Motherwell, W.B., Bingham, M.J. and Six, Y. (2001) Recent progress in the design and synthesis of artificial enzymes. *Tetrahedron*, **57**, 4663–4686.

6 Breslow, R. (1995) Biomimetic chemistry and artifical enzymes – catalysis by design. *Acc. Chem. Res.*, **28**, 146–153.

7 Murakami, Y., Kikuchi, J., Hisaeda, Y. and Hayashida, O. (1996) Artifical enzymes. *Chem. Rev.*, **96**, 721–758.

8 Rowan, A.E., Elemans, J.A.A.W. and Nolte, R.J.M. (1999) Molecular and supramolecular objects from glycoluril. *Acc. Chem. Res.*, **32**, 995–1006.

9 Sijbesma, R.P., Kentgens, A.P.M., Lutz, E.T.G., van der Maas, J.H. and Nolte, R.J.M. (1993) Binding features of molecular clips derived from diphenylglycoluril. *J. Am. Chem. Soc.*, **115**, 8999–9005.

10 Reek, J.N.H., Priem, A.H., Engelkamp, H., Rowan, A.E., Elemans, J.A.A.W. and Nolte, R.J.M. (1997) Binding features of molecular clips. Separation of the effects of hydrogen bonding and pi-pi interactions. *J. Am. Chem. Soc.*, **119**, 9956–9964.

11 Smeets, J.W.H., van Dalen, L., Kaats-Richters, V.E.M. and Nolte, R.J.M. (1990) Functionalized basket-shaped hosts. Synthesis and complexation studies with (alkali) metal and ammonium and diammonium ions. *J. Org. Chem.*, **55**, 454–461.

12 Schenning, A.P.H.J., de Bruin, B., Rowan, A.E., Kooijman, H., Spek, A.L. and Nolte, R.J.M. (1995) Strong binding of paraquat and polymeric paraquat derivatives by basket-shaped hosts. *Angew. Chem., Int. Ed. Engl.*, **34**, 2132–2134.

13 Coolen, H.K.A.C., van Leeuwen, P.W.N.M. and Nolte, R.J.M. (1992) A RhI-centered cage compound with selective catalytic properties. *Angew. Chem., Int. Ed. Engl.*, **31**, 905–907.

14 Coolen, H.K.A.C., Meeuwis, J.A.M., van Leeuwen, P.W.N.M. and Nolte, R.J.M. (1995) Substrate selective catalysis by rhodium metallohosts. *J. Am. Chem. Soc.*, **117**, 11906–11913.

15 Martens, C.F., Klein Gebbink, R.J.M., Feiters, M.C. and Nolte, R.J.M. (1994) Shape-selective oxidation of benzylic alcohols by a receptor functionalized with a dicopper(II) pyrazole complex. *J. Am. Chem. Soc.*, **116**, 5667–5670.

16 Klein Gebbink, R.J.M., Martens, C.F., Feiters, M.C., Karlin, K.D. and Nolte, R.J.M. (1997) Novel molecular receptors capable of forming Cu-2-O-2 complexes. Effect of preorganization on O-2 binding. *Chem. Commun.*, 389–390.

17 Sprakel, V.S.I., Feiters, M.C., Meyer-Klaucke, W., Klopstra, M., Brinksma, J., Feringa, B.L., Karlin, K.D. and Nolte, R.J.M. (2005) Oxygen binding and activation by the complexes of PY2- and TPA-appended diphenylglycoluril receptors with copper and other metals. *Dalton Trans.*, 3522–3534.

18 Rowan, A.E., Aarts, P.P.M. and Koutstaal, K.W.M. (1998) Novel porphyrin-viologen rotaxanes. *Chem. Commun.*, 611–612.

19 Elemans, J.A.A.W., Claase, M.B., Aarts, P.P.M., Rowan, A.E., Schenning, A.P.H.J. and Nolte, R.J.M. (1999) Porphyrin clips derived from diphenylglycoluril. Synthesis, conformational analysis, and binding properties. *J. Org. Chem.*, **64**, 7009–7016.

20 Elemans, J.A.A.W., Bijsterveld, E.J.A., Rowan, A.E. and Nolte, R.J.M. (2000) A host–guest epoxidation catalyst with enhanced activity and stability. *Chem. Commun.*, 2443–2444.

21 Elemans, J.A.A.W., Bijsterveld, E.J.A., Rowan, A.E. and Nolte, R.J.M. (2007) Manganese porphyrin hosts as epoxidation catalysts – Activity and stability control by axial ligand effects. *Eur. J. Org. Chem.*, 751–757.

22 Thordarson, P., Bijsterveld, E.J.A., Rowan, A.E. and Nolte, R.J.M. (2003) Epoxidation of polybutadiene by a topologically linked catalyst. *Nature*, **424**, 915–918.

23 Coumans, R.G.E., Elemans, J.A.A.W., Nolte, R.J.M. and Rowan, A.E. (2006) Processive enzyme mimic: Kinetics and thermodynamics of the threading and sliding process. *Proc. Natl. Acad. Sci. U.S.A.*, **103**, 19647–19651.

24 Muthukumar, M. (2001) Translocation of a confined polymer through a hole. *Phys. Rev. Lett.*, **82**, 3188–3191.

25 Hidalgo Ramos, P., Coumans, R.G.E., Deutman, A.B.C., Smits, J.M.M., de Gelder, R., Elemans, J.A.A.W., Nolte, R.J.M. and Rowan, A.E. (2007) Processive rotaxane systems. Studies on the mechanism and control of the threading process. *J. Am. Chem. Soc.*, **129**, 5699–5702.

26 Oritz de Montellano, P.R. (ed.) (1995) *Cytochrome P450: Structure, Mechanism and Biochemistry*, 2nd edn, Plenum Press, New York.

27 van Esch, J., Roks, M.F.M. and Nolte, R.J.M. (1986) Membrane-bound cytochrome P-450 mimic. Polymerized vesicles as microreactors. *J. Am. Chem. Soc.*, **108**, 6093–6094.

28 Gosling, P.A., Hoffmann, M.A.M., van Esch, J.H. and Nolte, R.J.M. (1993) Reductive activation of dioxygen by a manganese(III) porphyrin-rhodium(III)-formate catalytic-system. *J. Chem. Soc., Chem. Commun.*, 472–473.

29 Schenning, A.P.H.J., Hubert, D.H.W., van Esch, J.H., Feiters, M.C. and Nolte, R.J.M. (1994) Novel bimetallic model system for cytochrome P450: Effect of membrane environment on the catalytic oxidation. *Angew. Chem., Int. Ed. Engl.*, **33**, 2468–2470.

30 Schenning, A.P.H.J., Lutje Spelberg, J.H., Hubert, D.H.W., Feiters, M.C. and Nolte, R.J.M. (1998) A supramolecular cytochrome P450 mimic. *Chem.–Eur. J.*, **4**, 871–880.

31 Schenning, A.P.H.J., Lutje Spelberg, J.H., Driessen, M.C.P.F., Hauser, M.J.B., Feiters, M.C. and Nolte, R.J.M. (1995) Enzyme mimic displaying oscillatory

behavior. Oscillating reduction of manganese(III) porphyrin in a membrane-bound cytochrome P-450 model system. *J. Am. Chem. Soc.*, **117**, 12655–12656.

32 Hulsken, B., van Hameren, R., Gerritsen, J.W., Khoury, T., Thordarson, P., Crossley, M.J., Rowan, A.E., Nolte, R.J.M., Elemans, J.A.A.W. and Speller, S. (2007) Real-time single-molecule imaging of oxidation catalysis at a liquid-solid interface. *Nat. Nanotech.*, **2**, 285–289.

33 Velonia, K., Rowan, A.E. and Nolte, R.J.M. (2002) Lipase polystyrene giant amphiphiles. *J. Am. Chem. Soc.*, **124**, 4224–4225.

34 Boerakker, M.J., Hannink, J.M., Bomans, P.H., Frederik, P.M., Nolte, R.J.M., Meijer, E.M. and Sommerdijk, N.A.J.M. (2002) Giant amphiphiles by cofactor reconstitution. *Angew. Chem., Int. Ed.*, **41**, 4239–4241.

35 Reynhout, I.C., Cornelissen, J.J.L.M. and Nolte, R.J.M. (2007) Self-assembled architectures from biohybrid triblock copolymers. *J. Am. Chem. Soc.*, **129**, 2327–2332.

36 Hannink, J.M., Cornelissen, J.J.L.M., Farrera, J.A., Foubert, P., de Schryver, F.C., Sommerdijk, N.A.J.M. and Nolte, R.J.M. (2001) Protein-polymer hybrid amphiphiles. *Angew. Chem., Int. Ed.*, **40**, 4732–4734.

37 Engelkamp, H., Hatzakis, N.S., Hofkens, J., De Schryver, F.C., Nolte, R.J.M. and Rowan, A.E. (2006) Do enzymes sleep and work? *Chem. Commun.*, 935–940.

38 Velonia, K., Flomenbom, O., Loos, D., Masuo, S., Cotlet, M., Engelborghs, Y., Hofkens, J., Rowan, A.E., Klafter, J., Nolte, R.J.M. and de Schryver, F.C. (2005) Single-enzyme kinetics of CALB-catalyzed hydrolysis. *Angew. Chem., Int. Ed.*, **44**, 560–564.

39 Flomenbom, O., Velonia, K., Loos, D., Masuo, S., Cotlet, M., Engelborghs, Y., Hofkens, J., Rowan, A.E., Nolte, R.J.M., de Schyver, F.C. and Klafter, J. (2005) Stretched exponential decay and correlations in the catalytic activity of fluctuating single lipase molecules. *Proc. Natl. Acad. Sci. U.S.A.*, **102**, 2368–2372.

40 Hatzakis, N.S., Engelkamp, H., Velonia, K., Hofkens, J., Christianen, P.C.M., Svendsen, A., Patkar, S.A., Vind, J., Maan, J.C., Rowan, A.E. and Nolte, R.J.M. (2006) Synthesis and single enzyme activity of a clicked lipase-BSA hetero-dimer. *Chem. Commun.*, 2012–2014.

7
Selective Stoichiometric and Catalytic Reactivity in the Confines of a Chiral Supramolecular Assembly

Michael D. Pluth, Robert G. Bergman, and Kenneth N. Raymond

7.1
Introduction

Nature uses enzymes to activate otherwise unreactive compounds in remarkable ways. For example, DNases are capable of hydrolyzing phosphate diester bonds in DNA within seconds [1–3] – a reaction with an estimated half-life of 200 million years without an enzyme [4]. The fundamental features of enzyme catalysis have been much discussed over the last 60 years in an effort to explain the dramatic rate increases and high selectivities of enzymes. As early as 1946, Linus Pauling suggested that enzymes must preferentially recognize and stabilize the transition state over the ground state of a substrate [5]. Despite the intense study of enzymatic selectivity and ability to catalyze chemical reactions, the entire nature of enzyme-based catalysis is still poorly understood. For example, Houk and coworkers reported a survey of binding affinities in a wide variety of enzyme–ligand, enzyme–transition-state, and synthetic host–guest complexes and found that the average binding affinities were insufficient to generate many of the rate accelerations observed in biological systems [6]. Therefore, transition-state stabilization cannot be the sole contributor to the high reactivity and selectivity of enzymes, but, rather, other forces must contribute to the activation of substrate molecules.

Inspired by the efficiency and selectivity of Nature, synthetic chemists have admired the ability of enzymes to activate otherwise unreactive molecules in the confines of an active site. Although much less complex than the evolved active sites of enzymes, synthetic host molecules have been developed that can carry out complex reactions within their cavities. While progress has been made towards highly efficient and selective reactivity inside synthetic hosts, the lofty goal of duplicating enzymes' specificity remains [7–9]. Pioneered by Lehn, Cram, Pedersen, and Breslow, supramolecular chemistry has evolved well beyond the crown ethers and cryptands originally studied [10–12]. Despite the increased complexity of synthetic host molecules, most assembly conditions utilize self-assembly to form complex highly-symmetric

Supramolecular Catalysis. Edited by Piet W. N. M. van Leeuwen
Copyright © 2008 WILEY-VCH Verlag GmbH & Co. KGaA, Weinheim
ISBN: 978-3-527-32191-9

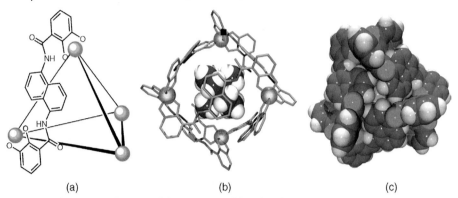

Figure 7.1 (a) Schematic of the M_4L_6 assembly with only one ligand shown for clarity. (b) Model of **1** with encapsulated NEt_4^+. (c) Space-filling model of **1** as viewed down the aperture coincident with the three-fold axis.

structures from relatively simple subunits. For supramolecular assemblies able to encapsulate guest molecules, the chemical environment in each assembly – defined by the size, shape, charge, and functional group availability – greatly influences the guest-binding characteristics [6,13–17].

Over the last decade, the Raymond group has made efforts toward understanding how encapsulation of molecules within a synthetic host affects the reactivity of the guest. Several host molecules of stoichiometry M_4L_6 [M = Ga^{III} (**1**), Al^{III}, In^{III}, Fe^{III}, Ti^{IV}, or Ge^{IV}, L = N,N'-bis(2,3-dihydroxybenzoyl)-1,5-diaminonaphthalene] (Figure 7.1) have been developed [18–21], and the strategy for carrying out reactions in **1** is discussed in this chapter. The M_4L_6 assembly is a well-defined, self-assembling tetrahedron formed from metal–ligand interactions with the ligands spanning each edge and the metal ions occupying the vertices. The tris-bidentate coordination of the catechol amides at the metal vertices makes each vertex a stereocenter and the rigid ligands transfer the chirality of one metal vertex to the others, thereby forming the homochiral ΔΔΔΔ or ΛΛΛΛ configurations [22,23]. While the −12 overall charge imparts water solubility, the interior cavity is defined by the naphthalene walls, thereby providing a hydrophobic environment that is isolated from the bulk aqueous solution. Initial studies of host formation and guest encapsulation focused on small tetraalkyl-ammonium cations such as NEt_4^+. Making use of the hydrophobicity and polyanionic charge of **1**, several highly reactive cations have been kinetically stabilized by encapsulation. These include tropylium [24], iminium [25], diazonium [26], and reactive phosphonium species [27], all of which decompose rapidly in water and are normally stable only under anhydrous or highly acidic aqueous conditions.

Although thermodynamically stable within **1**, encapsulated guests are able to freely exchange with other guests in solution. The kinetics and thermodynamics of guest encapsulation and exchange have been studied and the mechanism of guest exchange determined [28,29]. The activation barrier for guest ejection is dependent on the size of the guest. Despite the hemi-labile coordination of the catechol oxygens at the metal vertices, the assembly remains intact during the guest exchange process.

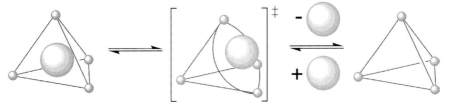

Figure 7.2 Guest exchange schematic for **1**. Dilation of the apertures allows for guest ingress and guest egress.

During guest exchange, the apertures coincident with the three-fold axis of **1** dilate to allow for guest ingress and egress (Figure 7.2). This exchange mechanism is consistent with the constrictive binding model proposed by Cram from work on carcerands and hemicarcerands [30].

As discussed in this chapter, the fundamental host–guest chemistry of **1** has been elaborated to include both stoichiometric and catalytic reactions. The constrained interior and chirality of **1** allows for both size- and stereo-selectivity [31–35]. Additionally, **1** itself has been used as a catalyst for the sigmatropic rearrangement of enammonium cations [36,37] and the hydrolysis of acid-labile orthoformates and acetals [38,39]. Our approach to using **1** to mediate chemical reactivity has been twofold: First, the chiral environment of **1** is explored as a source of asymmetry for encapsulated achiral catalysts. Second, the assembly itself is used to catalyze reactions that either require preorganization of the substrate or contain high energy intermediates or transition states that can be stabilized in **1**.

7.2
Chemistry of Organometallic Guests

Asymmetric catalysis is one of the fastest-growing and rapidly evolving fields in chemistry [40–44]. With the recent expanding search for both new asymmetric methodologies as well as the drive to synthesize newly isolated and often biologically-active molecules, understanding the stereocontrol of reactions is of the utmost importance. The stereocontrol of catalyzed chemical reactions is often achieved by incorporating a source of chirality into the reaction by means of a chiral catalyst or chiral auxiliary on the reactant. Many transition metal catalysts, for example, employ chiral ligands to affect the stereoselectivity of a catalyzed reaction [45–50]. Similarly, the rapidly developing field of organocatalysis often employs chiral compounds as catalysts to carry out a wide variety of transformations [51–56]. A related strategy for introducing asymmetry to a reaction is the use of a chiral medium such as a chiral solvent, chiral counterian, or chiral ionic liquid [57–61]. All of these methods seek to provide the reactive site of the reaction with a unique environment that allows for one enantiomer or transition state to be favored due to steric or functional group interactions. To further the understanding of close steric proximity on the stereocontrol of reactions, the innate chirality of **1** was explored as a substitute for a chiral ligand for chemical reactions occurring inside **1**. Ideally, active catalysts containing bulky and complex ligands could be simplified to their achiral analogues, and the chirality of **1** would furnish asymmetric induction. To investigate this hypothesis, simple organometallic guests were encapsulated in **1** to confirm that **1** is a viable host for such molecules.

Initially, monocations of the form CpRu(η^6-arene)$^+$ (Cp = η^5-cyclopentadienyl) were explored as guests in **1** [31]. Although these complexes are not catalytically active, they do provide a shape and size similar to that of many organometallic catalysts. Addition of CpRu(η^6-C$_6$H$_6$)$^+$ (**2**) to **1** formed the host–guest complex [**2** ⊂ **1**]$^{11-}$ (⊂ denotes encapsulation), with the guest resonances in the ^{1}H NMR spectrum characteristically shifted upfield by 2–3 ppm due to the magnetic anisotropy of the naphthalene walls of **1**. A similar but larger guest CpRu(p-cymene)$^+$ (**3**) was also encapsulated. Although **3** itself is achiral, the isopropyl methyl groups on the arene are enantiotopic and, when encapsulated in the purely-rotational T-symmetric host, were rendered diastereotopic. This confirmed the hypothesis that encapsulated guests are influenced by the chiral environment of **1**.

Expanding on the encapsulation of organometallic guests, half-sandwich complexes of the form CpRu(η^4-diene)(H$_2$O)$^+$ were encapsulated in **1** [32]. When the diene portion of the half-sandwich complex is unsymmetrically substituted, the ruthenium atom becomes a chiral center. Addition of CpRu(2-ethylbutadiene)(H$_2$O)$^+$ (**4**) to **1** revealed the existence of two diastereomers. Encapsulation of these racemic ruthenium complexes in racemic **1** leads to diastereomeric pairs of enantiomeric host–guest complexes (Δ/R, Δ/S, Λ/R, Λ/S) (Figure 7.3). However, chiral discrimination was not observed with the diastereomeric ratio (d.r.) being 50 : 50.

By increasing the steric bulk of the organometallic guest, and thereby increasing the steric interaction of the guest with **1**, it was hoped that higher diastereoselectivity would be observed. Replacement of the Cp ring of **4** with the more sterically demanding Cp* (Cp* = η^5-pentamethylcyclopentadienyl) ligand to form Cp*Ru(2-ethylbutadiene)(H$_2$O)$^+$ (**6**) again produced diastereomers, as observed by ^{1}H NMR spectroscopy, but provided a d.r. of 85 : 15. To probe the generality and extent of this diastereoselectivity, a series of 1- and 2-substituted diene complexes were investigated (Table 7.1). A small change in substitution, such as moving a methyl group from the 1- to the 2-position of the diene (e.g., **6** and **10** or **8** and **11**), greatly impacts the selectivity of **1**, suggesting that the chiral induction by **1** is sensitive to both the shape and the size of the encapsulated guests.

After establishing that **1** can project its chirality to encapsulated guests, the next step toward catalysis was the investigation and mechanistic study of stoichiometric reactions in **1**. A more chemically reactive organometallic guest was sought that could react with a substrate to form a chiral product. A suitable candidate was the complex [Cp*(PMe$_3$)Ir(Me)OTf] (**12**), which was developed and studied by the Bergman group [62–66]. This complex thermally activates C—H bonds in various molecules such as aldehydes, ethers, and hydrocarbons, including methane. Dissociation of the labile triflate ligand from **12** affords the reactive monocationic intermediate [Cp*(PMe$_3$)Ir(Me)]$^+$ (Scheme 7.1). This cationic species or its solvent adduct should be an ideal candidate for encapsulation in **1**. However, addition of **12** to an aqueous solution of **1** did not afford a host–guest complex, presumably because the aquo species Cp*(PMe$_3$)Ir(Me)(OH$_2$)$^+$ is too highly solvated. To circumvent this problem,

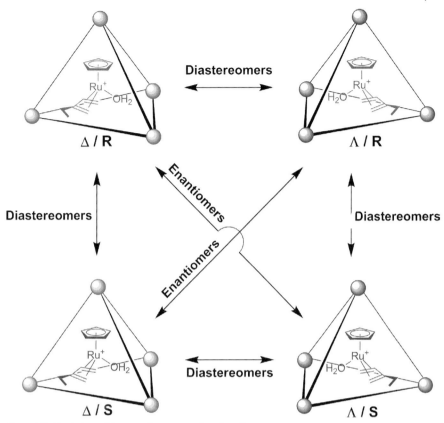

Figure 7.3 Depiction of the two diastereomeric pairs of enantiomers forming four possible stereoisomers of racemic 1 and racemic ruthenium half-sandwich compound 4.

Table 7.1 Observed diastereomeric ratios (d.r.s) of host–guest complexes of the form $K_{11}[Cp^*Ru(\eta^4\text{-diene})(H_2O) \subset 1]$.

Entry	R_1	R_2	Compound	D.r.
1	H	Me	5	52:48
2	H	Et	6	85:15
3	H	i-Pr	7	63:37
4	H	n-Pr	8	82:18
5	Me	H	9	58:42
6	Et	H	10	59:41
7	n-Pr	H	11	59:17

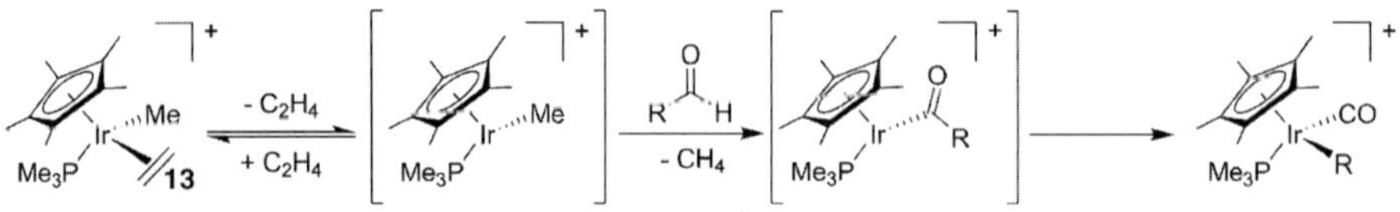

Scheme 7.1 Mechanism for C–H bond activation of aldehydes by **13**.

the more hydrophobic olefin species Cp*(PMe$_3$)Ir(Me)(η^2-olefin)$^+$ [olefin = ethylene (**13**), cis-2-butene (**14**)] were prepared and introduced to **1**. Both of these species formed host–guest complexes, [**13** ⊂ **1**]$^{11-}$ (**15**) and [**14** ⊂ **1**]$^{11-}$ (**16**), which likely benefited from the increased guest hydrophobicity and the potential π–π interactions between the coordinated olefin and the π-basic naphthalene walls of **1**. While **15** provided a modest d.r. of 55 : 45, the more sterically bulky **16** provided a higher d.r. of 70 : 30.

To generate the active iridium species, dissociation of the coordinated olefin was required. Gentle heating of the host–guest complex (45 °C for **16**, 75 °C for **15**) facilitated olefin dissociation and allowed C–H bond activation of the substrates to occur. Upon addition of acetaldehyde to the iridium host–guest complex, new resonances corresponding to encapsulated [Cp*(PMe$_3$)Ir(CO)(Me)]$^+$ (**17**) were observed. Various aldehydes were added to **16** to probe its reactivity. Interestingly, selectivity for both size and shape was observed. Small aldehydes, such as acetaldehyde, are readily C–H activated whereas large aldehydes, such as benzaldehyde, are not. Although in the absence of **1** both acetaldehyde and benzaldehyde undergo C–H bond activation in the presence of **14**, in a competition experiment with the two substrates in the presence of **16**, only acetaldehyde is C–H activated, suggesting that benzaldehyde is too large to enter the assembly. Table 7.2 shows the scope of aldehydes activated by **16** [33,34].

Small changes in the shape of the aldehydes have dramatic effects on reactivity with the encapsulated host–guest complex. For example, the host–guest complex reacts with isobutyraldehyde (entry 5) with a lower diastereoselectivity than with n-butyraldehyde (entry 3). This may be because the isobutyraldehyde complex is more spherical than the n-butyraldehyde complex. When comparing the five-carbon aldehydes (entries 6–8), only isovaleraldehyde undergoes C–H bond activation in **1**.

With the knowledge that **14** can activate aldehydes in **1**, the role of **1** in the reaction was explored further. Specifically, the relative rates of C–H bond activation and guest ejection, and the possibility of ion association with **1**, were investigated. The hydrophobic nature of **14** could allow for ion association on the exterior of **1**, which would be both enthalpically favorable due to the cation–π interaction, and entropically favorable due to the partial desolvation of **14**. To explore these questions, **14** was irreversibly trapped in solution by a large phosphine, which coordinates to the iridium complex and thereby inhibits encapsulation. Two different trapping phosphines were used. The first, triphenylphosphine tris-sulfonate sodium salt (TPPTS), is a trianionic water-soluble phosphine and should not be able to approach the highly anionic **1**, thereby only trapping the iridium complex that has diffused away from **1**. The second phosphine, 1,3,5-triaza-7-phosphaadamantane (PTA), is a water-soluble neutral phosphine that should be able to intercept an ion-associated iridium complex.

Table 7.2 Diastereoselectivities for C–H bond activation of aldehydes by **16**.

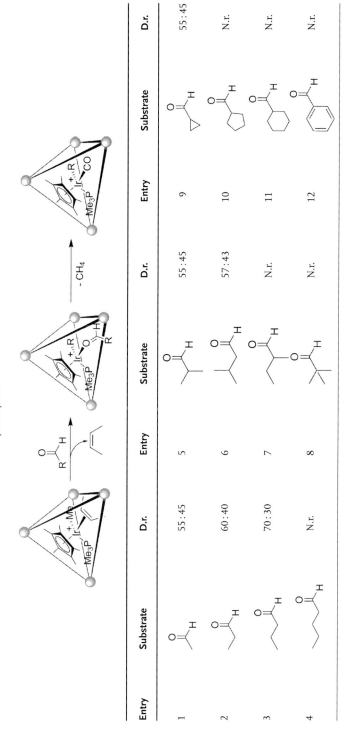

Entry	Substrate	D.r.	Entry	Substrate	D.r.	Entry	Substrate	D.r.
1	acetaldehyde	55:45	5	isobutyraldehyde	55:45	9	cyclopropanecarboxaldehyde	55:45
2	propionaldehyde	60:40	6	3-methylbutanal	57:43	10	cyclopentanecarboxaldehyde	N.r.
3	butyraldehyde	70:30	7	2-ethylbutanal	N.r.	11	cyclohexanecarboxaldehyde	N.r.
4	pentanal	N.r.	8	pivaldehyde	N.r.	12	benzaldehyde	N.r.

7 Selective Stoichiometric and Catalytic Reactivity

TPPTS: P(−C₆H₄−SO₃Na)₃ structure shown

PTA: 1,3,5-triaza-7-phosphaadamantane structure shown

Significantly, the neutral phosphine PTA trapped ejected **14** at a much faster rate than the trianionic phosphine TPPTS, suggesting the presence of an ion associated intermediate of **14** on the exterior of **1**. Upon addition of Na^+ or K^+, the rate of capture by TPPTS increases, suggesting that the added cations could facilitate dissociation of the ion associated complex. Furthermore, the exterior ion associated $Cp^*(PMe_3)Ir(Me)(PTA)^+$ was observed in solution by 2D NOESY spectrum with the Cp^* showing close proximity to the catechol resonances of **1**. From a series of kinetic and thermodynamic analyses, the guest dissociation mechanism in Figure 7.4 was proposed. This mechanism involves the formation of an ion associated complex prior to complete guest dissociation. Since the rate of C−H bond activation was considerably faster than that of guest ejection, these studies confirmed that the reactivity was taking place inside of the assembly, rather than in bulk solution.

With a greater understanding of the chemistry and mechanism of stoichiometric organometallic reactivity inside of **1**, catalytic systems were investigated. The prevalence of monocationic rhodium catalysts in the literature, and the water solubility of

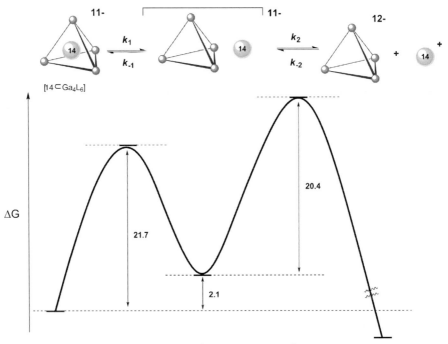

Figure 7.4 Energy coordinate diagram for the dissociation of **14** from **1**. The initial dissociation (k_1) produces an ion associated intermediate that can further dissociate (k_2) to the free guest.

7.2 Chemistry of Organometallic Guests

Table 7.3 Binding constants for square planar rhodium-phosphine complexes in **1**.

Entry	Compound	Guest	$10^{-2} K_a$ (M^{-1})
1	18	(PMe$_3$)$_2$Rh(COD)$^+$	5.2
2	19	(dmpe)Rh(COD)$^+$	5.7
3	20	(PMe$_3$)$_2$Rh(NBD)$^+$	1.2
4	21	(dmpe)Rh(NBD)$^+$	1.2
5	22	(PEt$_3$)$_2$Rh(COD)$^+$	Not encapsulated

these catalysts, prompted the investigation of such species as guests for **1**. Small bis-phosphine complexes of the form [(P-P)Rh(diene)][BF$_4$] [P-P = 2PR$_3$ or 1,2-bis(dimethylphosphino)ethane (dmpe), diene = 1,5-cyclooctadiene (COD) or norbornadiene (NBD)] were prepared and investigated as potential guests [67]. The cationic complexes catalyze various chemical transformations, including the isomerization of allylic alcohols, which is an ideal system for mediation by **1** since most allylic alcohols are readily water-soluble. While much work concerning rhodium complexes bearing large bis-phosphine ligands has appeared in the literature, the reactivity of the smaller phosphine derivatives remained unexplored. Treatment of **1** with the small rhodium cations provided evidence for encapsulation by ^{1}H NMR spectroscopy. Despite the small size of these complexes, a sharp size cutoff for encapsulation was observed, as the complex formed with triethylphosphine was not encapsulated. To favor catalysis occurring inside of **1**, the binding constants for the different rhodium complexes were determined so that a catalyst with a suitable affinity for **1** could be used (Table 7.3). Despite the appropriate size and hydrophobicity of these complexes, the binding affinities in **1** were substantially lower than for the iridium complexes studied in C–H bond activation. This discrepancy may be due to the less complementary shape of the square planar rhodium complexes and the possibility of solvation at the vacant axial coordination sites on the metal.

To generate the catalytically active species, the diene ligand of **18** was removed by hydrogenation with 1 atm H$_2$ to afford the bis-aquo species (PMe$_3$)$_2$Rh(H$_2$O)$_2$$^+$ (**23**). In the absence of **1**, addition of small allyl alcohols quickly yielded the isomerized product. However, the reaction did not tolerate terminal substitution, and isomerization of crotyl alcohol was not observed (Table 7.4). Significantly, when both crotyl alcohol and allyl alcohol were added to **23**, neither substrate underwent isomerization, suggesting that crotyl alcohol inhibits the reaction.

Having established that **23** is catalytically active in aqueous solution, the reactivity of **23** inside of **1** was investigated. Starting with [**18** ⊂ **1**]$^{11-}$, hydrogenation of the diene ligand with 1 atm H$_2$ afforded the catalytically active [**23** ⊂ **1**]$^{11-}$. In contrast to the results of the unencapsulated complex, selective isomerization of specific allylic alcohols was observed. Only substrates of a suitable shape or size were able to enter the host cavity and undergo isomerization (Table 7.5). Substrates with branching did not undergo isomerization in **1** in contrast to their reactivity in solution. Also of interest is the resistance of the encapsulated rhodium complex to degradation by crotyl alcohol. If both crotyl alcohol and allyl alcohol are added to the rhodium host–guest complex, isomerization of allyl alcohol is observed. This suggests that the

Table 7.4 Isomerization of allylic substrates by **23**.

Entry	Substrate	Yield %[a]	Entry	Substrate	Yield %[a]
1	⌇⌇OH (allyl)	95	5	(methallyl-type)OH	N.r.
2	⌇⌇OH (methyl-substituted)	95	6	Ph-allyl-OH	N.r.
3	⌇⌇OH (dimethyl)	95	7	allyl ether	95[b]
4	⌇⌇OH (crotyl)	N.r.	8	allyl ether	95[b]

Reaction: allyl-OH → propanal; 10% **23**, D_2O, 25 °C, 0.5 h

[a]Determined by ^1H NMR spectroscopy.
[b]1 : 1 E : Z enol ether was obtained.

host molecule prevents crotyl alcohol from entering the assembly and essentially protects the active catalyst from this detrimental substrate.

Despite the selective catalysis of this isomerization exhibited by [23 ⊂ 1]$^{11-}$, a larger substrate scope was desirable. This objective can be approached from two distinct directions, either of which requires a larger volume available for substrates to react inside of **1**. The first approach requires a larger molecular host so that the encapsulated catalysts could have a larger volume available for substrate binding. Such complexes have been prepared in the Raymond group [68–70] and the reactivity of these complexes is an active area of research. A second strategy for increased substrate scope is to use the assembly itself as a catalyst rather than encapsulate a catalyst inside the assembly. Numerous examples of synthetic host molecules catalyzing reactions have been observed, with often astonishing selectivity and rate acceleration.

Table 7.5 Isomerization of allylic substrates by [23 ⊂ 1]$^{11-}$.

Entry	Substrate	Yield %[a]	Entry	Substrate	Yield %[a]
1	⌇⌇OH (allyl)	95	5	(methallyl-type)OH	N.r.
2	⌇⌇OH (methyl-substituted)	N.r.	6	Ph-allyl-OH	N.r.
3	⌇⌇OH (dimethyl)	N.r.	7	allyl ether	95[b]
4	⌇⌇OH (crotyl)	N.r.	8	allyl ether	N.r.

Reaction: allyl-OH → propanal; 10% [23⊂1], D_2O, 25 °C, 0.5 h

[a]Determined by ^1H NMR spectroscopy.
[b]1 : 1 E : Z enol ether was obtained.

7.3
The Assembly as a Catalyst

7.3.1
Electrocyclic Rearrangements

The strategy of using a synthetic host molecule as a catalyst draws direct inspiration from enzyme catalysis. Enzymes often contain strategically placed functional groups in the active site, and while synthetic host cavities generally do not bear such complexity, the constrained binding environment can provide catalytic reactivity. One benefit of binding substrates in a finite cavity is the increased encounter frequency of the bound molecules, which may also be thought of as an increased local concentration. For example, Rebek and coworkers have observed a 200-fold rate acceleration through encapsulation in the Diels–Alder reaction of benzoquinone with cyclohexadiene mediated by a hydrogen bonded, self-assembled "softball" [71,72]. Unfortunately, such systems are often plagued by the high binding affinity of the product, which prevents catalytic turnover. In such cases where product inhibition is observed, choosing different reactants can often lower the binding affinity of the product. For example, in the Rebek system, the use of a different dienophile, 2,5-dimethylthiophene dioxide, provided a product with a lower binding affinity than the substrate, thereby allowing for catalytic turnover [73]. Similarly, Fujita and coworkers have used organopalladium cages to change the reactivity and selectivity of Diels–Alder reactions occurring within the molecular host [74,75]. Interestingly, usually unreactive aromatic dienes, such as triphenylene, perylene and peryene, reacted with N-cyclohexylmaleimide when trapped inside molecular hosts. Similarly, a related molecular cage has been used to change the regioselectivity of the Diels–Alder cycloaddition of anthracene and phthalimide guests such that the terminal rather than the central anthracene ring acts as the dienophile.

To use **1** itself as a catalyst, a chemical transformation with a monocationic substrate that is compatible with the supramolecular host needed to be identified. Ideally, the reaction would either produce a weakly bound product or a product that could undergo further reaction in solution to prevent its re-encapsulation in **1**. The utility of tetraalkylammonium cations as guests prompted a search for similar but more chemically reactive guests. Enammonium cations, associated with the 3-aza-Cope rearrangement, provided an attractive class of candidates [76–78]. The 3-aza-Cope (or aza-Claisen) reaction is a member of the [3,3] class of sigmatropic rearrangements and occurs thermally in N-allyl enamine systems with varying degrees of ease. Neutral allylic enamines thermally rearrange to δ-ene imines at elevated temperatures (170–250 °C); however, the corresponding quaternized molecules require much milder conditions (20–120 °C) [79–81]. The subsequent iminium product is hydrolyzed in water to the corresponding γ,δ-unsaturated aldehyde. Since neutral molecules are only very weakly bound by **1**, hydrolysis of the iminium product should circumvent product inhibition and allow for catalytic turnover (Scheme 7.2).

To determine if encapsulation in **1** affected the rate of the unimolecular rearrangement, various enammonium cations were prepared and the rates of rearrangement

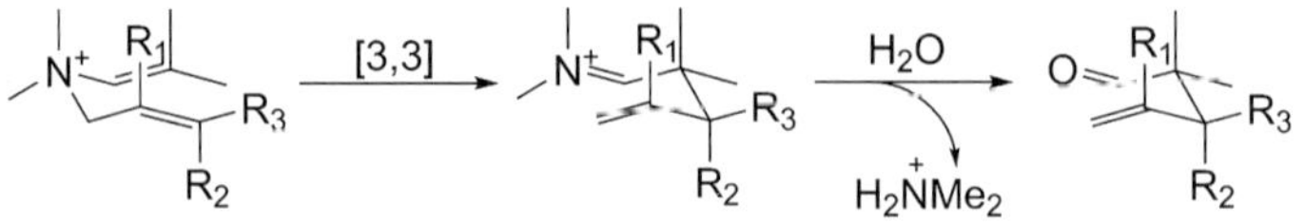

Scheme 7.2 General scheme for the 3-aza-Cope rearrangement. The enammonium cation undergoes a [3,3] sigmatropic rearrangement to form an iminium cation that can be hydrolyzed in water to give the associated aldehyde and dimethyl ammonium.

were measured for the free and encapsulated reactions. Encouragingly, in all cases, the encapsulated substrates rearranged faster than in the un-encapsulated reaction, with the largest rate acceleration being almost three orders of magnitude (Table 7.6) [36,37]. Interestingly, intermediately sized substrates appear to be an "optimal fit" in **1** and show the largest rate accelerations. Larger or smaller substrates are still accelerated by **1**, but to a lesser extent. As was also observed in the C–H bond activation of aldehydes, both shape and size selectivity are observed. For example, comparing the cis and trans substitution isomers (**26, 27** and **28, 29**) shows an increased acceleration for the trans isomers.

Having established that **1** catalyzes the unimolecular rearrangement, the origin of this acceleration was investigated. Addition of a strongly-binding guest, such as NEt_4^+, to **1** inhibited the catalysis, suggesting that the interior cavity of **1** was responsible for catalysis. Control studies of the rearrangement in different solvents showed no dependence on solvent polarity, suggesting that the hydrophobic interior of **1** was not the primary contributor to the acceleration. The prospect that the high negative charge of **1** was causing the rate acceleration was ruled out as adding salt (2 M KCl) in the absence of the assembly did not result in a notable increase in rate for the free rearrangement.

To probe the energetics of the reaction in **1**, the activation parameters were measured for substrates **26,27,30** for both the free and the encapsulated rearrangements (Table 7.7). The obtained parameters for the free rearrangement of substrate **26**, for example, are [$\Delta H^{\ddagger} = 23.1(8)$ kcal mol^{-1} and $\Delta S^{\ddagger} = -8(2)$ cal mol^{-1} K^{-1}] and are similar to those reported in the literature for related systems. This negative entropy of activation suggests that an organized transition state is required for the rearrangement. To ensure that this negative entropy of activation was not an artifact of solvation changes specific to the aqueous medium, the activation parameters for **26** were also measured in C_6D_5Cl, again revealing a negative entropy of activation. The encapsulated reaction of $[26 \subset 1]^{11-}$ in water gave an identical enthalpy of activation [$\Delta H^{\ddagger} = 23.0(9)$ kcal mol^{-1}]; however, the entropy of activation differed remarkably by almost 10 eu [$\Delta S^{\ddagger} = +2(3)$ cal mol^{-1} K^{-1}], suggesting preorganization of the encapsulated substrate by **1**.

Analysis of the activation parameters for the different encapsulated substrates reveals that the source of catalysis is more complex than simply a reduction of the entropy of activation, since different effects are observed for substrates **26,27,30**. While the rate acceleration for the encapsulated **26** was exclusively due to lowering the entropic barrier, for **27** and **30** a decrease in the enthalpic barrier for rearrangement is observed in addition. It is possible that, for **27** and **30** binding into the narrow confines of the metal–ligand assembly induces some strain on the bound molecules, thereby raising their ground-state energies compared to those of the unbound

Table 7.6 Substrate scope and rate constants for the free (k_{free}) and encapsulated (k_{encaps}) rearrangements.

Entry	Substrate	Product	$10^5 k_{free}$ (s^{-1})	$10^5 k_{encaps}$ (s^{-1})	Acceleration
1	24		3.49	16.3	5
2	25		7.61	198	26
3	26		3.17	446	141
4	27		4.04	135	90
5	28		1.69	74.2	150
6	29		0.37	316	44

(Continued)

Table 7.6 (Continued)

Entry	Substrate	Product	$10^5 k_{free}$ (s^{-1})	$10^5 k_{encaps}$ (s^{-1})	Acceleration
7	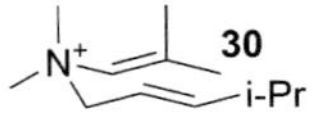 30, i-Pr	 i-Pr aldehyde	3.97	222	854
8	 31, Bu	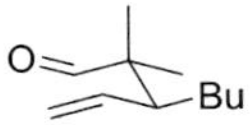 Bu aldehyde	0.033	1.17	56
9	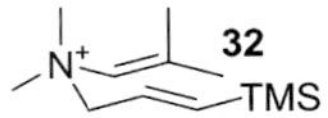 32, TMS	 TMS aldehyde	6.3	331	35
10	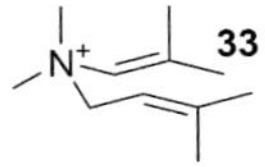 33	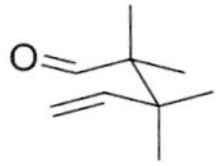 aldehyde	3.49	16.3	53

Table 7.7 Summary of activation parameters for the sigmatropic rearrangement of free and encapsulated substrates.

Entry	Substrate	Solvent	$\Delta H^{\ddagger}$ (kcal mol^{-1})	$\Delta S^{\ddagger}$ (cal mol^{-1} K^{-1})
1	26	D$_2$O	23.1(8)	−8(2)
2		C$_6$D$_5$Cl	23.4(5)	−5(2)
3		Encaps.	23.0(9)	+2(3)
4	27	D$_2$O	23.0(4)	−10(1)
5		Encaps.	21.8(7)	−5(2)
6	28	D$_2$O	23.6(3)	−11(1)
7		Encaps.	22.6(9)	−1(2)

substrates. The changes in $\Delta S^{\ddagger}$ suggest that encapsulation selects a preorganized conformation of the substrate which facilitates the rearrangement as shown in the mechanism for rearrangement and hydrolysis in **1** (Scheme 7.3). The space-restrictive host cavity allows for encapsulation of only tightly packed conformers that closely resemble the conformations of the transition states. The predisposed conformers, which have already lost several rotational degrees of freedom, are selected from an equilibrium mixture of all possible conformers, causing the entropic barrier for rearrangement to decrease. The lower enthalpic barrier for rearrangement in **1** is realized by the added strain that is induced by squeezing the ground state into the tight cavity. The strain becomes more significant for larger substrates, allowing for a noticeable decrease in $\Delta H^{\ddagger}$ when the optimal fit of the reactant transition state in the host cavity is exceeded, and the rate accelerations become attenuated, as seen with substrates **31** and **32**.

Analysis of 2D NOESY spectra of encapsulated enammonium substrates also suggests that the host assembly can selectively bind preorganized, reactive conformations of the substrates. The hypothesis of substrate preorganization upon encapsulation was further investigated using quantitative NOE growth rate experiments that allowed the conformation of the encapsulated substrates to be determined [82]. These studies were carried out for $[26 \subset 1]^{11-}$ and $[27 \subset 1]^{11-}$ and revealed that the ground state conformations of the substrate in **1** resembled the chair-like transition state for the rearrangement (Figure 7.5), supporting the hypothesis that the lowered entropic activation barrier for the rearrangement of the encapsulated substrate is due to preorganization of the substrate upon encapsulation.

Although the [3,3] sigmatropic rearrangement was occurring inside **1**, the question of the nature of the hydrolysis remained – does water enter the host cavity and hydrolyze the iminium inside of the assembly or is the iminium cation ejected and hydrolyzed in solution? To further understand the hydrolysis mechanism, the

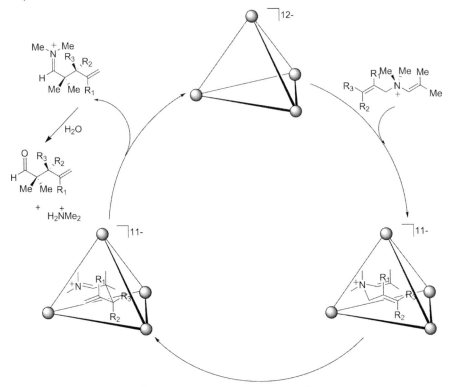

Scheme 7.3 Mechanism for the [3,3] aza-Cope rearrangement of enammonium substrates in **1**. Hydrolysis of the iminium product leads to catalytic turnover.

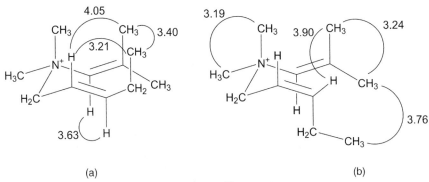

Figure 7.5 Intramolecular distances for [26 ⊂ 1]$^{11-}$ (a) and [27 ⊂ 1]$^{11-}$ (b) as determined by NOE buildup studies. Distances to methyl groups refer to a pseudoatom located at the average location of the three hydrogen atoms.

reaction mechanism was probed using **33** because, in this case, the starting enammonium cation is almost quantitatively converted into the iminium cation before hydrolysis, thereby allowing both species to be monitored by ^{1}H NMR spectroscopy. The product iminium cation can be hydrolyzed by water at neutral pH but is more quickly hydrolyzed by hydroxide. If hydrolysis of the iminium cation occurs exclusively inside the cavity of **1**, the rate of hydrolysis should be independent of hydroxide concentration since the only nucleophile in solution that can enter the cavity of **1** is water. If, however, the iminium cation is hydrolyzed outside of **1**, the rate of hydrolysis would be expected to change with pH.

From a series of experiments in solutions of pD varying from 6.5 to 12.8 and six equivalents of NMe_4^+, the buildup and hydrolysis of the iminium cationic intermediate was studied in the rearrangement and hydrolysis of $[33 \subset 1]^{11-}$. In neutral pD (pD 6.5–8), the rates of iminium hydrolysis are essentially constant, with water acting as the nucleophile. However, in more basic solution, a dependence on $[OD^-]$ is observed until approximately pD 11, at which point saturation is observed. The observed linear first-order dependence on hydroxide concentration from the pD rage of 9–10.5 supports the mechanistic model where the iminium cation is ejected from the assembly and then hydrolyzed in solution. The presence of saturation implies that, after pD 11, the rate of iminium dissociation from **1** becomes rate limiting because hydrolysis becomes faster than the re-encapsulation process.

To further investigate the dissociation of the iminium cation from **1**, the rearrangement of $[33 \subset 1]^{11-}$ was studied while varying the concentration of NMe_4^+ at high pD. Prior kinetic experiments demonstrated that the rates of guest exchange were sensitive to the concentration of other cations in solution such as NMe_4^+ [29]. Increasing $[NMe_4^+]$ led to faster hydrolysis with a first order rate dependence on $[NMe_4^+]$; however, at high concentrations of NMe_4^+ saturation was observed.

The dependence of the observed rate on $[NMe_4^+]$ can be interpreted in two ways. If an S_N2-type bimolecular pathway for guest exchange is present, the addition of NMe_4^+ may assist guest egress, thereby accelerating release of the iminium cation. However, NMe_4^+ is a very weakly binding guest, which makes this explanation unlikely. A second possibility is that ion association of NMe_4^+ with the exterior of **1** facilitates the displacement of ion associated iminium cations. This ion association would explain the pD dependence on the rate of hydrolysis since the anionic hydroxide could not approach the highly-charged **1** to intercept the ion associated complex. Only when the concentration of NMe_4^+ is increased does the ion associated intermediate get released into solution where it can be hydrolyzed by hydroxide.

To further probe this pathway, NBu_4^+, a cation too large to enter **1**, was used to displace the ion associated iminium product. As with NMe_4^+, increasing the concentration of NBu_4^+ led to faster rates of hydrolysis with a first order dependence on $[NBu_4^+]$. With this information, the mechanistic model was proposed that incorporates an intermediate ion associated iminium cation complex (Scheme 7.4). The encapsulated enammonium substrate rearranges inside of **1** to form the iminium cation. The rearrangement step, as anticipated, is independent of $[OD^-]$ or $[NMe_4^+]$. The iminium product can reversibly dissociate from **1** to the exterior of the assembly where it is tightly ion associated. In the presence of a suitable ion

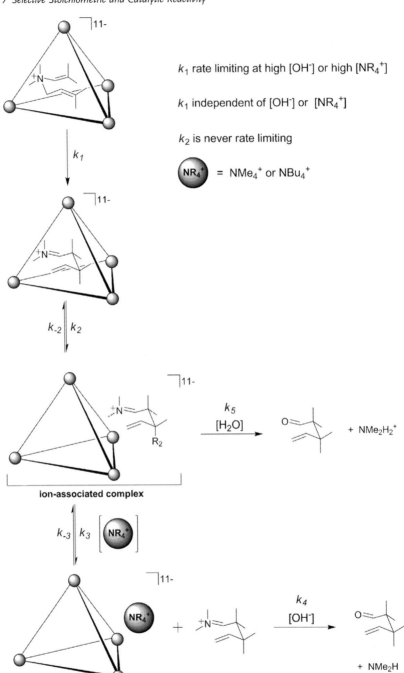

Scheme 7.4 Mechanism for hydrolysis of the iminium product in neutral or basic solution.

associating cation in solution, such as NMe$_4^+$ or NBu$_4^+$, the exterior ion associated iminium cation can be displaced from the exterior and released into the bulk solution, thereby allowing hydroxide to act as the nucleophile. In the absence of ion associating counterions, dissociation of the ion associated iminium is slow; in this case, the slow disappearance of the ion associated iminium is due to the background reaction of its water-mediated hydrolysis. This mechanism is supported by the fact that the rates of hydrolysis are independent of pD in the absence of ion associating counterions.

7.3.2
Acid-Catalyzed Reactions

Following the successful use of **1** as a catalyst for the unimolecular rearrangement of enammonium substrates, the further potential of **1** as a catalyst was explored. Given the propensity of **1** to preferentially bind cations over neutral guests, it was hoped that **1** could catalyze reactions that contained a cationic transition state. An ideal candidate for this type of reaction is the class of hydrolysis reactions that occur through an acid-catalyzed pathway. The subsequent protonated substrate or high-energy monocationic species on the reaction coordinate should be stabilized by **1**, hopefully leading to catalysis. Extension to this class of reactions would be significant because it would allow for catalysis of neutral substrates, thereby greatly increasing the potential scope of possible substrate for catalysis.

A common method used by nature to activate otherwise unreactive compounds is the precise arrangement of hydrogen-bonding networks and electrostatic interactions between the substrate and adjacent residues of the protein [83]. Electrostatic interactions alone can greatly favor charged states and have been responsible for large pK_a shifts of up to 5 pK_a units, as seen in acetoacetate decarboxylase [84]. Several reports in the literature have documented synthetic chemists' approaches to mimicking such pK_a shifts. Synthetic host molecules such as cyclodextrins and cucurbiturils have produced pK_a shifts of up to two units [85–88]. The breadth of work utilizing monocations as guests prompted our investigation of the ability of **1** to encapsulate protonated guest molecules.

To test the hypothesis that neutral guests can be protonated to allow for encapsulation, bis(dimethylphosphino)methane (**35**) was added to **1** and new upfield resonances corresponding to encapsulated **35** were observed both in the ^{1}H and ^{31}P NMR spectra. A ^{31}P{^{1}H} NMR spectrum in H$_2$O revealed a singlet and an un-decoupled spectrum gave $^1J_{HP} = 490$ Hz, corresponding to a one-bond P–H coupling, thus confirming protonation. In D$_2$O, $^1J_{DP} = 74$ Hz was observed, which confirmed deuteration. After establishing that protonation of phosphines allows for encapsulation in **1**, several potential amine guests were screened (Figure 7.6) [89].

Chelating tertiary diamines (**48–52, 55, 56**) are good guests even with large chelate rings. When the alkyl chain between the amines in increased to eight or more carbons, however, encapsulation is no longer observed. Diamines unable to chelate, such as DABCO (**63**), are not encapsulated. Sufficiently hydrophobic tertiary-monoamines are encapsulated. For example, triethylamine (**38**) is encapsulated, but the resonances appear to be broad, which suggests fast guest exchange. In contrast, the

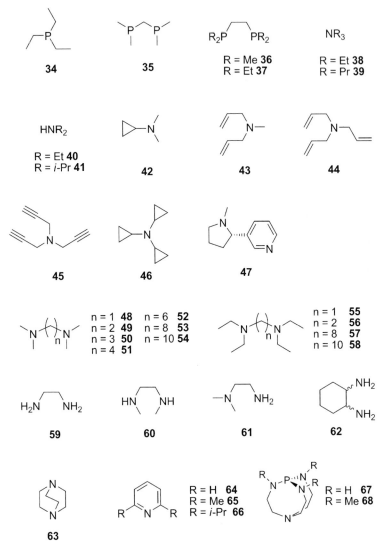

Figure 7.6 Scope of protonated amine and phosphine guests screened with **1**.

more bulky and hydrophobic amine tripropylamine (**39**) forms a clean host–guest complex. Similar trends are observed for diamines: diethylamine (**40**) is not encapsulated but the more hydrophobic diisopropylamine (**41**) is. Primary amines, either monodentate or chelating (**59,61,62**), are not encapsulated. This is presumably because primary amines are more highly solvated in water and the enthalpic loss upon desolvation is disfavored. Similarly, pyridine-based amines (**64–66**) are not encapsulated, which is likely due either to their inherently low basicity or to shape incompatibility with **1**. More exotic guests such as pro-azaphosphatrane superbases [90–93] (**67, 68**) can also be encapsulated in **1**.

To probe the thermodynamics of amine encapsulation, the binding affinities for different protonated amines for **1** were investigated. By studying the stabilization of the protonated form of encapsulated amines, the feasibility of stabilizing protonated intermediates in chemical reactions could be assessed. The thermodynamic cycle for encapsulation of a hypothetical substrate (S) is shown in Scheme 7.5. The acid–base equilibrium of the substrate is defined by K_1 and the binding constant of the protonated substrate in **1** is defined by K_2. Previous work has shown that neutral substrates can enter **1** [94]; however, the magnitude of this affinity (K_4) remains unexplored. Although neutral encapsulated amines were not observable in the study of protonated substrates, the thermodynamic cycle can be completed with K_3, which is essentially the acid–base equilibrium inside **1**.

All of the protonated amines encapsulated in **1** remained encapsulated even when the pH of the bulk solution was higher than the pK_a of the protonated amine, suggesting that **1** significantly stabilizes the encapsulated guest. To confirm that **1** was not acting as a kinetic trap for the encapsulated guests, the self-exchange rates were measured for the protonated amines using the selective inversion recovery (SIR) method [95–97]. All of the protonated amines encapsulated in **1** were found to exchange quickly on the NMR time scale, confirming that the stabilization of the encapsulated substrates was thermodynamic rather than kinetic (Table 7.8). To determine the magnitude of the stabilization of the protonated amines in **1**, guest encapsulation was monitored as a function of pH, allowing determination of the binding constants (K_{eff}). The product of the amine pK_a and the binding constant of the amine in **1** gives the effective basicity of the encapsulated amine (pK_{eff}) (Table 7.8). The pK_a shifts observed for these amines are the largest pK_a shifts observed in synthetic host molecules and approach those observed in enzymes.

Nature often exploits large pK_a shifts in enzymes to effect chemical catalysis; similarly, we hoped to apply the large shifts in the effective basicities of encapsulated guests to reaction chemistry. Initial studies focused on the hydrolysis of orthoformates, a class of molecules responsible for much of the formulation of the Brønsted theory of acids almost a century ago [98]. While orthoformates are readily hydrolyzed in acidic solution, they are exceedingly stable in neutral or basic solution [99]. However, in the presence of a catalytic amount of **1** in basic solution, small orthoformates are quickly hydrolyzed to the corresponding formate ester [38]. Addition of NEt_4^+ to the reaction inhibited the catalysis but did not affect the hydrolysis rate measured in the absence of **1**. With a limited volume in the cavity of **1**, substantial size selectivity was observed in the orthoformate hydrolysis. Orthoformates smaller than tripentyl

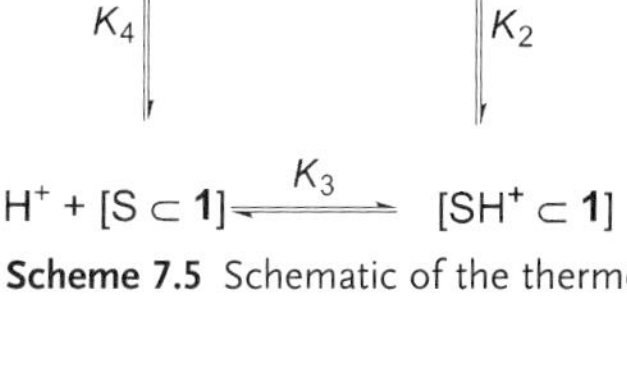

$$S + H^+ + 1 \xrightleftharpoons{K_1} SH^+ + 1$$

$$\Big\updownarrow K_4 \qquad\qquad \Big\updownarrow K_2$$

$$H^+ + [S \subset 1] \xrightleftharpoons{K_3} [SH^+ \subset 1]$$

Scheme 7.5 Schematic of the thermodynamic cycle for encapsulation of protonated guests in **1**.

Table 7.8 Kinetics and thermodynamics of protonated amine encapsulation in **1**. The self-exchange rates (k_{277}) of the protonated guests were measured at 277 K. $K_{eff(298)}$ is the binding constant of the encapsulated protonated amine and has an estimated uncertainty of 10%.

Amine	pK_a	k_{277} (s^{-1})	$K_{eff(298)}$ (M^{-1})	Log (K_{eff})	$-\Delta G°$ (kcal mol^{-1})	Effective Basicity (pK_{eff})
38	10.7	46(9)	130	2.1	2.9	12.8
39	10.7	0.31(4)	500	2.7	3.7	13.4
41	10.8	17(3)	2510	3.4	4.6	14.2
42	8.5	5.3(3)	2080	3.3	4.5	11.8
44	8.3	5.4(6)	25100	3.3	6.0	12.7
45	8.1	4.4(6)	31600	4.4	6.1	12.6
46	6.4	1.1(1)	1590	4.5	4.4	9.6
47	10.6	1.0(5)	650	3.2	3.8	13.4
49	9.1	47(9)	1260	2.8	4.2	12.2
50	9.8	1.1(2)	6310	3.1	5.2	13.6
51	9.8	0.24(3)	6450	3.8	5.2	13.6
52	9.8	1.9(3)	12600	4.1	5.6	13.9
54	10.8	1.13(2)	3160	3.5	4.8	14.3

orthoformate are readily hydrolyzed with 1 mol.% **1**, while larger substrates remain unreacted (Scheme 7.6).

Having established that **1** catalyzes the hydrolysis of orthoformates in basic solution, the reaction mechanism was probed. Mechanistic studies were performed using triethyl orthoformate (**70**) as the substrate at pH 11.0 and 50 °C. First-order substrate consumption was observed under stoichiometric conditions. Working under saturation conditions (pseudo-0th order in substrate), kinetic studies revealed that the reaction is also first order in [H$^+$] and in [**1**]. When combined, these mechanistic studies establish that the rate law for this catalytic hydrolysis of orthoformates by host **1** obeys the overall termolecular rate law: rate = k[H$^+$][Substrate][**1**], which reduces to rate = k'[H$^+$][**1**] at saturation.

We conclude that the neutral substrate enters **1** to form a host–guest complex, leading to the observed substrate saturation. The encapsulated substrate then undergoes encapsulation-driven protonation, presumably by deprotonation of water, followed by acid-catalyzed hydrolysis inside **1**, during which two equivalents of the corresponding alcohol are released. Finally, the protonated formate ester is ejected from **1** and further hydrolyzed by base in solution. The reaction mechanism (Scheme 7.7) shows direct parallels to enzymes that obey Michaelis–Menten kinetics due to the initial pre-equilibrium followed by a first-order rate-limiting step.

Lineweaver–Burk analysis using the substrate saturation curves afforded the corresponding Michaelis–Menten kinetic parameters of the reaction: $V_{max} = 1.79 \times 10^{-5}$ M s^{-1}, $K_M = 21.5$ mM, $k_{cat} = 8.06 \times 10^{-3}$ s^{-1} for **69**, and $V_{max} = 9.22 \times 10^{-6}$ M s^{-1}, $K_M = 7.69$ mM, $k_{cat} = 3.86 \times 10^{-3}$ s^{-1} for **72**. These parameters demonstrate substantial rate acceleration over the background reaction with k_{cat}/k_{uncat} for triethyl orthoformate and triisopropyl orthoformate being 560 and 890, respectively.

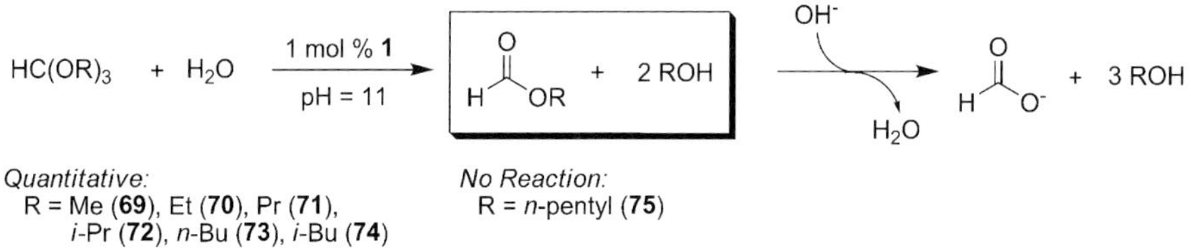

7.3 The Assembly as a Catalyst | 187

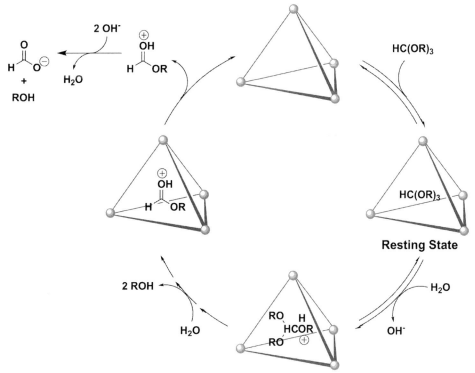

Scheme 7.6 Scope of orthoformates hydrolyzed in **1** under basic conditions.

Assuming a fast pre-equilibrium with respect to k_{cat}, K_M is essentially the dissociation constant of the encapsulated neutral substrate. The specificity factor k_{cat}/K_M can be used to compare the efficiency of hydrolysis by **1** for the two substrates. This constant corresponds to the second-order proportionality constant for the rate of conversion of the pre-formed host–guest complex to the product. Interestingly, **69** and **72** have specificity factors of 0.37 and 0.50 $M^{-1}s^{-1}$, respectively, showing that the more hydrophobic **72** is more efficiently hydrolyzed by **1**.

Also characteristic of enzymes that obey Michaelis–Menten kinetics is that suitable inhibitors can compete with the substrate for the enzyme active site, thus impeding the reaction. If the inhibitor binds reversibly to the enzyme active site, then the substrate can compete for the active site leading to competitive inhibition. To test for

Scheme 7.7 Mechanism for hydrolysis of orthoformates by **1**. The formate ester product is further hydrolyzed by base to formate anion and the corresponding alcohol.

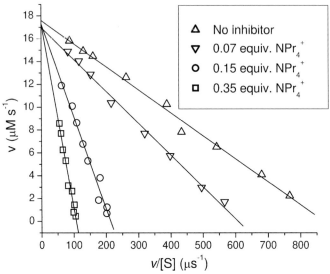

Figure 7.7 Eadie-Hofstee plot for the hydrolysis of triethyl orthoformate in **1**, pH 11, 100 mM K_2CO_3, 50 °C, using NPr_4^+ as a competitive inhibitor.

competitive inhibition for the hydrolysis of orthoformates by **1**, the rates of hydrolysis of triethyl orthoformate were measured in the presence of a varying amount of the strongly-binding inhibitor NPr_4^+ ($K_a = 10^{2.0(2)}\ M^{-1}$). By varying the concentration of substrate for each amount of inhibitor, the resulting saturation curves were compared using an Eadie–Hofstee plot (Figure 7.7) [100,101]. The saturation curves intersect on the y-axis, signifying that at infinite substrate concentration the maximum reaction velocity is independent of the amount of inhibitor, which confirms that competitive inhibition is indeed operating.

To expand the substrate scope for hydrolysis reactions catalyzed by **1**, the deprotection of acetals was investigated. Acetals are among the most commonly used protecting groups for aldehydes and ketones in organic synthesis due to their ease of installation and resistance to cleavage in neutral or basic solution [102]. Traditionally, aqueous acids, organic solutions acidified with organic or inorganic acids, or Lewis acids have been used for the reconversion of the acetal to carbonyl functionality [103–107]. However, several reports have documented various strategies for acetal cleavage under mild conditions [108–117], including the first acetal deprotection in basic solution using cerium ammonium nitrate at pH 8 in a water-acetonitrile solution [118].

Addition of 2,2-dimethoxypropane (**76**) to a solution of **1** in H_2O at pH 10 quickly yielded the hydrolysis products (acetone and methanol). To examine the reaction scope, various alkyl acetals and ketals were screened (Table 7.9). The hydrolysis reactions were screened by mild heating (50 °C) of 5 mol.% of **1** with respect to the acetal substrate at pH 10 in H_2O. Smaller substrates, which are able to fit into the cavity of **1**, are readily hydrolyzed. However, larger substrates, such as 2,2-dimethox-

7.3 The Assembly as a Catalyst

Table 7.9 Scope of acetals and ketals hydrolyzed by **1** in basic solution.

$$\underset{R^1 \;\; R^2}{\text{MeO} \;\; \text{OMe}} \xrightarrow[\text{pH 10, 50 °C, 6 h.}]{\text{5 mol\% } \mathbf{1} \,/\, H_2O} \underset{R^1 \;\; R^2}{\overset{O}{\|}} + \text{2 MeOH}$$

Entry	Substrate	Yield (%)	Entry	Substrate	Yield (%)
1	MeO OMe **76**	>95	9	2,2-dimethoxyadamantane **84**	87
2	MeO OMe **77**	>95	10	cyclohexyl C(OMe)₂ **85**	>95
3	MeO OMe **78**	>95	11	CH(OMe)₂ alkyl **86**	>95
4	MeO OMe **79**	>95	12	(CH₂)₅ CH(OMe)₂ **87**	>95
5	MeO OMe (CH₂)₂(CH₂)₅ **80**	>95	13	(CH₂)₇ CH(OMe)₂ **88**	<5
6	MeO OMe (CH₂)₈ **81**	<5	14	cyclopentyl C(OMe)₂ **89**	>95
7	cyclopentyl(OMe)₂ **82**	>95	15	cyclohexyl C(OMe)₂ **90**	>95
8	cyclohexyl(OMe)₂ **83**	>95			

yundecane (**81**, entry 6) or 1,1-dimethoxynonane (**88**, entry 13), remain unreacted, suggesting that they are too large to enter the interior cavity of **1**. In all cases, addition of a strongly binding inhibitor for the interior cavity of **1**, such as NEt_4^+, inhibits the overall reaction, confirming that **1** is the active catalyst.

For smaller acetals, such as **76**, the encapsulated substrate is not observed although the substrate resonances broaden, suggesting that the substrates are exchanging quickly on the NMR time scale. However, for larger acetals, broad guest resonances are observed upfield, suggesting a more slowly exchanging guest. For very bulky substrates, such as 2,2-dimethoxyadamantane (**84**, entry 9), the substrate is observed to be cleanly encapsulated in a 1 : 1 host–guest complex, indicating slow guest ingress and egress on the NMR time scale (Figure 7.8). By monitoring the ^1H NMR spectrum of **84** during the reaction, new peaks corresponding to the encapsulated product, 2-adamantanone, were observed.

With the observation that both the substrate and product were encapsulated, the binding affinities of both molecules within **1** were investigated to help explain the catalytic turnover. The total substrate, both free in solution and encapsulated, was monitored as a function of the concentration of **1**. The concentration of free substrate in solution was kept constant by always maintaining the presence of solid or liquid substrate in the system, which insured a uniform activity of the substrate throughout

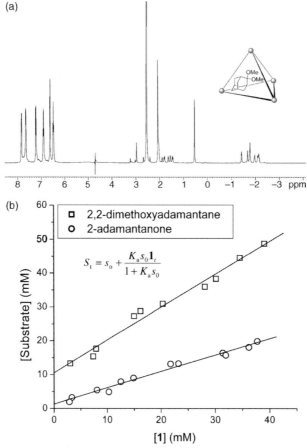

Figure 7.8 (a) ^{1}H NMR spectrum of encapsulated 2,2-dimethoxyadamantane in **1**. (b) Binding constant determination from the equation shown for 2,2-dimethoxyadamantane and 2-adamantanone in **1** in a 25 : 1 $H_2O : D_2O$ solution buffered to pH 10 with 100 mM carbonate, measured at 298 K.

the experiments. The total amount of substrate in solution can be defined as shown in the equation in Figure 7.8, where S_t is the total substrate concentration, s_0 is the constant concentration of free substrate in solution, 1_t is the total concentration of **1** and K_a is the association constant for the host–guest complex [119].

Using this equation, the binding constants, K_a, for the substrate **84** and its hydrolysis product 2-adamantanone (**91**) were determined from the data (Figure 7.8). Monitoring the encapsulation of both **84** and **91** over a concentration range from 2.8 to 40 mM **1**, in a 25 : 1 $H_2O : D_2O$ solution buffered to pH 10 with 100 mM carbonate, yielded binding constants of 3100 and 700 M^{-1} for 2,2-dimethoxyadamantane and 2-adamantanone, respectively. As expected, the hydrolysis product is bound less tightly by **1** and is less soluble in water than the substrate, which allows the observed catalytic turnover.

7.4
Conclusions and Outlook

The chemistry of a water-soluble, chiral supramolecular assembly has been explored over the last decade. Understanding the fundamental host–guest chemistry of the assembly **1**, such as the mechanism of guest exchange and the preference of monocationic guests, has allowed the chemistry of **1** to be expanded into the field of catalysis. In the hope of using the chirality of **1** as a chiral environment for encapsulated guests, a series of monocationic organometallic guests, both reactive and inert, have been encapsulated in **1**. Half-sandwich ruthenium complexes were encapsulated with diastereoselectivities of up to 85 : 15. Chiral-at-metal reactive iridium cations were encapsulated, and the C–H bond activation of aldehydes was carried out with diastereoselective product formation of up to 70 : 30. Furthermore, **1** itself was used as a catalyst for the [3,3] sigmatropic rearrangement of enammonium cations with rate accelerations of up to 10^3. Encapsulation of these cations in **1** locks the substrate in a reactive conformation, thereby reducing the entropic penalty in the transition state of the rearrangement. The preference for cationic substrates was exploited by using **1** to stabilize the cationic intermediate species, allowing for the catalysis of neutral substrates, as shown by the hydrolysis of orthoformates and acetals in basic solution.

As the field of supramolecular chemistry grows and the complexity of synthetic structures increases, the basic understanding of the host–guest chemistry is of utmost importance in the development of new chemistry. As synthetic chemists begin to emulate Nature's ability to carry out complex reactions in the confined cavities of enzymes, a fundamental understanding of the contributing forces to such reactivity is paramount. Understanding of the solvation effects, both upon encapsulation and in the self-assembly process of host molecules themselves, as well as the contributions of encapsulation to entropic concerns of the reaction are all important frontiers that remain underexplored. The field of supramolecular chemistry allows chemists to uniquely examine how weak forces can interact to produce spectacular results and is poised to contribute to our understanding of enzyme mimicry and catalysis as a whole.

References

1 Chin, J. (1991) Developing artificial hydrolytic metalloenzymes by a unified mechanistic approach. *Acc. Chem. Res.*, **24**, 145–152.

2 Uhlenbeck, O.C. (1987) A small catalytic oligoribonucleotide. *Nature*, **328**, 596–600.

3 Zaug, A.J., Michael, D.B. and Cech, T.R. (1986) The Tetrahymena ribozyme acts like an RNA restriction endonuclease. *Nature*, **324**, 429–433.

4 Chin, J. and Banaszczyk, M. (1989) Highly efficient hydrolytic cleavage of adenosine monophosphate resulting in a binuclear cobalt(III) complex with a novel doubly bidentate. mu. 4-phosphato bridge. *J. Am. Chem. Soc.*, **111**, 4103–4105.

5 Pauling, L. (1946) Molecular architecture and biological reactions. *Chem. Eng. News*, **24**, 1375–1377.

6 Houk, K.N., Leach, A.G., Kim, S.P. and Zhang, X. (2003) Binding affinities of host–guest, protein-ligand, and protein-transition-state complexes. *Angew. Chem., Int. Ed.*, **42**, 4872–4897.

7 Murakami, Y., Kikuchi, J., Hisaeda, Y. and Hayashida, O. (1996) Artificial enzymes. *Chem. Rev.*, **96** (2), 721–758.

8 Kennan, A.J. and Whitlock, H.W. (1996) Host-catalyzed isoxazole ring opening: A rationally designed artificial enzyme. *J. Am. Chem. Soc.*, **118** (12), 3027–3028.

9 Suh, J. (2000) Designing active sites of synthetic artificial enzymes. *Adv. Supramol. Chem.*, **6**, 245–286.

10 Breslow, R. and Dong, S.D. (1998) Biomimetic reactions catalyzed by cyclodextrins and their derivatives. *Chem. Rev.*, **98** (5), 1997–2012.

11 Cram, D.J. (1988) The design of molecular hosts, guests, and their complexes (Nobel Lecture). *Angew. Chem., Int. Ed.*, **27** (8), 1009–1020.

12 Lehn, J.-M. (1988) Supramolecular chemistry – scope and perspectives molecules, supermolecules, and molecular devices (Nobel Lecture). *Angew. Chem., Int. Ed.*, **27** (1), 89–112.

13 Biros, S.M. and Rebek, J. Jr (2007) *Chem. Soc. Rev.*, **36** (1), 93–104.

14 Oshovsky, G.V., Reinhoudt, D.N. and Verboom, W. (2007) Supramolecular chemistry in water. *Angew. Chem., Int. Ed.*, **46** (14), 2366–2393.

15 Pluth, M.D. and Raymond, K.N. (2007) Reversible guest exchange mechanisms in supramolecular host–guest assemblies. *Chem. Soc. Rev.*, **36** (2), 161–171.

16 Schmuck, C. (2007) Guest encapsulation within self-assembled molecular containers. *Angew. Chem., Int. Ed.*, **46** (31), 5830–5833.

17 Yoshizawa, M. and Fujita, M. (2005) Self-assembled coordination cage as a molecular flask. *Pure Appl. Chem.*, **77** (7), 1107–1112.

18 Caulder, D.L., Bruckner, C., Powers, R.E., Konig, S., Parac, T.N., Leary, J.A. and Raymond, K.N. (2001) Design, formation and properties of tetrahedral M4L4 and M4L6 supramolecular clusters1. *J. Am. Chem. Soc.*, **123** (37), 8923–8938.

19 Caulder, D.L., Powers, R.E., Parac, T.N. and Raymond, K.N. (1998) The self-assembly of a predesigned tetrahedral M4L6 supramolecular cluster. *Angew. Chem. Int. Ed.*, **37** (13–14), 1840–1843.

20 Caulder, D.L. and Raymond, K.N. (1999) The rational design of high symmetry coordination clusters. *J. Chem., Soc. Dalton Trans.*, 1185–1200.

21 Caulder, D.L. and Raymond, K.N. (1999) Supermolecules by design. *Acc. Chem. Res.*, **32** (11), 975–982.

22 Terpin, A.J., Ziegler, M., Johnson, D.W. and Raymond, K.N. (2001) Resolution and kinetic stability of a chiral supramolecular assembly made of labile components. *Angew. Chem., Int. Ed.*, **40** (1), 157–160.

23 Ziegler, M., Davis, A.V., Johnson, D.W. and Raymond, K.N. (2003) Supra-molecular chirality: A reporter of structural memory. *Angew. Chem., Int. Ed.*, **42** (6), 665–668.

24 Brumaghim, J.L., Michels, M., Pagliero, D. and Raymond, K.N. (2004) Encap-sulation and stabilization of reactive aromatic diazonium ions and the tropylium ion within a supramolecular host. *Eur. J. Org. Chem.*, 5115–5118.

25 Dong, V.M., Fiedler, D., Carl, B., Bergman, R.G. and Raymond, K.N. (2006) Molecular recognition and stabi-lization of iminium ions in water. *J. Am. Chem. Soc.*, **128** (45), 14464–14465.

26 Brumaghim, J.L., Michels, M. and Raymond, K.N. (2004) Hydrophobic chemistry in aqueous solution: Stabi-lization and stereoselective encapsulation of phosphonium guests in a supra-molecular host. *Eur. J. Org. Chem.*, 4552–4559.

27 Ziegler, M., Brumaghim, J.L. and Raymond, K.N. (2000) Stabilization of a reactive cationic species by supra-molecular encapsulation. *Angew. Chem., Int. Ed.*, **39** (22), 4119–4121.

28 Davis, A.V., Fiedler, D., Seeber, G., Zahl, A., Eldik van, R. and Raymond, K.N. (2006) Guest exchange dynamics in an M4L6 tetrahedral host. *J. Am. Chem. Soc.*, **128** (4), 1324–1333.

29 Davis, A.V. and Raymond, K.N. (2005) The big squeeze: Guest exchange in an M4L6 supramolecular host. *J. Am. Chem. Soc.*, **127** (21), 7912–7919.

30 Cram, D.J., Tanner, M.E. and Knobler, C.B. (1991) Host–guest complexation. 58. Guest release and capture by hemicarcerands introduces the phenomenon of constrictive binding. *J. Am. Chem. Soc.*, **113** (20), 7717–7727.

31 Fiedler, D., Pagliero, D., Brumaghim, J.L., Bergman, R.G. and Raymond, K.N. (2004) Encapsulation of cationic ruthenium complexes into a chiral self-assembled cage. *Inorg. Chem.*, **43** (3), 846–848.

32 Fiedler, D., Leung, D.H., Bergman, R.G. and Raymond, K.N. (2004) Enantioselective guest binding and dynamic resolution of cationic ruthenium complexes by a chiral metal–ligand assembly. *J. Am. Chem. Soc.*, **126** (12), 3674–3675.

33 Leung, D.H., Fiedler, D., Bergman, R.G. and Raymond, K.N. (2004) Selective C–H bond activation by a supramolecular host–guest assembly. *Angew. Chem., Int. Ed.*, **43** (8), 963–966.

34 Leung, D.H., Bergman, R.G. and Raymond, K.N. (2006) Scope and mechanism of the C–H bond activation reactivity within a supramolecular host by an iridium guest: A stepwise ion pair guest dissociation mechanism. *J. Am. Chem. Soc.*, **128** (30), 9781–9797.

35 Fiedler, D., Leung, D.H., Bergman, R.G. and Raymond, K.N. (2005) Selective molecular recognition, C–H bond activation, catalysis in nanoscale reaction vessels. *Acc. Chem. Res.*, **38** (4), 351–358.

36 Fiedler, D., Bergman, R.G. and Raymond, K.N. (2004) Supramolecular catalysis of a unimolecular transformation: Aza-Cope rearrangement within a self-assembled host. *Angew. Chem., Int. Ed.*, **43** (48), 6748–6751.

37 Fiedler, D., van Halbeek, H., Bergman, R.G. and Raymond, K.N. (2006) Supramolecular catalysis of unimolecular rearrangements: Substrate scope and mechanistic insights. *J. Am. Chem. Soc.*, **128** (31), 10240–10252.

38 Pluth, M.D., Bergman, R.G. and Raymond, K.N. (2007) Acid catalysis in basic solution: A supramolecular host promotes orthoformate hydrolysis. *Science*, **316**, 85–88.

39 Pluth, M.D., Bergman, R.G. and Raymond, K.N. (2007) Catalytic deprotection of acetals in basic solution using a self-assembled supramolecular "nanozyme". *Angew. Chem., Int. Ed.*, **119**, 8741–8743.

40 Walsh, P.J., Li, H. and de Parrodi, C.A. (2007) A Green chemistry approach to asymmetric catalysis: Solvent-free and highly concentrated reactions. *Chem. Rev.*, **107** (7), 2503–2545.

41 Dagorne, S., Bellemin-Laponnaz, S. and Maisse-Francois, A. (2007) Metal complexes incorporating monoanionic bisoxazolinate ligands: Synthesis, structures, reactivity and applications in asymmetric catalysis. *Eur. J. Inorg. Chem.*, 913–925.

42 Kobayashi, S. (2007) Symmetric catalysis in aqueous media. *Pure Appl. Chem.*, **79** (2), 235–245.

43 Gade, L.H. and Bellemin-Laponnaz, S. (2007) Chiral N-heterocyclic carbenes as stereodirecting ligands in asymmetric catalysis. *Top. Organomet. Chem.*, **21**, 117–157.

44 Chelucci, G. (2006) Synthesis and application in asymmetric catalysis of camphor-based pyridine ligands. *Chem. Soc. Rev.*, **35** (12), 1230–1243.

45 Burk, M.J. and Ramsden, J.A. (2006) *Handbook of Chiral Chemicals*, Vol. 2nd, pp. 249. CRC Press, Taylor & Francis, Boca Raton Florida.

46 Fu, G.C. (2006) Applications of planar-chiral heterocycles as ligands in asymmetric catalysis. *Acc. Chem. Res.*, **38** (11), 853–860.

47 Desimoni, G., Faita, G. and Jorgensen, K.A. (2006) C2-Symmetric chiral bis (oxazoline) ligands in asymmetric catalysis. *Chem. Rev.*, **106** (9), 3561–3651.
48 Wu, J. and Chan, A.S.C. (2006) P-Phos: A family of versatile and effective atropisomeric dipyridylphosphine ligands in asymmetric catalysis. *Acc. Chem. Res.*, **39** (10), 711–720.
49 Fu, G.C. (2004) Asymmetric catalysis with "planar-chiral" derivatives of 4-(dimethylamino)pyridine. *Acc. Chem. Res.*, **37** (8), 542–547.
50 Dieguez, M., Pamies, O. and Claver, C. (2004) Ligands derived from carbohydrates for asymmetric catalysis. *Chem. Rev.*, **104** (6), 3189–3216.
51 List, B. and Yang, J.W. (2006) CHEMISTRY: The organic approach to asymmetric catalysis. *Science*, **313**, 1584–1586.
52 Houk, K.N. and List, B. (2004) Asymmetric organocatalysis. *Acc. Chem. Res.*, **37** (8), 487–487.
53 Enders, D. and Balensiefer, T. (2004) Nucleophilic carbenes in asymmetric organocatalysis. *Acc. Chem. Res.*, **37** (8), 534–541.
54 Dalko, P.I. and Moisan, L. (2004) In the golden age of organocatalysis. *Angew. Chem., Int. Ed.*, **43** (39), 5138–5175.
55 Dalko, P.I. and Moisan, L. (2001) Enantioselective organocatalysis. *Angew. Chem., Int. Ed.*, **40** (20), 3726–3748.
56 List, B. (2001) Asymmetric aminocatalysis. *Synlett*, 1675–1686.
57 Ni, B., Zhang, Q. and Headley, A.D. (2007) Functionalized chiral ionic liquid as recyclable organocatalyst for asymmetric Michael addition to nitrostyrenes. *Green Chem.*, **9** (7), 737–739.
58 Schulz, P.S., Müller, N., Bösmann, A. and Wasserscheid, P. (2007) Effective chirality transfer in ionic liquids through ion-pairing effects. *Angew. Chem., Int. Ed.*, **46** (8), 1293–1295.
59 Gausepohl, R., Buskens, P., Kleinen, J., Bruckmann, A., Lehmann, C.W., Klankermayer, J. and Leitner, W. (2006) Highly enantioselective aza-Baylis–Hillman reaction in a chiral reaction medium. *Angew. Chem., Int. Ed.*, **45** (22), 3689–3692.
60 Luo, S., Mi, X., Zhang, L., Liu, S., Xu, H. and Cheng, J.P. (2006) Functionalized chiral ionic liquids as highly efficient asymmetric organocatalysts for Michael addition to nitroolefins. *Angew. Chem., Int. Ed.*, **45** (19), 3093–3097.
61 Zhou, L. and Wang, L. (2007) Chiral ionic liquid containing L-proline unit as a highly efficient and recyclable asymmetric organocatalyst for aldol reaction. *Chem. Lett.*, **36** (5), 628–629.
62 Burger, P. and Bergman, R.G. (1993) Facile intermolecular activation of carbon–hydrogen bonds in methane and other hydrocarbons and silicon–hydrogen bonds in silanes with the iridium(III) complex Cp*(PMe3)Ir(CH3)(OTf). *J. Am. Chem. Soc.*, **115** (22), 10462–10463.
63 Arndtsen, B.A. and Bergman, R.G. (1995) Unusually mild and selective hydrocarbon C–H bond activation with positively charged iridium(III) complexes. *Science*, **270**, 1970–1973.
64 Luecke, H.F. and Bergman, R.G. (1997) Synthesis and C–H activation reactions of cyclometalated complexes of Ir(III): Cp*(PMe3)Ir(CH3)+ does not undergo intermolecular C–H activation in solution via a cyclometalated intermediate. *J. Am. Chem. Soc.*, **119** (47), 11538–11539.
65 Alaimo, P.J., Arndtsen, B.A. and Bergman, R.G. (2000) Alkylation of iridium via tandem carbon–hydrogen bond activation/decarbonylation of aldehydes. Access to complexes with tertiary and highly hindered metal-carbon bonds. *Organometallics*, **19** (11), 2130–2143.
66 Klei, S.R., Golden, J.T., Burger, P. and Bergman, R.G. (2002) Cationic Ir(III) alkyl and hydride complexes: Stoichiometric and catalytic C–H activation by Cp*(PMe3)Ir(R)(X) in

homogeneous solution. *J. Mol. Catal. A: Chem.*, **189** (1), 79–94.

67 Leung, D.H., Bergman, R.G. and Raymond, K.N. (2007) Highly selective supramolecular catalyzed allylic alcohol isomerization. *J. Am. Chem. Soc.*, **129** (10), 2746–2747.

68 Yeh, R.M., Xu, J., Seeber, G. and Raymond, K.N. (2005) Large M4L4 (M = Al(III), Ga(III), In(III), Ti(IV)) tetrahedral coordination cages: An extension of symmetry-based design. *Inorg. Chem.*, **44** (18), 6228–6228.

69 Johnson, D.W. and Raymond, K.N. (2001) The self-assembly of a [Ga4L6]12- tetrahedral cluster thermodynamically driven by host–guest interactions. *Inorg. Chem.*, **40** (20), 5157–5161.

70 Scherer, M., Caulder, D.L., Johnson, D.W. and Raymond, K.N. (1999) Triple helicate – tetrahedral cluster interconversion controlled by host–guest interactions. *Angew. Chem., Int. Ed.*, **38** (11), 1587–1592.

71 Kang, J.M. and Rebek, J. Jr (1997) Acceleration of a Diels–Alder reaction by a self-assembled molecular capsule. *Nature*, **385**, 50–52.

72 Kang, J.M., Hilmersson, G., Santamaria, J. and Rebek, J. Jr (1998) Diels–Alder reactions through reversible encapsulation. *J. Am. Chem. Soc.*, **120** (15), 3650–3656.

73 Kang, J.M., Santamaria, J., Hilmersson, G. and Rebek, J. Jr (1998) Self-assembled molecular capsule catalyzes a Diels–Alder reaction. *J. Am. Chem. Soc.*, **120** (29), 7389–7390.

74 Yoshizawa, M., Tamura, M. and Fujita, M. (2006) Diels–Alder in aqueous molecular hosts: Unusual regioselectivity and efficient catalysis. *Science*, **312**, 251–254.

75 Nishioka, Y., Yamaguchi, T., Yoshizawa, M. and Fujita, M. (2007) Unusual [2 + 4] and [2 + 2] cycloadditions of arenes in the confined cavity of self-assembled cages. *J. Am. Chem. Soc.*, **129** (22), 7000–7001.

76 Walters, M.A. (1996) Ab initio investigation of the 3-aza-Cope reaction. *J. Org. Chem.*, **61** (3), 978–983.

77 Przheval, N.M. and Grandberg, I.I. (1987) The aza-Cope rearrangement in organic synthesis. *Uspekhi Khim.*, **56** (5), 814–843.

78 Elkik, E. and Francesch, C. (1968) Enamine alkylation mechanism. *Bull. Soc. Chim. Fr.*, **3**, 903–910.

79 Opitz, G. (1961) Enamines. VII. Course of the allyl- and propargyl-allenyl rearrangements in the alkylation of enamines. *Justus Liebigs Ann. Chem.*, 122–132.

80 Blechert, S. (1989) The hetero-Cope rearrangement in organic synthesis. *Synthesis*, 71–82.

81 Nubbemeyer, U. (2005) Recent advances in charge-accelerated aza-Claisen rearrangements. *Top. Curr. Chem.*, **244**, 149–213.

82 Neuhaus, D. and Williamson, M.P. (2001) *The Nuclear Overhauser Effect in Structural and Conformational Analysis*, 2nd edn, VCH Publishers, New York.

83 Ha, N.-C., Kim, M.-S., Lee, W., Choi, K.Y. and Oh, B.-H. (2000) Detection of large pKa perturbations of an inhibitor and a catalytic group at an enzyme active site, a mechanistic basis for catalytic power of many enzymes. *J. Biol. Chem.*, **275**, 41100–41106.

84 Westheimer, F.H. (1995) Coincidences, decarboxylation, and electrostatic effects. *Tetrahedron*, **51** (1), 3–20.

85 Bakirci, H., Koner, A.L., Schwarzlose, T. and Nau, W.M. (2006) Analysis of host-assisted guest protonation exemplified for p-sulfonatocalix[4]arene – Towards enzyme-mimetic pKa shifts. *Chem.–Eur. J.*, **12** (18), 4799–4807.

86 Marquez, C. and Nau, W.M. (2001) Two mechanisms of slow host–guest complexation between cucurbit[6]uril and cyclohexylmethylamine: pH-Responsive supramolecular kinetics. *Angew. Chem. Int. Ed.*, **40**, 3155–3160.

87 Mohanty, J., Bhasikuttan, A.C., Nau, W.M. and Pal, H. (2006) Host–guest

complexation of neutral red with macrocyclic host molecules: Contrasting pKa shifts and binding affinities for cucurbit[7]uril and -cyclodextrin. *J. Phys. Chem. B*, **110** (10), 5132–5138.

88 Zhang, X., Gramlich, G., Wang, X. and Nau, W.M. (2002) A joint structural, kinetic, and thermodynamic investigation of substituent effects on host–guest complexation of bicyclic azoalkanes by -cyclodextrin. *J. Am. Chem. Soc.*, **124** (2), 254–263.

89 Pluth, M.D., Bergman, R.G. and Raymond, K.N. (2007) Making amines strong bases: Thermodynamic stabilization of protonated guests in a highly-charged supramolecular host. *J. Am. Chem. Soc.*, **129** (37), 11459–11467.

90 Kisanga, P.B. and Verkade, J.G. (2001) Synthesis of new proazaphosphatranes and their application in organic synthesis. *Tetrahedron*, **57** (3), 467–475.

91 Laramay, M.A.H. and Verkade, J.G. (1990) The "anomalous" basicity of P(NHCH2CH2)3N relative to P(NMeCH2CH2)3N and p(NBzCH2CH2)3N: a chemical consequence of orbital charge balance? *J. Am. Chem. Soc.*, **112** (25), 9421–9422.

92 Lensink, C., Xi, S.K., Daniels, L.M. and Verkade, J.G. (1989) The unusually robust phosphorus-hydrogen bond in the novel cation [cyclic] HP(NMeCH2CH2)3N +. *J. Am. Chem. Soc.*, **111** (9), 3478–3479.

93 Verkade, J.G. (1993) Atranes: New examples with unexpected properties. *Acc. Chem. Res.*, **26** (9), 483–489.

94 Biros, S.M., Bergman, R.G. and Raymond, K.N. (2007) The hydrophobic effect drives the recognition of hydrocarbons by an anionic metal–ligand cluster. *J. Am. Chem. Soc.*, **129**, 12094–12095.

95 Bain, A. and Cramer, J.A. (1993) A method for optimizing the study of slow chemical exchange by NMR spin-relaxation measurements. Application to tripodal carbonyl rotation in a metal complex. *J. Magn. Reson. Ser. A*, **103** (2), 217–222.

96 Bain, A.D. and Cramer, J.A. (1996) Slow chemical exchange in an eight-coordinated bicentered ruthenium complex studied by one-dimensional methods. Data fitting and error analysis. *J. Magn. Reson. Ser. A*, **118** (1), 21–27.

97 Perrin, C.L. and Dwyer, T.J. (1990) Application of two-dimensional NMR to kinetics of chemical exchange. *Chem. Rev.*, **90** (6), 935–967.

98 Bronsted, J.N. and Wynne-Jones, W.F.K. (1929) Acid catalysis in hydrolytic reactions. *Trans. Faraday Soc.*, **25**, 59–77.

99 Cordes, E.H. and Bull, H.G. (1974) Mechanism and catalysis for hydrolysis of acetals, ketals, and ortho esters. *Chem. Rev.*, **74** (5), 581–603.

100 Eadie, G.S. (1942) The inhibition of cholinesterase by physostigmine and prostigmine. *J. Biol. Chem.*, **146**, 85.

101 Hofstee, B.J.H. (1952) On the evaluation of the constants Vm and KM in enzyme reactions. *Science*, **116**, 329–331.

102 Greene, T.W. and Wuts, P.G.M. (1978) *Protective Groups in Organic Synthesis*, 2nd edn, John Wiley & Sons, New York.

103 Deslongchamps, P., Dory, Y.L. and Li, S. (1994) *Can. J. Chem.*, **72** (10), 2021–2027.

104 Kirby, A.J. (1984) *Acc. Chem. Res.*, **17** (9), 305–311.

105 Knowles, J.P. and Whiting, A. (2007) *Eur. J. Org. Chem.*, **20**, 3365–3368.

106 Nakamura, M., Isobe, H. and Nakamura, E. (2003) Cyclopropenone acetals-synthesis and reactions. *Chem. Rev.*, **103** (4), 1295–1326.

107 Pchelintsev, V.V., Sokolov, A.Y. and Zaikov, G.E. (1988) Polymer degradation and stability, **21** (4), 285–310.

108 Ates, A., Gautier, A., Leroy, B., Plancher, J.-M., Quesnel, Y., Vanherck, J.-C. and Marko, I.E. (2003) Mild and chemoselective catalytic deprotection of ketals and acetals using cerium(IV) ammonium nitrate. *Tetrahedron*, **59** (45), 8989–8999.

109 Carrigan, M.D., Sarapa, D., Smith, R.C., Wieland, L.C. and Mohan, R.S. (2002) A simple and efficient chemoselective

method for the catalytic deprotection of acetals and ketals using bismuth triflate. *J. Org. Chem.*, **67** (3), 1027–1030.
110 Dalpozzo, R., Nino De, A., Maiuolo, L., Procopio, A., Tagarelli, A., Sindona, G. and Bartoli, G. (2002) Simple and efficient chemoselective mild deprotection of acetals and ketals using cerium (III) triflate. *J. Org. Chem.*, **67** (25), 9093–9095.
111 Eash, K.J., Pulia, M.S., Wieland, L.C. and Mohan, R.S. (2000) A simple chemoselective method for the deprotection of acetals and ketals using bismuth nitrate pentahydrate. *J. Org. Chem.*, **65** (24), 8399–8401.
112 Fujioka, H., Okitsu, T., Sawama, Y., Murata, N., Li, R. and Kita, Y. (2006) Reaction of the acetals with TESOTf-base combination; speculation of the intermediates and efficient mixed acetal formation. *J. Am. Chem. Soc.*, **128** (17), 5930–5938.
113 He, Y., Johansson, M. and Sterner, O. (2004) Mild microwave-assisted hydrolysis of acetals under solvent-free conditions. *Synth. Commun.*, **34** (22), 4153–4158.
114 Komatsu, N., Taniguchi, A., Wada, S. and Suzuki, H. (2001) A catalytic deprotection of S,S- S,O- and O,O-acetals using Bi(NO3)3.5H2O under air. *Adv. Synth. Catal.*, **343** (5), 473–480.
115 Krishnaveni, N.S., Surendra, K., Reddy, M.A., Nageswar, Y.V.D. and Rao, K.R. (2003) Highly efficient deprotection of aromatic acetals under neutral conditions using -cyclodextrin in water. *J. Org. Chem.*, **68** (5), 2018–2019.
116 Mirjalili, B.F., Zolfigol, M.A. and Bamoniri, A. (2002) Deprotection of acetals and ketals by silica sulfuric acid and wet SiO2. *Molecules*, **7** (10), 751–755.
117 Sabitha, G., Babu, R.S., Reddy, E.V. and Yadav, J.S. (2000) A novel, efficient, and selective cleavage of acetals using bismuth(III) chloride. *Chem. Lett.*, **29** (9), 1074–1075.
118 Marko, I.E., Ates, A., Gautier, A., Leroy, B., Plancher, J.-M., Quesnel, Y. and Vanherck, J.-C. (1999) Cerium(IV)-catalyzed deprotection of acetals and ketals under mildly basic conditions. *Angew. Chem., Int. Ed.*, **38** (21), 3207–3209.
119 Conners, K.A. (1987) *Binding Constants The Measurement of Molecular Complex Stability*, John Wiley & Sons, New York.

8
New Supramolecular Approaches in Transition Metal Catalysis; Template-Ligand Assisted Catalyst Encapsulation, Self-Assembled Ligands and Supramolecular Catalyst Immobilization

Joost N.H. Reek

8.1
Introduction

Traditionally, supramolecular catalysis involves the combination of host–guest chemistry with catalysis [1]. Inspired by the high efficiency displayed by natural catalysts, enzymes, many scientist started in this research area by connecting known active sites (metal centers and other active groups) to known binding sites, anticipating that bringing them together would provide better catalysts in terms of selectivity and activity. This resulted in many novel catalyst systems that displayed better properties than the parent compounds, but none of them approach the rate accelerations reported for enzymes. As most of these systems just bring the components together, the rate enhancements can be understood in terms of effective molarity [2], and indeed very large effects are not expected. Better results are obtained if the transition state of the reaction is also stabilized to some extent, in analogy to enzymatic reactions [3], leading to reduced free energies of activation.

The approach we have been following in the past decade is different from the traditional supramolecular approach. We have *implemented supramolecular technologies in traditional transition metal catalysis*. In the field of transition metal catalysis most effort has been devoted to ligand development and strategies to recover the catalyst from the product stream, and in these approaches we have introduced supramolecular strategies. Instead of using functional groups to bring the catalyst and the substrate together, we have also used them to:

1. control of the second coordination sphere by encapsulation of the transition metal catalyst (Figure 8.1A);
2. bring ligand building blocks together to form multidentate ligands for transition metal catalysis by assembly (Figure 8.1B);
3. reversibly anchor ligands and their transition metal complexes to a soluble or insoluble support (Figure 8.1C).

Supramolecular Catalysis. Edited by Piet W. N. M. van Leeuwen
Copyright © 2008 WILEY-VCH Verlag GmbH & Co. KGaA, Weinheim
ISBN: 978-3-527-32191-9

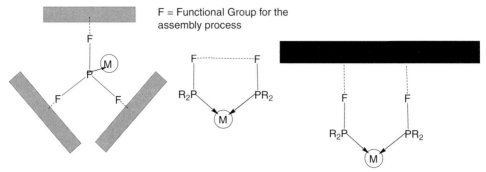

Figure 8.1 Use of functional groups to form encapsulated catalysts by assembly (A), to form bidentate ligands by assembly (B), and to support the catalyst by supramolecular interactions (C).

In Section 8.2 we elaborate on catalyst encapsulation, and in Section 8.3 we discuss the application of supramolecular ligands in transition metal catalysis. Finally, in Section 8.4 we show that the issue of catalyst recovery can also be addressed by the supramolecular anchoring of ligands and their metal complexes to support. As we will see there is a great similarity in building blocks that have been applied for both strategies, and the functional groups are essentially the same. Throughout, we focus on our own achievements in this area.

8.2
Template-Ligand Assisted Catalyst Encapsulation

We began our supramolecular approach to transition metal catalysis with the fundamental question if we could encapsulate a transition metal catalyst, thereby putting it in a controlled local environment (for a different approach see Chapter 7). Porphyrins appeared to be good building blocks as they were frequently used to make various supramolecular assemblies [4]. In particular, the pyridine-zinc(II)porphyrin motif has been extensively exploited. The binding is dynamic (rapid exchange on NMR time scale) and in apolar solvents such as toluene the association constant is around 10^3–10^4 M^{-1} [5]. The amine-zinc(II)porphyrin interaction was previously used to afford molecular capsules in which the guest that was encapsulated functions as the template for the encapsulation process. In the first example reported that used this templated approach to encapsulation, a first generation DAB dendrimer containing four amine groups had been used as a template-guest molecule. Upon addition of two equivalents of shape self-complementary bis(zinc)porphyrins, a closed molecular capsule was formed that is held together by the template enclosed inside (Figure 8.2) [6]. Interestingly, the guest-molecule is enclosed in the capsule formed by the two building blocks, but the assembly process is fast, enabling rapid exchange processes to occur, which is interesting if such supramolecular structures are applied in catalysis.

Figure 8.2 Templated approach to molecular encapsulation. The meso-phenyl groups of the porphyrin building blocks have been omitted for clarity.

We anticipated that the use of template-ligands, molecules that can coordinate to zinc(II)porphyrins via nitrogen donor atoms and have soft-donor atoms such as phosphines that can coordinate to catalytically active transition metal fragments, would lead to a new and general strategy to encapsulate transition metal catalysts. This new generation of transition metal catalysts has a second coordination sphere that is controlled by the capsule. Molecular modeling studies indeed show that the assembly of three porphyrin building blocks to tris(meta-pyridyl)phosphine results in complete encapsulation of the ligand (Figures 8.3 and 8.4).

We started this templated approach to catalyst encapsulation using tris(meta-pyridyl)phosphine and zinc(II)TPP (TPP = tetraphenylporphyrin) [7]. From NMR and UV/Vis titration experiments we found a selective assembly process via coordination of the nitrogen to the zinc(II)TPP, rendering the phosphine donor atom completely encapsulated by the three porphyrin components (Figure 8.4). The phosphine center is still available for coordination to transition metals, providing

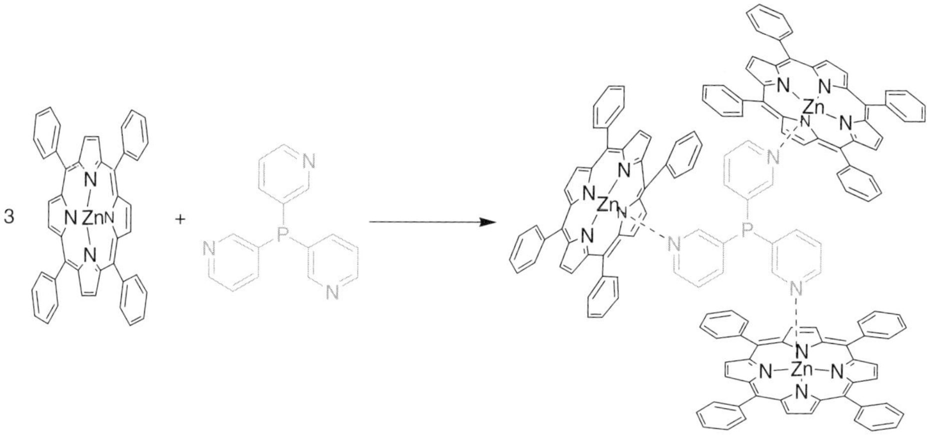

Figure 8.3 Templated approach to ligand encapsulation. Molecular modeling demonstrates that the phosphine ligand is completely encapsulated (see also Figure 8.4).

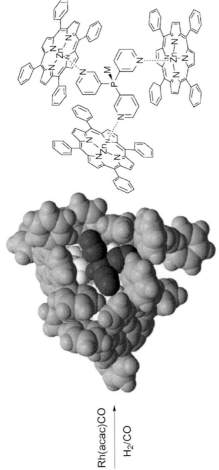
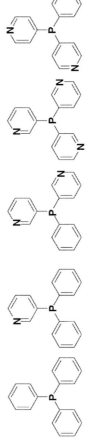

Encapsulated Catalyst Assembly: M=HRh(CO)₃

Other Phosphine Templates

Figure 8.4 Modeled structure of an encapsulated transition metal catalyst, consisting of zinc(II)TP) (gray), tris(*meta*-pyridyl)phosphine (**3**) (white) and M = [HRh(CO)₃] (dark-gray). The different pyridylphosphine template-ligands used for assembly processes are also shown.

readily access to hemi-spherically, encapsulated transition metal complexes. Later we found that coordination of zinc(II)TPP to tris(*meta*-pyridyl)phosphine is cooperative, i.e., the last zinc(II)TPP binds three times more strongly than the first, probably because of π–π interactions between the meso-phenyl groups of two adjacent porphyrin units [8]. Importantly, in the presence of a Rh(acac)(CO)$_2$ or PdCl$_2$ as metal precursors only monophosphine coordinated metal complexes were formed when the template-ligand was encapsulated by zinc(II)TPP, indicating that the coordination of the nitrogen to the zinc(II)TPP is sufficiently strong to enforce one of the phosphine ligands to dissociate (or prevent coordination) from (to) the metal complex. This shows that the strategy indeed leads to catalyst encapsulation and, in addition, to a controlled second coordination sphere around the catalytically active transition metal center.

Initial studies showed that the encapsulated palladium catalyst based on the assembly outperformed its non-encapsulated analogue by far in the Heck coupling of iodobenzene with styrene [7]. This was attributed to the fact that the active species consist of a monophosphine-palladium complex. The product distribution was not changed by encapsulation of the catalyst. A similar rate enhancement was observed in the rhodium-catalyzed hydroformylation of 1-octene (Scheme 8.1). At room temperature, the catalyst was 10 times more active. For this reaction a completely different product distribution was observed. The encapsulated rhodium catalyst formed preferentially the branched aldehyde (L/B ratio 0.6), whereas usually the linear aldehyde is formed as the main product (L/B > 2 in control experiments). These effects are partly attributed to geometry around the metal complex: monophosphine coordinated rhodium complexes are the active species, which was also confirmed by high-pressure IR and NMR techniques.

The origin of the encapsulation effects on catalysis has been studied in more detail (at 25 °C) by changing the template. The supramolecular catalyst derived from mono-pyridylphosphine template **1** and zinc(II)TPP gives similar results to **1** in the absence of zinc(II)TPP and the model PPh$_3$-catalyst; the regioselectivity (L/B = 2.8) is comparable to what is usually obtained for this reaction. The use of bis(*meta*-pyridyl)phosphine (**2**) as a template gives rise to a 2:1 assembly and therefore the catalyst is only partly encapsulated. This partly encapsulated catalyst is three times more active and leads to an increase in the selectivity for the branched product **B** (L/B = 1.1). According to high-pressure IR measurements this assembly also leads to the exclusive formation of monophosphine coordinated rhodium complexes. The effect of complete catalyst encapsulation becomes evident if this result is compared to that of catalyst assemblies obtained with the tris(*meta*-pyridyl)

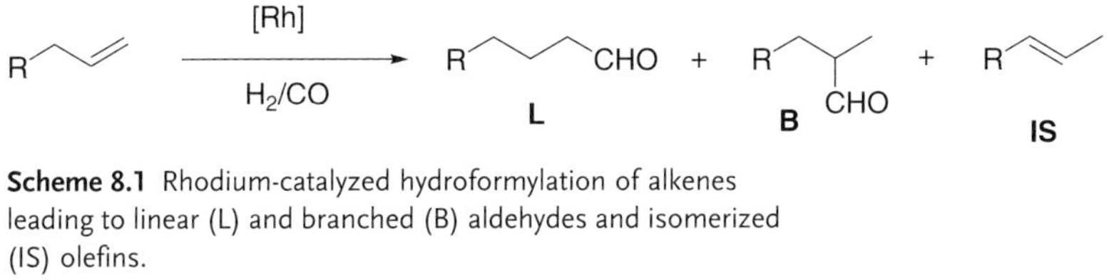

Scheme 8.1 Rhodium-catalyzed hydroformylation of alkenes leading to linear (L) and branched (B) aldehydes and isomerized (IS) olefins.

phosphine template (3) and zinc(II)TPP: a ten-fold increase in activity and a higher preference for the branched product was obtained (L/B = 0.6, 63% **B**). When 3 was combined with three equivalents of ruthenium(II)TPP(CO), a similar increase in activity (eight-fold) and high selectivity for the branched product **B** was observed (L/B = 0.4, 67% **B**). The optimal ratio between the pyridylphosphine and metalloporphyrin was determined and for templates 2 and 3 the highest **B**-selectivity was achieved at stoichiometric amounts (or excess) of porphyrin with respect to the number of pyridine donor atoms. Higher porphyrin concentrations do not result in an increase of selectivity for the branched product, whereas lowering the porphyrin/phosphine ratio gives rise to increasing amounts of **L** product. In these cases a mixture of catalysts is present, fully and partly encapsulated, which explains these results. The encapsulation effects observed in catalysis at 80 °C are essentially the same, although the differences between the parent complexes based on templates 1–3 and their encapsulated analogues are smaller. The approach was extended by using different substituted porphyrins and trispyridyl phosphite building blocks, giving rise to a small library of encapsulated catalysts, of which the archetypical encapsulated system (Figure 8.4) gives rise to the highest selectivity for the branched aldehyde.

The geometry of the template-ligand is of crucial importance and small changes results in significantly different capsules and consequently in different catalytic behavior. The nitrogen donor atoms of template 3 point in almost the same direction as the lone pair of the phosphorus atom, which is an important prerequisite for the assembly of enclosed phosphines. In contrast, those of template ligand 4 point away with an angle of approximately 109 °. As a result, the tris-zinc(II)TPP-4 assembly has a much more open structure, which was evident from molecular modeling studies and later confirmed by -ray structure analysis (Figure 8.5) [9]. Interestingly, the structure formed in the solid state did not have the expected 1 : 3 ratio of building blocks as found in solution, but instead a 2 : 5 stoichiometry was observed. One of the zinc(II) porphyrin units is part of two adjacent capsule assemblies as a result of a rather

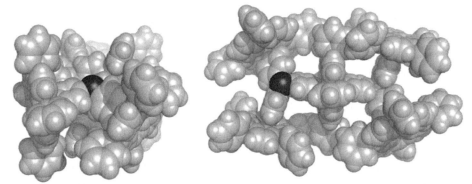

Figure 8.5 Molecular crystal structure of [4]$_2$·[zinc(II)TPP]$_5$ found in the solid state. The front view shows that the phosphine (black) of the template-ligand in this complex is still rather accessible.

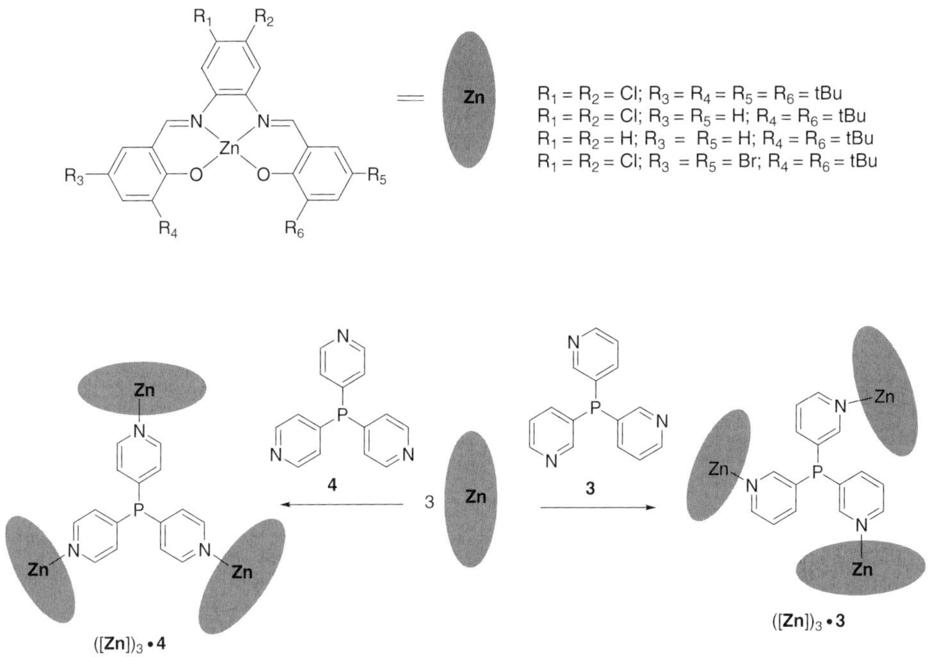

Scheme 8.2 Formation of catalyst assemblies by selective pyridine-zinc(II) coordinative motifs using zinc(II)salphen complexes and different pyridylphosphine templates.

unusual hexacoordinated zinc metal, leading to assembly $[4]_2 \cdot [\text{zinc(II)TPP}]_5$. The presence of both hexa- and pentacoordinate zinc(II)porphyrins within one supramolecular structure is very rare. Importantly, from the front view, the phosphine is clearly far more accessible than the phosphine of assembly tris-zinc(II)TPP-**3** and, as a consequence, the formation of bisphosphine ligated metal complexes is possible with these assemblies. This is reflected in catalysis, since the assembly tris-zinc(II)TPP-**4** gave identical activity and selectivity in the hydroformylation of 1-octene as the rhodium catalyst based on **4** [in the absence of zinc(II)TPP] [9].

More recently, this approach was further extended to the formation of catalyst assemblies consisting of zinc(II)salphen building blocks using the tris(pyridyl) phosphine templates **3** and **4** (Scheme 8.2) [10]. We found a high binding constant for the pyridine association to zinc(II)salphen in toluene ($K_{ass} = 10^5 – 10^6 \text{ M}^{-1}$), typically two orders of magnitude higher than the analogous binding of a pyridine to zinc(II)TPP, which is attributed to the higher Lewis acidity of the zinc atom in the salphen framework. The difference in coordination chemistry between template ligands **3** and **4** with zinc(II)salphens was studied by IR and NMR spectroscopic studies and X-ray crystallographic analyses. Figure 8.6 presents the molecular structure of a 3 : 1 zinc(II)salphen/**4** assembly. In these salphen based molecular capsules the phosphorus center is inside the capsule of zinc(II)salphen/**3**, and more accessible in that of zinc(II)salphen/**4**.

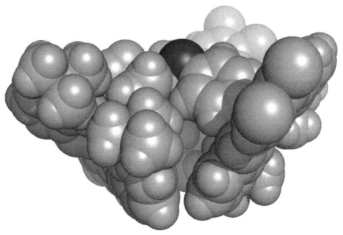

Figure 8.6 X-Ray (solid state) structure of the assembly formed by tris(*para*-pyridyl)phosphine (**4**) and a zinc(II)salphen. The phosphorus atom is shown in black.

A monophosphine complex is formed when **3** is mixed with three equivalents of a zinc(II)salphen complex and half an equivalent of Rh(acac)(CO)$_2$ (acac = acetyl acetonate), whereas the assembly based on template **4** and the zinc(II)salphen complexes forms a bis-phosphine rhodium species. In the latter case, the bis-phosphine rhodium complex is completely encapsulated by six salphen building blocks. This difference in mono- versus diphosphine ligation to the Rh^I-center and, to a lesser extent, the difference in electronic features (and thus donating properties of the phosphine) between template ligands **3** and **4**, can be used to induce a different catalytic behavior.

The encapsulation effects of pyridylphosphine templates **4** and **3** using these zinc (II)salphen have been examined in the hydroformylation of 1-octene. As for the porphyrin assemblies, higher activity and selectivity towards the branched aldehyde was found (up to 57%) for the capsules based on template **3** and (several differently substituted) zinc(II)-salphen complexes. In most cases the effects proved to be somewhat less pronounced than that found with the porphyrin capsules. Apparently, the precise structure of the capsule is of importance in steering the selectivity. This was also found for the various substituted porphyrins. Interestingly, the assemblies based on **4** were *less* active than the non-encapsulated parent derivative **4** and PPh$_3$, with the typical L/B ratio found for bisphosphine ligated rhodium catalysts. The lower activity is explained by a complete encapsulation of the catalyst, reducing the accessibility of the active center for substrates, which might be useful for size-selective catalysis. Catalyst assemblies based on **3** showed significantly *higher* activity (at least five-fold) than parent **3**, which is in line with the results obtained for the porphyrin based capsules.

More recently, the encapsulated catalyst HRh(CO)$_3$(**3·TPP**)$_3$) (Figure 8.4) was also used as a catalyst for the hydroformylation of 2-octene and 3-octene [11]. Hydroformylation of *trans*-2-octene at room temperature using a (non-encapsulated) rhodium catalyst based on tris(*meta*-pyridyl)phosphine provides, as expected, the 2-methyloctanal and 2-ethylheptanal as products in an almost 1 : 1 ratio, showing that there is no preference for either product. In contrast, the encapsulated rhodium catalyst forms 2-ethylheptanal with an outstanding selectivity of 88%. Notably, nonanal and 2-propylhexanal are only formed in trace amounts, indicating that under these conditions no isomerization takes place. In addition to this astonishingly high selectivity, the conversion is roughly doubled, which is in line with the rate enhancements observed for 1-octene. Applying a reaction temperature of 40 °C resulted in a higher reaction rate for both the non-encapsulated and the encapsulated rhodium complexes. Interestingly, the regioselectivity induced by the encapsulated catalyst was still very high, producing 80% 2-ethylheptanal. At 80 °C all four possible aldehyde products are formed, indicating that isomerization occurs. The selectivity of the encapsulated catalyst is no longer retained because the isomerization reaction transforms the pure starting *trans*-2-octene into a mixture of alkenes.

The hydroformylation of *trans*-3-octene at room temperature using the (non-encapsulated) rhodium catalyst based on tris(*meta*-pyridyl)phosphine afforded 2-ethylheptanal and 2-propylhexanal in exactly a 1 : 1 ratio. The encapsulated catalyst provided an unprecedented selectivity for 2-propylhexanal of 75% (Scheme 8.3). Again the selectivity is largely retained at 40 °C whereas at 80 °C the isomerization side reaction prohibits the selective formation of aldehydes. Similar regioselectivities were obtained in the hydroformylation of *trans*-2-hexene, *trans*-2-nonene and *trans*-3-nonene at 25 °C.

Preliminary kinetic data suggest that, in line with literature, the rate-determining step in the hydroformylation reaction is either alkene coordination or migratory insertion of the hydride to the rhodium-alkene. As the isomerization is suppressed at room temperature, this step is likely irreversible, which is commonly observed at low temperatures and high CO pressure. After this point all steps are either fast or irreversible and so the regioselectivity is determined early in the cycle. Since, after alkene-coordination, both isomers can still be formed, we propose that the regioselectivity induced by the encapsulated catalysts is determined during the hydride-migration step. During this migration the substrate has to rotate, and the steric restrictions of the metal-olefin complex imposed by the innerside of the capsule reduce rotational freedom. Notably, this is the first catalyst system able to discriminate between carbon atoms C3 and C4 in *trans*-3-octene! The next logical step we are exploring is the use of chiral capsules to produce the branched aldehydes in enantiopure form. Initial experiments using zinc(II)salphen building blocks show

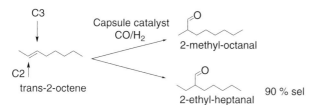

Scheme 8.3 Hydroformylation of internal octenes, aiming for selective formation of one of the aldehyde products.

that the selectivity effects for this reaction are exclusive for the porphyrin based capsules as in all experiments statistical product distributions are found. Thus, again, the structure of the zinc(II) building block is very important in creating capsules of the proper shape for selective catalysis.

So far the supramolecular ligands and templated encapsulation were based on the complementary use of donor atoms; zinc does not have an affinity for phosphine ligands, and the monopyridyl ligands have lower affinity for the catalytically active metals used. We recently demonstrated that a similar type of assemblies could be based on template ligands that only have nitrogen donor atoms, in which case selective coordination is based on steric differences [12]. In the 3-pyridyl-BIAN ligand (*m*Py-BIAN, see Scheme 8.4) encapsulation takes place via the pyridyl groups, whereas palladium complexes are, in the presence of zinc(II)salphen building blocks, coordinated to the bis-imine site. It was anticipated that coordination of the zinc(II)salphen to the imine-nitrogen would not be possible for steric reasons. Indeed, control experiments show that the Ph-BIAN ligand does not coordinate to zinc(II)salphens. In addition, Job-plot analysis showed that the *m*Py-BIAN ligand is encapsulated by two zinc(II)salphens that are coordinated via the pyridyl-nitrogen. The palladium precursor [(COD)PdMeCl] has a preference for the bis-imine ligand because of the chelate effect. In the absence of zinc(II)salphen building blocks a mixture of complexes is formed, indicating that under these conditions the pyridyl groups are also partly coordinated to palladium. Cationic palladium complexes could also be prepared and encapsulated in similar fashion, and these are known to give active catalysts for the CO/styrene polymerization reaction.

From *m*Py-BIAN and building blocks A–D (Scheme 8.4) four assembled active catalysts for the CO/TBS (TBS = 4-*tert*-butylstyrene) copolymerization are formed that give very different results. Activities vary, depending on the structure of the salphen building block, between 98-gCP per g-Pd to a maximum of 411-gCP per g-Pd. This activity is much higher than that of the non-encapsulated parent complex, which showed an activity of 284-gCP per g-Pd. Interestingly, the supramolecular catalysts *m*Py-BIAN·A$_2$–*m*Py-BIAN·D$_2$ all provide selective catalysts leading to highly stereoregular (syndiotactic) polymers, as is clear from the ^{1}H and ^{31}C NMR spectra, whereas the parent complex gives rise to atactic copolymers. The highest degree of syndiotacticity is displayed by complex *m*Py-BIAN·D$_2$, with a maximum percentage of *uu* triads of 87%, which is among the most selective catalysts known to date. In addition, the molecular weight of the copolymer formed can be tuned by small changes of the salphen building blocks and is much higher than the molecular weight of the polymer produced by the parent complex (average molecular weight of $35 \cdot \times 10^3$). The supramolecular catalysts formed polymers with molecular weights of between $44 \cdot \times 10^3$ (*m*Py-BIAN·A$_2$) and $118 \cdot \times 10^3$ (*m*Py-BIAN·C$_2$). These results demonstrate that the catalytic activity of the complex and the molecular weight of the copolymer obtained can be tuned by the properties of the zinc(II)salphens used for the assembly to the ligand of the catalyst. The catalyst activity and copolymer stereoregularity are amongst the best reported for CO/TBS copolymerization under similar conditions (no use of trifluoroethanol, room temperature), and the molecular weight of the copolymer produced under these conditions by *m*Py-BIAN·C$_2$ ($118 \cdot \times 10^3$) is

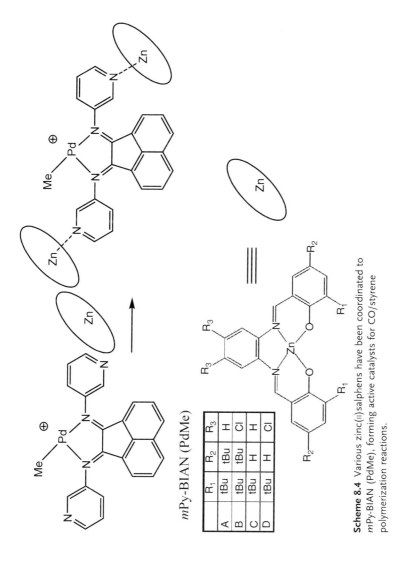

Scheme 8.4 Various zinc(II)salphens have been coordinated to mPy-BIAN (PdMe), forming active catalysts for CO/styrene polymerization reactions.

the highest reported thus far. More importantly, with this example we show that we can use nitrogen-only based template ligands, thereby expanding the scope of ligands that can be used in this supramolecular approach.

We have recently extended the ligand-template approach to systems in which the ligand-template is an integral part of the capsule [13]. A concave Xantphos-based diphosphine ligand, to which transition metal complexation can occur, was functionalized at the four phenyl groups of the two phosphines. Self-assembly of the tetracationic diphosphine ligand with a tetraanionic calix[4]arene leads to the formation of a reversible heterocapsule based on ionic interactions (**H**). Coordination of palladium to the template-ligand results in metal encapsulation [**H** ⊃ Pd-Ar] (Figure 8.7), and reactions can still take place, as demonstrated by stoichiometric CO-insertion reactions.

8.3
Self-Assembled Ligands in Transition Metal Catalysis

The properties of a homogeneous catalyst can be tailored to very specific conversions by a proper design of the ligand. Important parameters that describe ligand properties include the cone-angle (Θ) [14], desribing the steric properties of the ligand, and the X-parameter, describing the electronic properties of phosphine ligands [15]. The importance of the geometry of bidentate ligands was recognized much later and the "natural bite angle" was introduced as a way of describing the properties of bidentate chelating ligands [16]. Although these parameters are very helpful and computational techniques are becoming increasingly important, one cannot easily predict the activity and selectivity of a transition metal catalyst. This is especially true for chiral catalysts developed for asymmetric conversions and, therefore, catalyst development (for asymmetric reactions) relies to a great extent on trial-and-error and sophisticated guesses. As a consequence, combinatorial approaches and high-throughput experimentation have become increasingly important [17]. Several elegant screening techniques for combinatorial catalysis [18] have been developed, but there is only limited technology available to prepare large catalyst libraries with sufficient diversity. Recently, it has become clear that catalyst based on monodentate ligands can lead to selective catalysts [19–24]. This class of ligands is generally easier to prepare, and an automated protocol for the preparation of chiral phosphoramidite ligands has been reported. Bidentate ligands are much more difficult to prepare and new tools are required that lead to large libraries of sufficient diversity [25,26]. We anticipated that making bidentate by self-assembly would lead to such a new tool, and as such we [16,21,22,24,26,27,30], and others [28], have recently introduced supramolecular bidentate ligands as a new class. These bidentate ligands form via self-assembly of two monodentate ligands by simply mixing the proper monodentate ligands functionalized with complementary binding motifs. Especially for the class of heterobidentate ligands, which can be distinguished from homobidentate ligands in that they consist of two different donor atoms, the supramolecular approach could be very powerfull. Heterobidentate ligands offer a higher level of spacial control around the

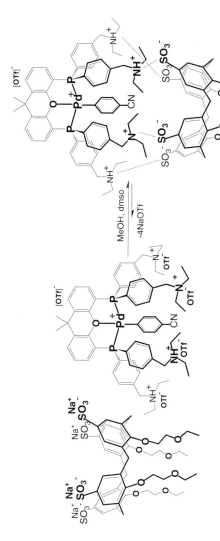

Figure 8.7 Encapsulation of palladium in a heterocapsule formed by ionic interactions between a concave Xantphos-based diphosphine ligand and a tetraanionic calix[4]arene.

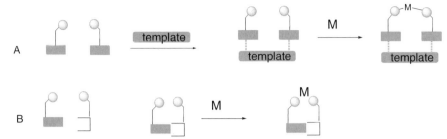

Scheme 8.5 Two strategies to make bidentate ligands by assembly: association of ligand building blocks to a template (A) or the direct approach by bringing ligand blocks together via noncovalent interactions (B).

metal center, which, for many reactions, results in higher selectivity; however, in general the synthesis of this class of ligands is more complicated. Two main strategies can be followed (Scheme 8.5) to form bidentate ligands by assembly:

(A) a template can be used that contains binding sites for the selective assembly of two (different) ligands,
(B) the ligands can be functionalized with complemetary binding motifs.

An important difference between these two strategies is the number of components involved. Approach A leads to a three-component-assembly and, as a consequence, the system is more complicated, but it does easily produce large ligand libraries ($10 \times 10 \times 10 = 1000$ members based on 30 compounds). Approach B is simpler, but more building blocks are required to arrive at large ligand libraries ($30 \times 30 = 900$ members based on 60 compounds).

8.3.1
Template Approach

With design principle A (Scheme 8.5) in mind we developed a prototype system based on pyridyl-zinc(II)porphyrin interactions. This binding motif has also been used for template ligand assisted catalyst encapsulation (Section 8.2). We anticipated that supramolecular bidentate ligands could be obtained by assembling two pyridyl containing phosphorus ligands (5–7) onto a bis-zinc(II)porphyrin template 8 [29,30] (Scheme 8.6). The supramolecular bidentate phosphorus ligand forms *in situ* by the selective coordination of the nitrogen donor atom of building blocks 5–7 to the zinc atoms of the porphyrin and the complex formation has been confirmed by UV/Vis spectroscopy experiments and high-pressure NMR spectroscopy.

Assemblies based on 8 and pyridine phosphorus ligands 5–7 were used as supramolecular ligands in the rhodium-catalyzed hydroformylation and showed typical bidentate behavior. The chelating bidentate assembly exhibited lower activities (a factor of three) than the monodentate analogue. Only a slightly higher selectivity for the linear aldehyde was observed. The chiral ligand assemblies based

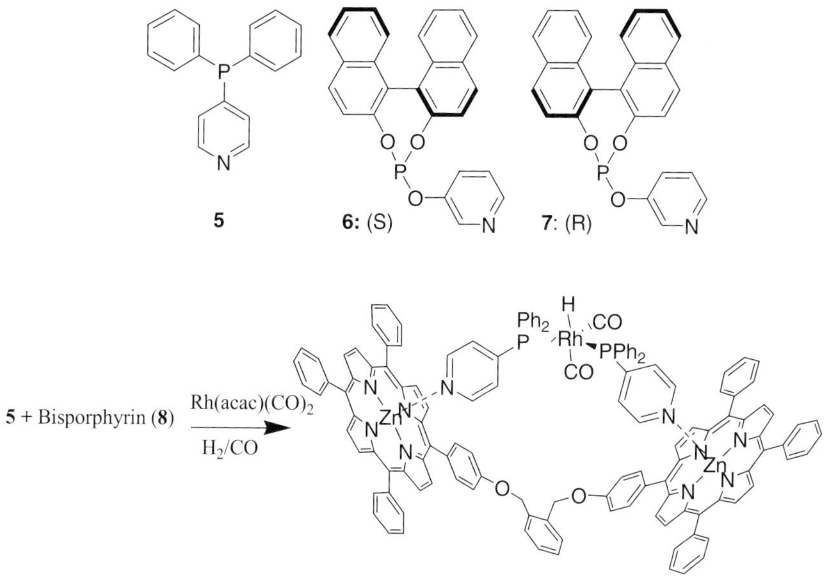

Scheme 8.6

on **6**(*S*) and **7**(*R*) were studied in the asymmetric rhodium-catalysed hydroformylation of styrene [31,32]. Rhodium complexes based on monodentates **6**(*S*) and **7**(*R*) resulted in low enantiomeric excess (approximately 7%), which is in line with previous results for monodentate ligands [33]. Interestingly, templated ligand assemblies **8**[**6**(*S*)]$_2$ and **8**[**7**(*R*)]$_2$ afforded significantly higher enantioselectivity (33%), along with an increase in activity. Similar ligand assemblies were prepared using tin–carboxylate interactions, showing that the strategy is certainly not limited to zinc(II)–nitrogen interactions [32].

We recently have extended this approach to more rigid bis-zinc(II)salphen [34] building blocks and we also applied more nitrogen-containing ligand building blocks [35]. The bis-zinc(II)salphen building block is more rigid and, consequently, higher selectivities were expected upon application as a template. The formation of a templated chelating bidentate ligand in the presence of a transition metal precursor was demonstrated by NMR experiments and unambiguously proven by X-ray crystallography. Crystals of Pt[**9**·(**1**)$_2$]Cl$_2$ were grown from a dichloromethane/acetonitrile solvent mixture, and the molecular structure clearly shows the formation of a chelating bidentate ligand; the *meta*-pyridyldiphenylphosphine ligands **1** are coordinated via the nitrogen donor atoms to the axial positions of the bis-zinc(II)salphen template and the phosphorus donor atoms are coordinated to platinum(II) in a cis fashion. Crystals of Pd[**9** (**5**)$_2$]MeCl grown from a dichloromethane/acetonitrile solvent mixture were also analyzed by X-ray crystallography. The solid-state structure clearly shows the formation of a chelating bidentate ligand [34], with the phosphorus donor atoms coordinated to palladium(II) in trans-disposition and the *para*-pyridyldiphenylphosphine ligands coordinated via the nitrogen donor atoms to the axial positions of the bis-zinc(II)salphen template. Application of these templated bidentate ligands in transition metal catalysis showed, in most cases, the typical bidentate character of the supramolecular ligand. Compared to the ligands based on the more flexible bis-zinc(II)porphyrin template the selectivities were not much improved, suggesting that, in contrast to what we expected, the rigidiy of the template is not an important factor for highly regio- and enantio-selective catalysts.

To extend the strategy to heterobidentate ligands we expected that templates with two different (orthogonal) binding sites were required. However, we unexpectedly found that the use of a template with two identical binding sites in the presence of two different ligand building blocks resulted in the exclusive formation of the hetero-bidentate ligand (Scheme 8.7). For example, if **1**, **10** and the template **9** are mixed in a 1 : 1 : 1 ratio in the presence of a metal-precursor two sets of double doublets appear in the ^{31}P NMR spectrum, which is consistent with the formation of the heterocomplex. In contrast, in the absence of template only the homocomplexes were formed. Initial computational analysis suggest that the orgin of this selective formation is a result of steric effects. The metal complexes with two rather bulky ligands has a rather large energy penalty for the preorganization of its pyridyl functions required to coordinate to the template. As a result, the mixture of homocomplexes rearrganges into the heterocomplex, which has a lower energy penalty. We have studied several different ligand combinations, and so far have always obtained the heterocomplex unless both ligand building blocks are very bulky. Interestingly, when we applied the hetero-complexes in the asymmertic hydroformylation of styrene, all heterocomplexes gave much higher ee's than the homobidentates, with the best catalyst (**1**, **10** on **9**) producing the aldehyde with relatively high ee (72%).

We also demonstrated that bidentate ligands can be constructed by using, simultenously, several templates. Relatively small ditopic nitrogen ligands (DABCO, 1,4-diaza-bicyclo[2.2.2]octane) have been used as templates, which are known to form 2 : 1 sandwich complexes with zinc(II)porphyrins. This assembly process was also used to form chelating bidentate ligands by the application of various phosphite functionalized porphyrins. Initial studies using these complexes in the hydroformy-lation of 1-octene indicated that the assembled bidentate only showed moderate bidentate character. The activity was still 80% of that of the catalyst without template while the selectivity changed only slightly. On the basis of this result it was anticipated that a more rigid assembly was required to express the chelate effect significantly. To this end bisphosphite assemblies were developed based on the trisporphyrin-phosphite ligand **11** and three equivalents of DABCO (**12**) [36]. In the presence of a metal precursor a multicomponent supramolecular complex was obtained in which two ligands are firmly fixed by three bridging DABCO ligands (Scheme 8.8). This was confirmed by UV/Vis and NMR spectroscopic studies. In comparison with the single porphyrin system this assembly expressed very strong chelate bidentate behavior in the hydroformylation reaction. The rhodium catalyst based on assembly [Rh(**11**)$_2$(**12**)$_3$] showed a high linear to branched ratio (L/B = 15.1 at 80 °C) and a lower activity than the catalyst based on a single porphyrin-DABCO sandwich. Furthermore, at a lower DABCO-porphyrin ratio the selectivity dropped, pointing at

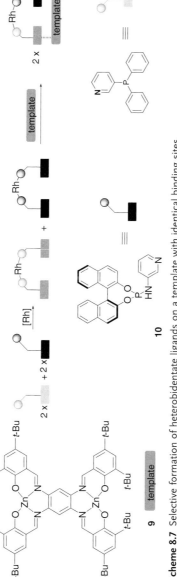

Scheme 8.7 Selective formation of heterobidentate ligands on a template with identical binding sites.

216 | *8 New Supramolecular Approaches in Transition Metal Catalysis*

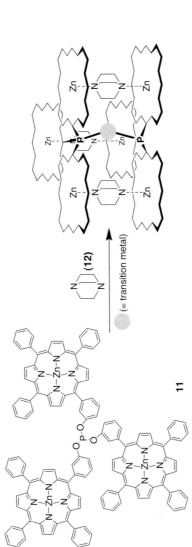

Scheme 8.8 Schematic representation of a multicomponent assembly fixing two phosphites in space to form a chelating bidentate ligand.

formation of more flexible ligands. These experiments unambiguously showed that multicomponent ligand assemblies based on noncovalent interactions can act as a rigid chelating bidentate ligand. Furthermore, it was concluded that the flexibility range can be fine-tuned with these systems by altering the number of bridging ditopic DABCO ligands. Upon lowering the temperature to 30 °C a more selective catalyst was obtained (the L/B ratio increased to 22.8) [37]. This suggests that the supramolecular catalyst system [Rh(**11**)$_2$(**12**)$_3$] exhibits less dynamic behavior at lower temperatures.

8.3.2
Direct Approach

We have investigated the assembly of ligand building blocks to form bidentate ligands using various interactions (Scheme 8.5, strategy B). We initially explored this strategy by using the zinc(II)porphyrin–pyridyl interaction, with one ligand building block carrying the zinc-porphyrin motif and the other containing a pyridyl motif (for a typical example see Scheme 8.9) [38].

The assembly formation of heterobidentate ligated complexes was studied in detail using bidentate ligand **13·b** as a typical example (Figure 8.8). UV/Vis titrations and NMR spectroscopy experiments showed that the pyridyl moiety of **b** selectively coordinates to the zinc(II)porphyrin **13** with a binding constant of $K_{(1.b)} = 3.8 \times 10^3 \, M^{-1}$. An increase in the K for **13·b** to $64.5 \times 10^3 \, M^{-1}$ was observed in the presence of metal precursor [HRh(**a**)$_3$(CO)]. This provided the formation of a bidentate chelating system with a corresponding chelate energy of 7 kJ mol^{-1}. This compares well with the classical textbook examples of chelate effects such as 1,2-ethylene diamine coordinating to Co(II), which proceeds with a chelate energy of 10 kJ mol^{-1}. Bidentate ligand assembly **14·b** gave similar results.

The chelating behavior was also evident from HP-NMR experiments. The addition of triphenylphosphine (**a**) to a catalyst solution of [HRh(CO)$_2$(**13·b**)] did not affect the complex. Moreover, the addition of 1 equiv. of **b** to a solution of HRh(CO)$_2$(**13**)PPh$_3$ resulted in the exclusive formation of [HRh(CO)$_2$(**13·b**)] upon release of free triphenylphosphine. The chelating effect of the supramolecular ligand assembly effectively competes with triphenylphosphine, leading to exclusive formation of the rhodium complex of **13·b**. In the complex [HRh(CO)$_2$(**13·b**)] the supramolecular ligand **13·b** coordinates in an equatorial–equatorial fashion to the rhodium metal center, whereas the HRh(CO)$_2$(**13**)PPh$_3$ exists in a mixture of complexes (ee and ea).

To demonstrate the power of the concept as a tool for high-throughput experimentation we studied a library of 48 supramolecular bidentate ligands based on six phosphite functionalized porphyrins (**13–20**) in combination with eight nitrogen-containing phosphorus ligand building blocks (**b–i**) (Figure 8.8).

Two series of control experiments were conducted. In the first column on the left-hand side in Figure 8.9 we display the results obtained when the porphyrin-phosphites are applied as monodentate ligands. In a second control experiment, ligand **a** (PPh$_3$) was used as the ligand next to the porphyrin-phosphites, since it can

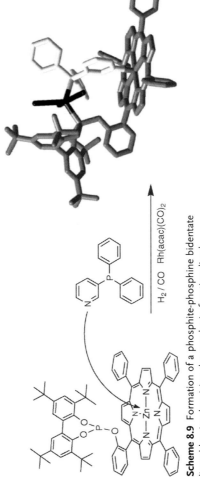

Scheme 8.9 Formation of a phosphite-phosphine bidentate ligand by simply mixing the porphyrin-functionalized phosphite and the 3-pyridyldiphenylphosphine ligand in the presence of a metal precursor.

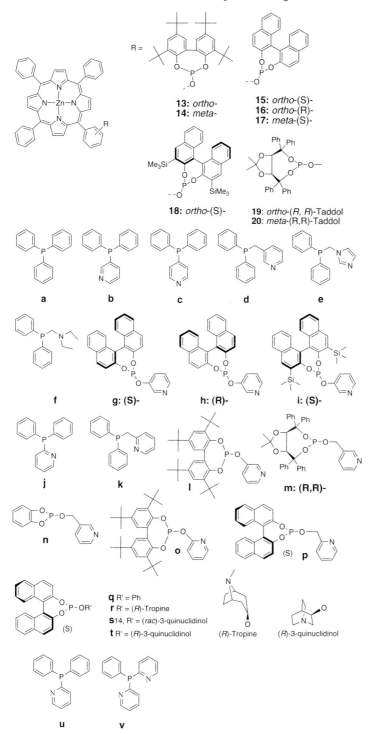

Figure 8.8 Ligand building blocks used to form bidentate ligands.

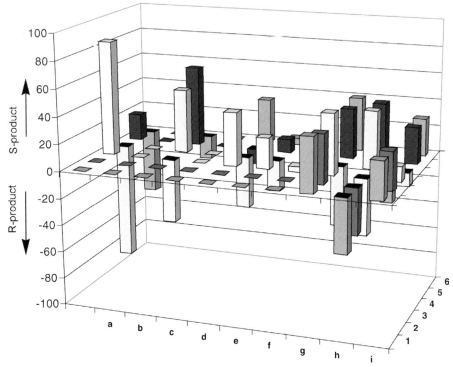

Figure 8.9 Results of palladium-catalyzed asymmetric alkylation of 1,3-diphenylallyl acetate and dimethyl malonate upon applying a ligand library of self-assembled bidentate ligands. The ligands **a–i** on the x-axis are defined in Figure 8.8.

not form a bond with the zinc(II)porphyrin (Figure 8.9, second row on the left). The library was studied in the palladium-catalyzed asymmetric allylic alkylation [39] of *rac*-1,3-diphenyl-2-propenyl acetate with dimethyl malonate at room temperature. The catalysts explored provided a wide scope of results, with enantiomeric excesses ranging from 85% (*S*) to 86% (*R*) (Figure 8.9). The ee of the product depends strongly on the ligand assembly used, with both components being important.

The control experiments in the absence of a second ligand include the most selective catalyst in this study; unexpectedly ligand **16** applied as a monodentate species provided 86% of the (*R*)-product. Notably, upon addition of a second, non-chiral, ligand (**b**) to this system the (*S*)-product was preferentially obtained in 60% ee. Control experiments including triphenylphosphine (**a**) as the second ligand next to one of the porphyrin-phosphites (**13–20**) demonstrated the importance of the presence of the binding motif. All control-reactions gave rise to racemic product, indicating that in these experiments the catalysis is dominated by palladium triphenylphosphine species. This extraordinary difference with the supramolecular ligands highlights the importance of the chelate effect. The catalyst based on ligand assembly **15·b** gave products with an ee of 60%, and also in the presence of a 30-fold

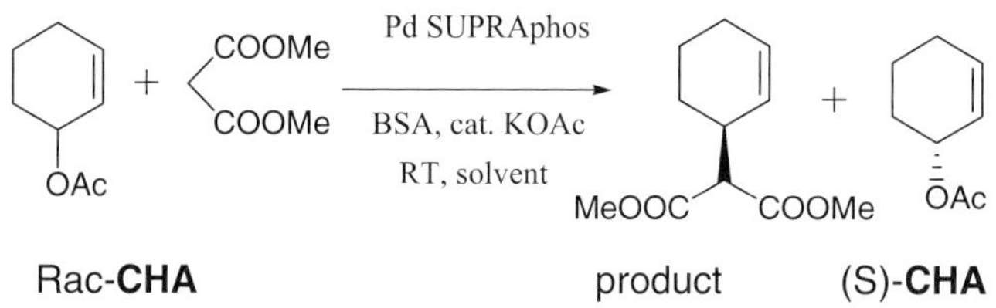

Scheme 8.10 Schematic representation of the palladium-catalyzed kinetic resolution of *rac*-**CHA** using SUPRAphos ligands.

excess of **b**. Under these conditions the catalysis is still dominated by the catalyst based on **15·b**, indicating a strong chelate effect of the supramolecular ligand system. Similar to a previous report for covalently linked phosphine-phosphite ligands [40], a small difference in the length of the bridge between the phosphine and phosphite results in a large difference in enantioselectivity [e.g., **10·b** 60% (*R*) and **10·d** 44% (*S*)].

Palladium catalysts can also be used for kinetic resolution, a process in which one of the enantiomers of the racemic starting material is converted into the desired product while the other enantiomer is recovered unchanged. Kinetic resolution can be very powerful because extremely high enantiopurities can be achieved, albeit at the cost of the yield in less favourable cases. In most examples in which metal catalysts are used both enantiomers of the substrate show some reactivity towards the catalyst. While the palladium-catalyzed allylic substitution reaction has been widely studied, relatively few catalyst have been reported in the literature for kinetic resolution. We also applied a combinatorial approach to search for novel catalysts for the kinetic resolution of cyclohexenyl acetate (*rac*-**CHA**, Scheme 8.10), using the SUPRAphos ligand library (building blocks **15**, **17**, **19**, **20**, and most of **b–v**) [41]. We screened the catalysts library based on SUPRAphos ligands in the palladium-catalyzed kinetic resolution of *rac*-**CHA**. We simultaneously monitored the conversion, the ee of substrate *rac*-**CHA**, and that of the product. Notably, for kinetic resolution we aim for 50% conversion (100% of one enantiomer). The SUPRAphos library contains phosphine-phosphite as well as bisphosphite type ligands and as control experiments we also used the building blocks as chiral, monodentate ligands. A few catalysts based on heterobidentate supramolecular ligands give rise to kinetic resolution of *rac*-**CHA**; four catalysts provide the substrate in 99% ee at a conversion of between 69 and 84%. As second best result we obtained 36% ee only for the substate. The four best catalysts had in common that the ligand consisted of **15** and an ortho-pyridyl functionalized phosphine (e.g., **u** and **v**). The kinetic resolution (S-factor 12) is acceptable, yielding high ee's of the (*S*)-enantiomer above conversions of 60% at high rates (TOF 450 mol mol^{-1} h^{-1}), but the ee of the product is low (30%) and opposite, albeit constant during the reaction. The combination of a high kinetic resolution and a low ee of the product is remarkable as, usually, a catalyst will lead to acceptable ee's of the product and a poor kinetic resolution. The currently accepted view is that the transition state for the oxidative addition is similar to that of the nucleophilic attack, explaining why, usually, high kinetic resolution is accompanied with high ee of the product. The main difference between the current system and those reported is the dynamic character of the ligand, enabling the catalyst to change its coordination sphere during the various reaction steps. For instance, a decoordination of the achiral pyridylphosphine ligands from the zinc is envisioned, which could either coordinate to palladium or cause deracemization of the substrate attached to the palladium.

Based on this proof of principle we decided to extend the ligand library (for some examples see entries included in Figure 8.8) and currently we have over 40 building blocks with which we can construct over 400 bidentate ligands by assembly. With the large library in hand it is interesting to move from benchmark reactions to more challenging catalytic processes. As a first example, the asymmetric hydrogenation of a (commercially interesting) tri-substituted cyclic enamide, N-(3,4-dihydronaphthalen-2-yl)acetamide, was targeted (Figure 8.10) [42].

Prior to our study there were only a few ruthenium-based catalysts known that provided high enantioselectivity (up to 90%) in this reaction, albeit at high hydrogen pressure, while rhodium catalysts can be used at much lower pressures but generally provide the product in low to moderate ees (<72%) [43,44]. We performed the screening experiments using high-throughput equipment, enabling 96 reactions to be performed in parallel (0.04-mmol scale) [10,45]. The results of these screening

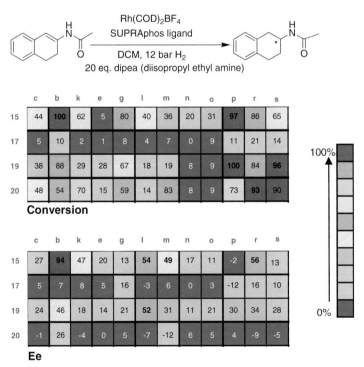

Figure 8.10 Conversions (%) (top) and ees (bottom) obtained in the rhodium-catalyzed hydrogenation of N-(3,4-dihydronaphthalen-2-yl)acetamide upon applying various SUPRAphos ligands.

experiments (Figure 8.10) show a large variety; the conversion ranged from 0 to 100% and enantioselectivities (ee) of the product obtained were between −12 to 94%.

From the 60 catalysts screened in the first round, four members displayed an enantioselectivity >50% (indicated green and red). Five catalysts of the library provided a conversion of >90% under the conditions applied (displayed in red). The catalyst based on supramolecular ligand **15·b** provided a true hit in this reaction (shown in red), showing an ee of 94% at 100% conversion. Control experiments showed that indeed the bidentate character is very important; if the interaction between the two building blocks is lacking the ee is very low. Some traditional bidentate phosphine-phosphite ligands (BINAPhos) were also studied for comparison, but the catalysts based on these ligands provided the product with a maximum of 60% ee. The catalyst based on supramolecular ligand **15·b** was also used at lower catalyst loadings (1 mol.%) and at larger scale, producing similarly high selectivities.

Importantly, the supramolecular strategy revealed a catalyst that outperforms all conventional catalysts known to date. This unambiguously shows that the supramolecular approach to create bidentate ligands is a very powerful tool that brings about new catalysts with properties that surpass those of catalysts already known. Moreover, it shows that these "hits" can be found in a very short time-frame. In addition, further catalyst optimization will be much faster using the supramolecular approach. The fact the only one hit was obtained from 60 catalysts clearly stresses the need for large numbers of catalysts to address catalytically challenging issues sufficiently.

In the examples shown so far in this chapter the assembly processes of the ligands are based on metal–ligand interactions. We [46] and others [47] have also utilized hydrogen bonds for this purpose. Initially, we focused on self-complementary urea-motifs to form supramolecular homo-bidentate ligands. Urea-appended phosphine ligands form palladium methyl chloride complexes in which the phosphorus ligands are coordinated in a trans disposition. From NMR and IR spectroscopic measurements it was concluded that the urea groups of the ligands are indeed hydrogen bonded (Figure 8.11).

Interestingly, the bis-urea pocket represents also a good host for anions and, indeed, upon the addition of ammonium chloride the pocket is occupied with the chloride anion. The binding pocket can change the reactivity of the metal center as it facilitates substitution processes at the palladium center. Upon the introduction of

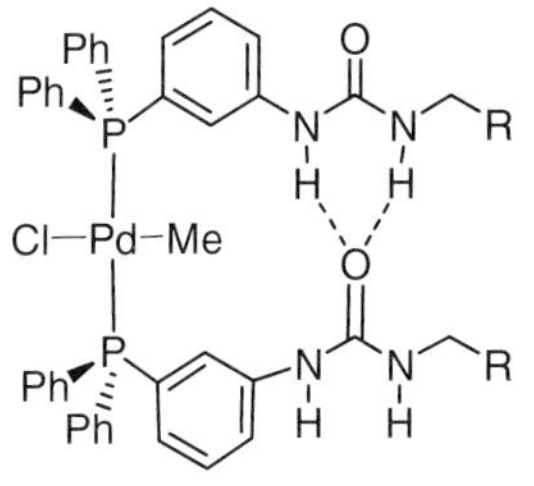

Figure 8.11 Bidentate ligands formed by urea-based hydrogen bonding motifs.

CO to the depicted complex the chloride is displaced by CO and moves to the bisurea binding pocket. After insertion of the CO into the palladium at higher temperatures the chloride moves back to the palladium. This shows that hydrogen bond motifs can also be used to construct bidentate ligands, but one should be aware of the non-innocent character.

After demonstrating the use of urea based ligands, we have introduced a new class of supramolecular bidentate phosphite ligands based on the urea-binding motif, and applied them successfully in the rhodium-catalyzed asymmetric hydrogenation of various substrates [48]. The urea-containing phosphite ligand building blocks (Figure 8.12, **21–26**) all have a similar urea-type hydrogen bond motif, a chiral phosphite ligand based on the bisnaphthol backbone, and a (chiral) spacer to connect these functions. Small differences in spacer and motif were applied to create a small series of ligand building blocks. Ligand **25** is the only one containing an (R)-bisnaphthol backbone, all others contain the (S)-bisnaphthol backbone. The spacers of building blocks **23–26** contain additional chirality (Figure 8.12). Ligand

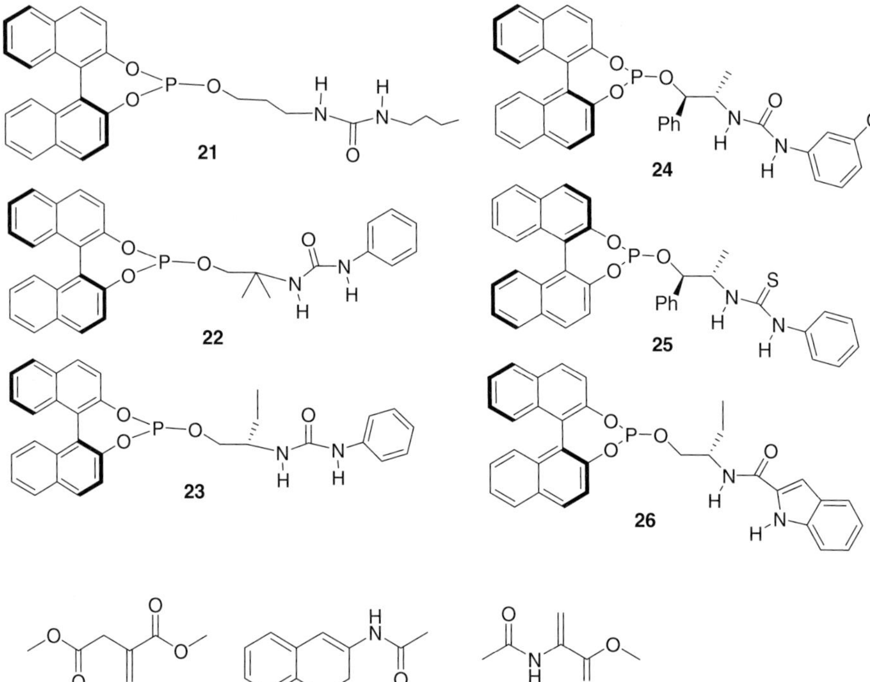

Figure 8.12 Urea functionalized phosphate ligands for asymmetric hydrogenation of dimethyl itaconate (DMI), N-(3,4-dihydronaphthalen-2-yl)acetamide (DNA) and methyl 2-acetamidoacrylate (MAA).

25 is the only one that contains a thiourea motif, while **26** contains an indole-amide binding motif whereas all other systems are functionalized with the urea binding motif. Notably, all ligand building blocks are easily accessible and are amendable to simple variation to facilitate the preparation of large, diverse libraries at a later stage. Indeed, all compounds were obtained via a simple two-step synthetic procedure, consisting of a coupling of an amino-alcohol with an iso(thio)cyanate to obtain the urea-alcohol that is subsequently reacted with 2,2′-bisnaphthol phosphorochloridite to obtain the urea-containing phosphites. Two of the six ligands provided rhodium complexes that hydrogenated dimethyl itaconate (DMI) in over 90% ee (highest 96% ee) and most catalysts gave 100% conversion of this substrate. We also studied the hydrogenation of N-(3,4-dihydronaphthalen-2-yl)acetamide (DNA), which is a much more challenging substrate both in terms of conversion and selectivity. Upon applying rhodium catalysts based on **21–26**, conversions were obtained between 0.4 to 100% and the selectivity ranged from 37.9% for the (*S*)-enantiomer to 76.5% for the (*R*)-enantiomer. In the hydrogenation of methyl 2-acetamidoacrylate (MAA), four catalysts provided full conversion while [Rh(**25**)$_2$] and [Rh(**21**)$_2$] showed low and no conversion, respectively. Both [Rh(**22**)$_2$] and [Rh(**23**)$_2$] provided the product in high selectivity, 94% and 92% ee, respectively. In general, the chirality of the bisnaphthol determines the chirality of the product while fine-tuning is established by the choice of the amino–alcohol spacer. The easy accessibility of these ligands and the huge potential for catalyst tuning makes them highly suitable for combinatorial approaches and high-throughput screening experimentation. We have extended the building block library using robotic synthesis strategies and we currently have about 120 building blocks available with which we can assemble 2500 bidentate ligands. The question remains how large and diverse a ligand library should be to provide a catalyst for every substrate. We hope to answer this question in due course.

8.4
Supramolecular Anchoring of Catalysts to Support

The advantages of homogeneous catalysis originate from the substrate and catalyst being all dissolved in the same phase. However, this implies a catalyst separation step to separate the product from the catalyst. As recycling of the catalyst could be interesting (or crucial) for economical reasons, many methodologies for separating catalysts from the products have been developed. These methodologies usually involve tedious synthetic procedures to create a complex supported catalyst that is optimized for only one process. We have introduced the concept of reversible noncovalent anchoring of catalysts (and excess ligand if applicable) to soluble and insoluble supports that are functionalized with binding sites tailored to bind the functionalized ligands. This provides simple and efficient recycling methods, while also allowing re-functionalization of the support and variation of the catalyst loading even during catalysis (Scheme 8.11).

The first examples of noncovalent anchoring of catalysts to soluble supports appeared in the literature in 2001. Concurrently, Mecking and ourselves were

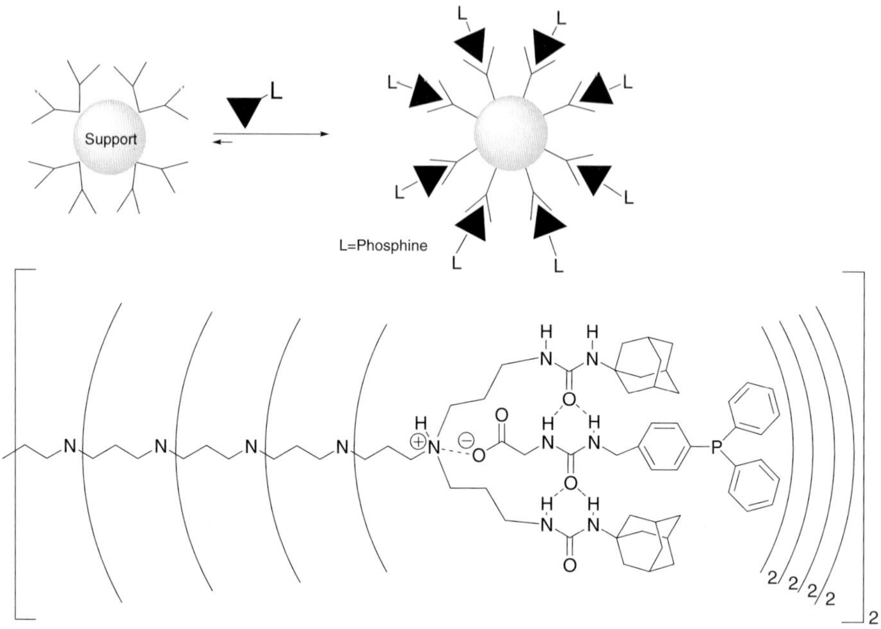

Scheme 8.11 Concept of reversible anchoring of catalysts to a (dendritic) support that is functionalized with a binding motif complementary to that of the catalyst, and a typical example used in our group.

investigating the reversible anchoring of catalysts to supports. Our approach involved the noncovalent anchoring of the catalyst to a dendrimer support using well-defined binding sites to keep control over the exact location of the catalyst on the support (Scheme 8.11). To this end we have utilized the fifth generation poly(propylene imine) dendrimer functionalized with urea-adamantyl units at the periphery [49], which allows the directional noncovalent anchoring of 32 guest molecules. For our purpose we used phosphine ligands with the complementary-binding motif (Scheme 8.11) [50]. The binding of guest molecules into the periphery of the dendrimer was studied in detail and appeared to be sufficiently high to warrant further study. The dendrimer complex with 16 metal complexes remained intact after size exclusion chromatography (SEC). The dendritic host containing 32 phosphine ligands assembled to the periphery of the dendrimer was used as a multidentate ligand in the Pd-catalyzed allylic amination of crotyl acetate by piperidine. High reaction rates were observed, similar to the unbound analogue, indicating that every active site on the dendrimer acts as an independent catalyst. The supramolecular anchoring of the catalysts does not decrease the activity or the selectivity, in contrast to the general observation for catalysts immobilized on insoluble supports. The noncovalently functionalized dendritic catalysts used in these batchwise reactions were recycled using SEC, but the catalyst was partly decomposed during such a procedure. The catalysts were also applied in a continuous-flow membrane reactor. The retention measured for the supramolecular acid-dendrimer complex was 99.4%, implying that 99.4% of the catalyst remains inside the reactor after each "cycle". Indeed, the dendritic supramolecular catalyst was successfully applied in the allylic amination reaction in the continuous-flow membrane reactor. The conversion remained fairly constant during the first 10 hours of the experiment, after which a small decrease was observed due to catalyst deactivation (not leaching!). These experiments clearly demonstrate that noncovalently supported catalysts can be used as recyclable catalysts. These supramolecular supports can be conveniently reloaded with new catalysts, in contrast to their covalent analogues. We recently extended the approach to hyperbranched polymers, which are cheaper, but less defined analogues of dendrimers [51]. The hyperbranched polymer contains different binding sites because of its less defined nature. Nonetheless, these materials are well suited for noncovalent catalyst immobilization. After optimizing the binding site attached to the ligands (we added an additional carboxyl group for additional interactions) we successfully applied the hyperbranched polymers containing urea-adamantyl groups as support in a continuous flow membrane reactor. In addition, we demonstrated that once the supramolecular system is in the reactor the support could be de-functionalized by washing with methanol and, subsequently, new catalyst for a different catalytic process could be uploaded to the system. The second reaction performed was an asymmetric hydrosilylation; the conversion was constant over prolonged reaction times.

We have extended the approach to dendrimers immobilized on silica [52]. For this purpose the first generation (and also larger generations) DAB dendrimer was functionalized with urea propyl-$(SiOR)_3$ groups that, via an easy process, were grafted onto silica. The catalyst was noncovalently anchored by simply adding a solution of a metal complex with ancillary ligands utilized with a typical complementary motif. The supramolecular interaction between the transition metal catalyst and the binding site is sufficiently strong to enable efficient catalyst recycling. In addition, the support can be readily re-functionalized with different catalyst systems, by washing with methanol to remove the first catalyst system and then attaching the new catalyst system by simply stirring in an apolar solvent such as toluene.

The resulting noncovalently immobilized complexes have been used as ligand systems for both the Pd-catalyzed allylic amination reaction and the Rh-catalyzed hydroformylation. A glycine-urea functionalized PPh_3 ligand was noncovalently attached to the immobilized dendritic support. Application of this system in the Pd-catalyzed allylic amination attains similar yields and product distributions as the homogeneous analogue for the Pd-catalyzed allylic amination of crotyl acetate by piperidine, while exhibiting a reduced rate, as is commonly observed, for heterogenized systems (90% conversion achieved after 30 min compared with 5 min for the homogeneous system). Interestingly, the catalyst could be recycled three times via a simple filtration step. Subsequently, the catalyst was separated from the support and the support was uploaded with hydroformylation catalysts.

A rhodium catalyst based on the glycine-urea functionalized Xantphos ligand was subsequently used in combination with the same support material, as a catalyst for the hydroformylation of 1-octene. In eleven consecutive reactions the catalyst did not show any deterioration or metal leaching. Similar to what was previously observed for covalently anchored systems a decrease in activity and selectivity is observed compared to the homogeneous system. Interestingly, higher activity and selectivity for the linear aldehyde are observed for the noncovalently anchored ligand compared to covalently anchored Nixantphos, while in the homogeneous phase these Nixantphos and Xantphos ligands show similar activity and selectivity.

These noncovalently anchored catalysts in general exhibit a behavior similar to their covalently bound analogues, but can now be separated from the support after the reactions by a simple filtration step. So far, these immobilized systems have not been used in continuous flow reactors.

8.5
Conclusion

We have explored various new supramolecular strategies in transition metal catalysis and demonstrate that new opportunities arise by doing so. We have encapsulated transition metal catalysts by assembly of building blocks such as zinc(II)porphyrins to ligand templates and found that this can lead to higher activity of the catalyst enclosed in this sphere. In addition, the encapsulated rhodium complexes show unusual selectivity, and with encapsulated catalyst internal alkenes can even be functionalized in a selective manner. This is not possible by any other catalyst (nor stoichiometric procedure) known so far. Supramolecular chemistry also provides a tool to make a new class bidentate ligands that has a high potential to become the next generation. The power of bidentate ligands formed by self-assembly is that a relatively small number of building blocks already result in a massive catalyst library. For example, a library of 1000 ligands based on a three-component self-assembled catalyst requires only 30 building blocks.

Supramolecular chemistry also provides new tools for catalyst anchoring. We have shown that catalysts can be noncovalently attached to various soluble and insoluble supports, affording recyclable catalysts. Interestingly, the reversible nature of the noncovalent bond gives rise to new opportunities. In the first instance, we foresee an important role for supramolecular bidentate ligands in combinatorial catalysis – but as a consequence of the entirely new properties many new applications are envisioned. We look forward to new developments and results in this exciting emerging area of supramolecular catalysis.

Acknowledgments

I am grateful to all undergraduate students, graduate students and post-docs (whose names appear in the references) that made valuable contributions to the work described in this chapter. I also would like to thank all academic colleagues with whom I have had valuable discussions on the various topics. Science lives only vividly by the grace of open discussions and it is these that I deeply appreciate.

References

1. (a) Sanders, J.K.M. (1998) Supramolecular catalysis in transition. *Chem.–Eur. J.*, **4**, 1378–1383. (b) Motherwell, W.B., Bingham, M.J. and Six, Y. (2001) Recent progress in the design and synthesis of artificial enzymes. *Tetrahedron*, **57**, 4663–4686.

2. Cacciapaglia, R., Di Stefano, S. and Mandolini, L. (2004) Effective molarities in supramolecular catalysis of two-substrate reactions. *Acc. Chem. Res.*, **37**, 113–122.

3. (a) Kirby, A.J. (1996) Enzyme mechanisms, models, and mimics. *Angew. Chem., Int. Ed. Engl.*, **35**, 707–724. (b) Gao, J., Ma, S., Major, D.T., Nam, K., Pu, J. and Truhlar, D.G. (2006) Mechanisms and free energies of enzymatic reactions. *Chem. Rev.*, **106**, 3188–3209. (c) Dewar, M.J.S. and Storch, D.M. (1985) Alternative view of enzyme reactions. *Proc. Natl. Acad. Sci. U.S.A.*, **82**, 2225–2229.

4. See for reviews: (a) Suslick, K.S., Rakow, N.A., Kosal, M.E. and Chou, J.-H. (2000) The materials chemistry of porphyrins and metalloporphyrins. *J. Porphyrins Phthalocyanines*, **4**, 407–413. (b) Bélanger, S., Keefe, M.H., Welch, J.L. and Hupp, J.T. (1999) Rapid derivatization of mesoporous thin-film materials based on Re(I) zinc-porphyrin "molecular squares": Selective modification of mesopore size and shape by binding of aromatic nitrogen donor ligands. *Coord. Chem. Rev.*, **190–192**, 29–45. (c) Toma, H.E. and Araki, K. (2000) Supramolecular assemblies of ruthenium complexes and porphyrins. *Coord. Chem. Rev.*, **196**, 307–329. (d) Vriezcma, D.M., Aragones, M.C., Elemans, J.A.A.W., Cornelissen, J.J.L.M., Rowan, A.E. and Nolte, R.J.M. (2005) Self-assembled nanoreactors. *Chem. Rev.*, **105**, 1445–1489.

5. Michelsen, U. and Hunter, C.A. (2000) Self-assembled porphyrin polymers. *Angew. Chem.*, **112**, 780–783; Self-assembled porphyrin polymers. *Angew. Chem., Int. Ed.*, **39**, 764–767; Mackay, L.G., Wylie, R.S. and Sanders, J.K.M. (1994) Catalytic acyl transfer by a cyclic porphyrin trimer – efficient turnover without product inhibition. *J. Am. Chem. Soc.*, **116**, 3141–3142; Okumura, A., Funatsu, K., Sasaki, Y. and Imamura, T. (1999) A novel porphyrin octamer with a cyclic tetramer core. *Chem. Lett.*, 779–780.

6. Reek, J.N.H., Schenning, A.P.H.J., Bosman, A.W., Meijer, E.W. and Crossley, M.J. (1998) Templated assembly of a molecular capsule. *Chem. Commun.*, 11–12.

7. Slagt, V.F., Kamer, P.C.J., van Leeuwen, P.W.N.M. and Reek, J.N.H. (2001) Assembly of encapsulated transition metal catalysts. *Angew. Chem.*, **113**, 4401–4404; Assembly of encapsulated transition metal catalysts. *Angew. Chem., Int. Ed.*, **40**, 4271–4274.

8. Slagt, V.F., Kamer, P.C.J., van Leeuwen, P.W.N.M. and Reek, J.N.H. (2004) Encapsulation of transition metal catalysts by ligand-template directed assembly. *J. Am. Chem. Soc.*, **126**, 1526–1536.

9. Kleij, A.W., Kuil, M., Tooke, D.M., Spek, A.L. and Reek, J.N.H. (2005) Template-assisted ligand encapsulation; the impact of an unusual coordination geometry on a supramolecular pyridylphosphine-Zn(II) porphyrin assembly. *Inorg. Chem.*, **44**, 7696–7698.

10. (a) Kleij, A.W., Kuil, M., Tooke, D.M., Lutz, M., Spek, A.L. and Reek, J.N.H. (2005) Zn-II-salphen complexes as versatile building blocks for the construction of supramolecular box assemblies. *Chem.–Eur. J.*, **11**, 4743–4750. (b) Kleij, A.W., Lutz, M., Spek, A.L. and Reek, J.N.H. (2005) Encapsulated transition metal catalysts comprising peripheral Zn(II) salen building blocks: Template-controlled reactivity and selectivity in hydroformylation catalysis. *Chem. Commun.* 3661–3663.

11 Kuil, M., Soltner, T., van Leeuwen, P.W.N.M. and Reek, J.N.H. (2006) High-precision catalysts: Regioselective hydroformylation of internal alkenes by encapsulated rhodium complexes. *J. Am. Chem. Soc.*, **128**, 11344–11345.

12 Flapper, J. and Reek, J.N.H. (2007) Templated Encapsulation of Pyridyl-Bian Palladium Complexes: Tunable Catalysts for Co/4-/tert/-Butylstyrene Copolymerisation. *Angew. Chem. Int. Ed.*, **46**, 5890.

13 Koblenz, T.S., Dekker, H.L., de Koster, C.G., van Leeuwen, P.W.N.M. and Reek, J.N.H. (2006) Bisphosphine based heterocapsules for the encapsulation of transition metals. *Chem. Commun.*, 1700–1702.

14 Tolman, C.A. (1977) Steric effects of phosphorus ligands in organometallic chemistry and homogeneous catalysis. *Chem. Rev.*, **77**, 313–348.

15 Tolman, C.A. (1970) Electron donor–acceptor properties of phosphorus ligands. Substituent additivity. *J. Am. Chem. Soc.*, **92**, 2953–2956.

16 (a) Casey, C.P. and Whiteker, G.T. (1990) *Isr. J. Chem.*, **30**, 299. (b) van Leeuwen, P.W.N.M., Kamer, P.C.J., Reek, J.N.H. and Dierkes, P. (2000) Ligand bite angle effects in metal-catalyzed C–C bond formation. *Chem. Rev.*, **100**, 2741–2769.

17 (a) Burgess, K., Lim, H.-J., Porte, A.M. and Sulikowski, G.A. (1996) New catalysts and conditions for a C–H insertion reaction identified by high throughput catalyst screening. *Angew. Chem., Int. Ed. Engl.*, **35**, 220–222. (b) Gennari, C. and Piarulli, U. (2003) Combinatorial libraries of chiral ligands for enantioselective catalysis. *Chem. Rev.*, **103**, 3071–3100. (c) Crabtree, R.H. (1999) Combinatorial and rapid screening approaches to homogeneous catalyst discovery and optimization. *Chem. Commun.*, 1611–1616. (d) Hagemeyer, A., Jandeleit, B., Liu, Y., Poojary, D.M., Turner, H.W., Volpe, A.F. Jr. and Weinberg, W.H. (2001) Applications of combinatorial methods in catalysis. *Appl. Catal. A. Gen.*, **221**, 23–43. (e) Dahmen, S. and Bräse, S. (2001) Combinatorial methods for the discovery and optimisation of homogeneous catalysts. *Synthesis*, 1431. (f) Murphy, V., Turner, H.W. and Weskamp, T. (2002) *Applied Homogeneous Catalysis with Organometallic Compound* 2nd edn., (eds B. Cornils and W.A. Herrmann), Wiley-VCH, Weinheim, vol 2, pp. 740. (g) de Vries, J.G. and de Vries, A.H.M. (2003) The power of high-throughput experimentation in homogeneous catalysis research for fine chemicals. *Eur. J. Org. Chem.*, **5**, 799–811. (h) Ding, K., Du, H., Yuan, Y. and Long, J. (2004) Combinatorial approach to chiral catalyst engineering and screening: Rational design and serendipity. *Chem.–Eur. J.*, **10**, 2872–2884.

18 (a) Hoveyda, A. (2002) *Handbook of Combinatorial Chemistry* (eds K.C. Nicolaou, R. Hanko and W. Hartwig), Wiley-VCH, Weinheim, vol 2, pp. 991–1016. (b) Taylor, S.J. and Morken, J.P. (1998) Thermographic selection of effective catalysts from an encoded polymer-bound library. *Science*, **280**, 267–270. (c) Chen, P. (2003) Electrospray ionization tandem mass spectrometry in high-throughput screening of homogeneous catalysts. *Angew. Chem., Int. Ed.*, **42**, 2832–2847. (d) Reetz, M.T. (2001) Combinatorial and evolution-based methods in the creation of enantioselective catalysts. *Angew. Chem., Int. Ed.*, **40**, 284–310.

19 Recent review about monodentate ligands, see: (a) Jerphagnon, T., Renaud, J.-L. and Bruneau, C. (2004) Chiral monodentate phosphorus ligands for rhodium-catalyzed asymmetric hydrogenation. *Tetrahedron: Asymmetry*, **15**, 2101–2111, and references cited therein. (b) de Vries, J.G. (2005) *Handbook of Chiral Chemicals*, 2nd edn. (eds D. Ager and S. Laneman), 2nd edn, CRC Press, New York.

20 Reetz, M.T. and Mehler, G. (2000) Highly enantioselective Rh-catalyzed hydrogenation reactions based on chiral monophosphite ligands. *Angew. Chem., Int. Ed.*, **39**, 3889–3890.

21 van den Berg, M., Minnaard, A.J., Schudde, E.P., van Esch, J., de Vries, A.H.M., de Vries, J.G. and Feringa, B.L. (2000) Highly enantioselective rhodium-catalyzed hydrogenation with monodentate ligands. *J. Am. Chem. Soc.*, **122**, 11539–11540.

22 Claver, C., Fernandez, E., Gillon, A., Heslop, K., Hyett, D.J., Martorell, A., Orpen, A.G. and Pringle, P.G. (2000) Biarylphosphonites: A class of monodentate phosphorus(III) ligands that outperform their chelating analogues in asymmetric hydrogenation catalysis. *Chem. Commun.* 961–962.

23 (a) Reetz, M.T. and Li, X. (2005) Mixtures of configurationally stable and fluxional atropisomeric monodentate P ligands in asymmetric Rh-catalyzed olefin hydrogenation. *Angew. Chem., Int. Ed.*, **44**, 2959–2962. (b) Reetz, M.T. and Li, X. (2004) Combinatorial approach to the asymmetric hydrogenation of beta-acylamino acrylates: Use of mixtures of chiral monodentate P-ligands. *Tetrahedron*, **60**, 9709–9714. (c) Reetz, M.T., Mehler, G. and Meiswinkel, A. (2004) Mixtures of chiral monodentate phosphites, phosphonites and phosphines as ligands in Rh-catalyzed hydrogenation of N-acyl enamines: Extension of the combinatorial approach. *Tetrahedron: Asymmetry*, **15**, 2165–2167. (d) Reetz, M.T., Sell, T., Meiswinkel, A. and Mehler, G. (2003) A new principle in combinatorial asymmetric transition-metal catalysis: Mixtures of chiral monodentate P ligands. *Angew. Chem., Int. Ed.*, **42**, 790–793.

24 (a) Monti, C., Gennari, C., Piarulli, U., de Vries, J.G., de Vries, A.H.M. and Lefort, L. (2005) Rh-catalyzed asymmetric hydrogenation of prochiral olefins with a dynamic library of chiral TROPOS phosphorus ligands. *Chem.–Eur. J.*, **11**, 6701–6717. (b) de Vries, A.H.M., Lefort, L., Boogers, J.A.F., de Vries, J.G. and Ager, D.J. (2005) Instant ligand libraries beat the time-to-market constraint. *Chim. Oggi.*, **23**, (Suppl. on Chiral Technol.), 18. (c) Hoen, R., Boogers, J.A.F., Bernsmann, H., Minnaard, A.J., Meetsma, A., Tiemersma-Wegman, T.D., de Vries, A.H.M., de Vries, J.G. and Feringa, B.L. (2005) Achiral ligands dramatically enhance rate and enantioselectivity in the Rh/phosphoramidite-catalyzed hydrogenation of α,β-disubstituted unsaturated acids. *Angew. Chem., Int. Ed.*, **44**, 4209–4212. (d) Peña, D., Minnaard, A.J., Boogers, J.A.F., de Vries, A.H.M., de Vries, J.G. and Feringa, B.L. (2003) Improving conversion and enantioselectivity in hydrogenation by combining different monodentate phosphoramidites; a new combinatorial approach in asymmetric catalysis. *Org. Biomol. Chem.*, **1**, 1087–1089.

25 (a) Jensen, J.F., Sotofte, I., Sorensen, H.O. and Johannsen, M. (2003) Modular approach to novel chiral aryl-ferrocenyl phosphines by Suzuki cross-coupling. *J. Org. Chem.*, **68**, 1258–1265. (b) Blume, F., Zemolka, S., Fey, T., Kranich, R. and Schmalz, H.-G. (2002) Identification of suitable ligands for a transition metal-catalyzed reaction: Screening of a modular ligand library in the enantioselective hydroboration of styrene. *Adv. Synth Catal.*, **344**, 868–883. (c) Bastero, A., Claver, C., Ruiz, A., Castillón, S., Daura, E., Bo, C. and Zangrando, E. (2004) Insights into CO/styrene copolymerization by using Pd-II catalysts containing modular pyridine-imidazoline ligands. *Chem.–Eur. J.*, **10**, 3747–3760. (d) Kranich, R., Eis, K., Geis, O., Muhle, S., Bats, J.W. and Schmalz, H.-G. (2000) A modular approach to structurally diverse bidentate chelate ligands for transition metal catalysis. *Chem.–Eur. J.*, **6**, 2874–2894. (e) Imbos, R., Minnaard, A.J. and Feringa, B.L. (2002) A highly enantioselective intramolecular Heck reaction with a monodentate ligand. *J. Am. Chem. Soc.*, **124**, 184–185.

26 For recent review about bidentate ligands in asymmetric hydrogenation, see: Tang,

W. and Zhang, X. (2003) New chiral phosphorus ligands for enantioselective hydrogenation. *Chem. Rev.*, **103**, 3029–3069, and references cited therein.

27 Wilkinson, M.J., van Leeuwen, P.W.N.M. and Reek, J.N.H. (2005) New directions in supramolecular transition metal catalysis. *Org. Biomol. Chem.*, **3**, 2371–2383.

28 See also Chapters 2 and 9. (a) Breit, B. and Seiche, W. (2003) Hydrogen bonding as a construction element for bidentate donor ligands in homogeneous catalysis: Regioselective hydroformylation of terminal alkenes. *J. Am. Chem. Soc.*, **125**, 6608–6609. (b) Breit, B. and Seiche, W. (2005) Self-assembly of bidentate ligands for combinatorial homogeneous catalysis based on an A-T base-pair model. *Angew. Chem., Int. Ed.*, **44**, 1640–1643. (c) Breit, B. and Seiche, W. (2003) Hydrogen bonding as a construction element for bidentate donor ligands in homogeneous catalysis: Regioselective hydroformylation of terminal alkenes. *J. Am. Chem. Soc.*, **125**, 6608–6609. (d) Seiche, W., Schuschkowski, A. and Breit, B. (2005) Bidentate ligands by self-assembly through hydrogen bonding: A general room temperature/ambient pressure regioselective hydroformylation of terminal alkenes. *Adv. Synth. Catal.*, **347**, 1488–1494. (e) Chevallier, F. and Breit, B. (2006) Self-assembled bidentate ligands for Ru-catalyzed anti-Markovnikov hydration of terminal alkynes. *Angew. Chem., Int. Ed.*, **45**, 1599–1602. (f) Weis, M., Waloch, C., Seiche, W. and Breit, B. (2006) Self-assembly of bidentate ligands for combinatorial homogeneous catalysis: Asymmetric rhodium-catalyzed hydrogenation. *J. Am. Chem. Soc.*, **128**, 4188–4189. (g) Breit, B. and Seiche, W. (2006) Self-assembly of bidentate ligands for combinatorial homogeneous catalysis based on an A-T base pair model. *Pure Appl. Chem.*, **78**, 249. (h) Takacs, J.M., Reddy, D.S., Moteki, S.A., Wu, D. and Palencia, H. (2004) Asymmetric catalysis using self-assembled chiral bidentate P,P-ligands. *J. Am. Chem. Soc.*, **126**, 4494. (i) Takacs, J.M., Chaiseeda, K. and Moteki, S.A. (2006) Rhodium-catalyzed asymmetric hydrogenation using self-assembled chiral bidentate ligands. *Pure Appl. Chem.*, **78**, 501. (j) Duckmanton, P.A., Blake, A.J. and Love, J.B. (2005) Palladium and rhodium ureaphosphine complexes: Exploring structural and catalytic consequences of anion binding. *Inorg. Chem.*, **44**, 7708–7710.

29 Slagt, V.F., van Leeuwen, P.W.N.M. and Reek, J.N.H. (2003) Bidentate ligands formed by self-assembly. *Chem. Commun.*, 2474–2475.

30 Slagt, V.F., van Leeuwen, P.W.N.M. and Reek, J.N.H. (2007) Supramolecular bidentate phosphorus ligands based on bis-zinc(II) and bis-tin(IV) porphyrin building blocks. *Dalton Trans.*, 2302–2310.

31 *Rhodium-Catalysed Hydroformylation*, (2000) (eds. P.W.N.M. van Leeuwen and C. Claver) Kluwer Academic Publishers, Dordrecht.

32 (a) Agbossou, F., Carpentier, J.-F. and Morteux, A. (1995) Asymmetric hydroformylation. *Chem. Rev.*, **95**, 2485–2506. (b) Gladiali, S., Bayon, J.C. and Claver, C. (1995) Recent advances in enantioselective hydroformylation. *Tetrahedron: Asymmetry*, **6**, 1453–1474.

33 Gladiali, S., Dore, A., Fabbri, D., De Lucchi, O. and Manassero, M. (1995) *Tetrahedron: Asymmetry*, **6**, 1453.

34 (a) Kleij, A.W., Kuil, M., Tooke, D.M., Lutz, M., Spek, A.L. and Reek, J.N.H. (2005) Zn-II-salphen complexes as versatile building blocks for the construction of supramolecular box assemblies. *Chem.–Eur. J.*, **11**, 4743–4750. (b) Kleij, A.W., Lutz, M., Spek, A.L., van Leeuwen, P.W.N.M. and Reek, J.N.H. (2005) Encapsulated transition metal catalysts comprising peripheral Zn(II)salen building blocks: Template-controlled reactivity and selectivity in hydroformylation catalysis. *Chem. Commun.*, 3661–3663.

35 (a) Kuil, M., Goudriaan, P.E., van Leeuwen, P.W.N.M. and Reek, J.N.H. (2006)

Template-induced formation of heterobidentate ligands and their application in the asymmetric hydroformylation of styrene. *Chem. Commun.*, 4679–4681. (b) Kuil, M., Goudriaan, P.E., Kleij, A.W., Tooke, D.M., Spek, A.L., van Leeuwen, P.W.N.M. and Reek, J.N.H. (2007) Rigid bis-zinc(II) salphen building blocks for the formation of template-assisted bidentate ligands and their application in catalysis. *Dalton Trans.*, 2311–2320.

36 Slagt, V.F., van Leeuwen, P.W.N.M. and Reek, J.N.H. (2003) Multicomponent porphyrin assemblies as functional bidentate phosphite ligands for regioselective rhodium-catalyzed hydroformylation. *Angew. Chem., Int. Ed.*, 42, 5619–5623.

37 Usually the l:b ratio decreases at lower temperatures: (a) Lazzaroni, R., Uccello-Barretta, G., Scamuzzi, S., Settambolo, R. and Caiazzo, A. (1996) H-2 NMR investigation of the rhodium-catalyzed deuterioformylation of 1,1-diphenylethene: Evidence for the formation of a tertiary alkyl-metal intermediate. *Organometallics*, 15, 4657–4659. (b) Botteghi, C., Cazzolato, L., Marchetti, M. and Paganelli, S. (1995) New synthetic route to pharmacologically active 1-(N,N-dialkylamino)-3,3-diarylpropanes via rhodium-catalyzed hydroformylation of 1,1-diaylethenes. *J. Org. Chem.*, 60, 6612–6615.

38 (a) Slagt, V.F., Röder, M., Kamer, P.C.J., van Leeuwen, P.W.N.M. and Reek, J.N.H. (2004) Supraphos: A supramolecular strategy to prepare bidentate ligands. *J. Am. Chem. Soc.*, 126, 4056–4057. (b) Reek, J.N.H., Röder, M., Goudriaan, P.E., Kamer, P.C.J., van Leeuwen, P.W.N.M. and Slagt, V.F. (2005) Supraphos: A supra-molecular strategy to prepare bidentate ligands. *J. Organometal. Chem.*, 605, 4505.

39 (a) Tsuji, J. (1986) New general synthetic methods involving π-allylpalladium complexes as intermediates and neutral reaction conditions. *Tetrahedron*, 42, 4361–4401. (b) Trost, B.M. and Van Vranken, D.L. (1996) Asymmetric transition metal-catalyzed allylic alkylations. *Chem. Rev.*, 96, 395–422. (c) Helmchen, G. and Pfaltz, A. (2000) Phosphinooxazolines – A new class of versatile, modular P,N-ligands for asymmetric catalysis. *Acc. Chem. Res.*, 33, 336–345. (d) Pfaltz, A. (1993) Chiral semicorrins and related nitrogen-heterocycles as ligands in asymmetric catalysis. *Acc. Chem. Res.*, 26, 339–345. (e) Trost, B.M. and Crawley, M.L. (2003) Asymmetric transition-metal-catalyzed allylic alkylations: Applications in total synthesis. *Chem. Rev.*, 103, 2921–2943.

40 Deerenberg, S., Schrekker, H.S., van Strijdonck, G.P.F., Kamer, P.C.J., van Leeuwen, P.W.N.M., Fraanje, J. and Goubitz, K. (2000) New chiral phosphine-phosphite ligands in the enantioselective palladium-catalyzed allylic alkylation. *J. Org. Chem.*, 65, 4810–4817.

41 (a) Jiang, X.-B., van Leeuwen, P.W.N.M. and Reek, J.N.H. (2007) SUPRAphos-based palladium catalysts for the kinetic resolution of racemic cyclohexenyl acetate. *Chem. Commun.*, 2287–2289.

42 Jiang, X.-B., Lefort, L., Goudriaan, P.E., de Vries, A.H.M., van Leeuwen, P.W.N.M., de Vries, J.G. and Reek, J.N.H. (2006) Screening of a supramolecular catalyst library in the search for selective catalysts for the asymmetric hydrogenation of a difficult enamide substrate. *Angew. Chem., Int. Ed.*, 45, 1223–1227.

43 For the examples of Rh-catalyzed hydrogenation of N-(3,4-dihydro-2-naphthalenyl)-acetamide, see: (a) Zhang, Z., Zhu, G., Jiang, Q., Xiao, D. and Zhang, X. (1999) *J. Org. Chem.*, 64, 1774. (b) Bernsmann, H., van den Berg, M., Hoen, R., Minnaard, A.J., Mehler, G., Reetz, M.T., de Vries, J.G. and Feringa, B.L. (2005) PipPhos and MorfPhos: Privileged monodentate phosphoramidite ligands for rhodium-catalyzed asymmetric hydrogenation. *J. Org. Chem.*, 70, 943–945. (c) Hoen, R., van den Berg, M., Bernsmann, H., Minnaard, A.J., de Vries, J.G. and

Feringa, B.L. (2004) Catechol-based phosphoramidites: A new class of chiral ligands for rhodium-catalyzed asymmetric hydrogenations. *Org. Lett.*, **6**, 1433–1436. (d) Guillen, F., Rivard, M., Toffano, M., Legros, J.Y., Daran, J.C. and Fiaud, J.-C. (2002) Synthesis and first applications of a new family of chiral monophosphine ligand: 2,5-diphenylphosphospholanes. *Tetrahedron*, **58**, 5895–5904. (e) Tang, W., Chi, Y. and Zhang, X. (2002) An ortho-substituted BIPHEP ligand and its applications in Rh-catalyzed hydrogenation of cyclic enamides. *Org. Lett.*, **4**, 1695–1698.

44 For examples of Ru-catalyzed hydrogenation of N-(3,4-dihydro-2-naphthalenyl)-acetamide see: (a) Renaud, J.L., Dupau, P., Hay, A.-E., Guingouain, M., Dixneuf, P.H. and Bruneau, C. (2003) Ruthenium-catalysed enantioselective hydrogenation of trisubstituted enamides derived from 2-tetralone and 3-chromanone: Influence of substitution on the amide arm and the aromatic ring. *Adv. Synth. Catal.*, **345**, 230–238. (b) Dupau, P., Le Gendre, P., Bruneau, C. and Dixneuf, P.H. (1999) Optically active amine derivatives: Ruthenium-catalyzed enantioselective hydrogenation of enamides. *Synlett*, **11**, 1832–1834. (c) Dupau, P., Bruneau, C. and Dixneuf, P.H. (2001) Enantioselective hydrogenation of the tetrasubstituted C=C bond of enamides catalyzed by a ruthenium catalyst generated *in situ*. *Adv. Synth. Catal.*, **343**, 331–334. (d) Devocelle, M., Mortreux, A., Agbossou, F. and Dormoy, J.-R. (1999) Alternative synthesis of the chiral atypical beta-adrenergic phenylethanolaminotetraline agonist SR58611A using enantioselective hydrogenation. *Tetrahedron Lett.*, **40**, 4551–4554. (e) Tschaen, D.M., Abramson, L., Cai, D., Desmond, R., Dolling, U.-H., Frey, L., Karady, S., Shi, Y.-J. and Verhoeven, T.R. (1995) Asymmetric-synthesis of MK-0499. *J. Org. Chem.*, **60**, 4324–4330.

45 In a full paper on this topic we will report the hydrogenation of a variety of substrates.

46 Knight, L.K., Freixa, Z., van Leeuwen, P.W.N.M. and Reek, J.N.H. (2006) Supramolecular trans-coordinating phosphine ligands. *Organometallics*, **24**, 954–960.

47 Lu, X.-X., Tang, H.-S., Ko, C.-C., Wong, J.K.-Y., Zhu, N. and Yam, V.W.-W. (2005) Anion-assisted trans-cis isomerization of palladium(II) phosphine complexes containing acetanilide functionalities through hydrogen bonding interactions. *Chem. Commun.*, 1572–1574.

48 Sandee, A.J., v.d. Burg, Lidy, and Reek, J.N.H. (2007) UREAphos: supramolecular bidentate ligands for asymmetric hydrogenation. *Chem. Commun.*, 864–866.

49 Baars, M.W.P.L., Karlsson, A.J., Sorokin, V. and Meijer, B.F.M. (2000) Supramolecular modification of the periphery of dendrimers resulting in rigidity and functionality. *Angew. Chem., Int. Ed.*, **39**, 4262–6265.

50 de Groot, D., de Waal, B.F.M., Reek, J.N.H., Schenning, A.P.H.J., Kamer, P.C.J., Meijer, E.W. and van Leeuwen, P.W.N.M. (2001) Noncovalently functionalized dendrimers as recyclable catalysts. *J. Am. Chem. Soc.*, **123**, 8453–8458.

51 Ribaudo, F. (2007) Thesis University of Amsterdam, The Netherlands, Manuscript in preparation.

52 Chen, R., Bronger, R.P.J., Kamer, P.C.J., van Leeuwen, P.W.N.M. and Reek, J.N.H. (2004) Noncovalent anchoring of homogeneous catalysts to silica supports with well-defined binding sites. *J. Am. Chem. Soc.*, **126**, 14557–14566.

9
Chirality-Directed Self-Assembly: An Enabling Strategy for Ligand Scaffold Optimization

James M. Takacs, Shin A. Moteki, and D. Sahadeva Reddy

9.1
Introduction

The design of new chiral ligands drives the development of asymmetric catalysts. Ultimately, the goal of all chiral ligand designs is to (a) create a "chiral catalytic pocket" (topography) around the metal to direct stereochemistry and (b) impart the appropriate electronic characteristics at the metal center for efficient catalysis. Most modular approaches start with one or a small set of "interesting scaffolds" and sequentially append the ligating groups. The challenge is to find the "right ligating group", the one that is properly oriented by the selected scaffold and thus makes the catalyst system efficient; the approach is largely hit or miss. Ligand scaffold optimization provides a novel alternative design strategy for improving catalyst efficiency. A modular ligand design is used, and the overall hypothesis is simple: use the nature of the ligating group to define a chiral surface and the electronic nature of the metal–ligand interaction, and then vary the ligand scaffold combinatorially to define its optimal orientation for asymmetric catalysis. Self-assembly is proving to be an efficient enabling strategy for ligand scaffold optimization in asymmetric catalysis.

9.2
The Need for New Catalyst Systems

Developing efficient methods for the reliable, stereocontrolled synthesis of small molecules via asymmetric catalysis is of both practical and fundamental importance in modern organic chemistry [1]. While tremendous advances have been realized, continued progress relies upon developing innovative, new approaches to the unsolved problems that remain. Among the most important of these problems is the need for more selective asymmetric catalysts, including ones that afford high enantioselectivity, better substrate generality, and/or more precisely substrate tunable

Supramolecular Catalysis. Edited by Piet W. N. M. van Leeuwen
Copyright © 2008 WILEY-VCH Verlag GmbH & Co. KGaA, Weinheim
ISBN: 978-3-527-32191-9

catalyst systems. We and others have been attracted to the use combinatorial approaches for ligand scaffold construction in the design and optimization of new ligands and catalyst systems; our interest has largely been directed toward asymmetric catalysis.

Combinatorial methods are often contrasted to those deemed "rational design" and draw the ire of some. The rational approach to reaction optimization typically examines the influence of one variable at a time. Of course, this approach is only truly valid when the variable being probed is an independent variable. When the reaction variables are highly interdependent, there is no reason to think that the effect of subtly changing the structure of a ligand, for example, only influences the presumed intermediate or transition structure of interest. Metal-catalyzed reactions inherently rely upon numerous intermediates in complex equilibria, only some of which contribute to the catalyzed reaction pathway of interest. In short, efficient combinatorial approaches to catalyst optimization allow one to generate data sets of much greater breadth – perhaps sufficient breadth to uncover trends that may be missed in a one variable at a time optimization algorithm.

9.3
A Typical Modular Approach to Chiral Bidentate Ligand Design

The need for advances in asymmetric catalysis fuels the search for new ligand and catalyst designs, especially those using combinatorial and modular approaches [2]. All chiral ligand designs ultimately aim to (a) create a "chiral pocket" (topography) around the metal to direct stereochemistry and (b) impart the appropriate electronic characteristics at the metal center for efficient catalysis. Most modular approaches start with one or a small set of "interesting scaffolds", usually relatively compact, rigid structures, to which the ligating groups are sequentially appended (Figure 9.1). The idea is to systematically vary the nature of the appended ligating groups (i.e., vary their elemental identity, shape, steric demand, electronic character) to simultaneously optimize or tune the asymmetric environment at the site of catalysis and define an appropriate metal–ligand interaction for catalysis. Anyone who has prepared a series of even 20–25 chiral ligands using this approach recognizes the synthetic challenge is usually quite significant. Furthermore, the challenge to find the "right ligating group", one that optimally "fits" (i.e., is properly oriented by) the selected scaffold and thus makes the catalyst system efficient, is also quite significant. For a given

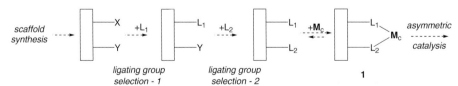

Figure 9.1 A typical strategy for ligand design: scaffold construction followed by attaching the ligating groups.

scaffold, even small changes in the ligating group(s) often have a big effect on the efficiency and enantioselectivity of the reaction. Similarly, even small changes in a rigid scaffold usually have a profound effect. There does not seem to be a simple way to fine tune the ligand using this approach. Despite advances in computational approaches to ligand/catalyst design, the typical modular approach to chiral bidentate ligand design is still largely a hit or miss proposition.

9.4 A Further Rationale for Developing Combinatorial Approaches to Scaffold Optimization

BINAP is regarded as a "privileged structure" for asymmetric catalysis, i.e., it has proven successful in a range of asymmetric reactions [3]. Nonetheless, there are problems that BINAP has not solved and efforts to improve upon its privileged shape continue. The biaryl scaffold is central to BINAP's success; Figure 9.2 illustrates a series of structures designed to further optimize essentially that single rotational degree of freedom, the biaryl axis. The examples shown represent but a fraction of those that have been individually synthesized, resolved and evaluated – a very substantial synthetic effort spread over many research groups and years. Each derivative shown in Figure 9.2 has the identical diphenylphosphino ligating groups and the rather subtle variations in the biaryl scaffold are largely intended to optimize the orientation of the diphenylphosphino moiety in the chiral catalyst. Each derivative exhibits some advantage over BINAP in a particular reaction, but the details are unimportant. The structures are shown here because they illustrate the substantial

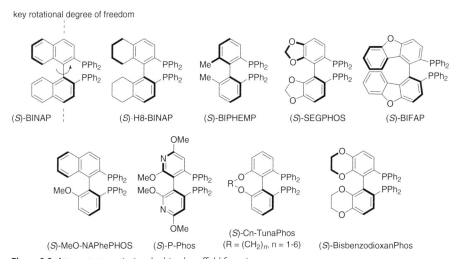

Figure 9.2 Attempts to optimize the biaryl scaffold focusing on a single rotational degree of freedom.

effort required to prepare even a small set of biaryl-scaffold analogs and highlight the relevance of combinatorial approaches that focus on exploring and optimizing scaffold structure as an integral element of ligand design.

9.5
Approaches to Scaffold Optimization

Gilbertson's polypeptide derived ligands constitute one of the earliest approaches explored for ligand scaffold optimization (Figure 9.3) [4]. Varying the amino acid sequence of the ligand scaffold is integral to his design. For example, a library of resin-bound polypeptides **4** bearing two unnatural phosphinoserine derivatives was prepared and used for the rhodium-catalyzed asymmetric hydrogenation of enamide **2**. The level of enantioselectivity varies as a function of the ligand scaffold; 38% ee was obtained in the best case. Unfortunately, space does not permit detailed discussion of the more recent work; however, Reek (Chapter 7) [5] and Lin (Chapter 4) [6] have made particularly insightful use of metal-directed self-assembly for the preparation of chiral self-assembled ligands (SALs). Breit's elegant use of complementary hydrogen-bonding patterns exploits a conceptually related approach to metal-directed self-assembly (Chapter 2) [7]. New systems now appear with increasing frequency [8–11]; several short reviews give excellent overviews of the current state-of-the-art [12].

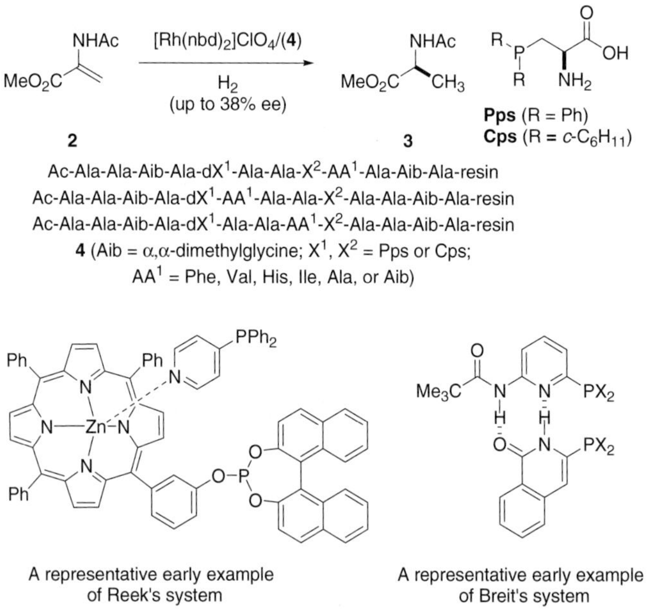

Ac-Ala-Ala-Aib-Ala-dX^1-Ala-Ala-X^2-AA^1-Ala-Aib-Ala-resin
Ac-Ala-Ala-Aib-Ala-dX^1-AA^1-Ala-Ala-X^2-Ala-Ala-Aib-Ala-resin
Ac-Ala-Ala-Aib-Ala-dX^1-Ala-Ala-AA^1-X^2-Ala-Ala-Aib-Ala-resin
4 (Aib = α,α-dimethylglycine; X^1, X^2 = Pps or Cps; AA^1 = Phe, Val, His, Ile, Ala, or Aib)

A representative early example of Reek's system

A representative early example of Breit's system

Figure 9.3 Representative approaches and structural motifs used for scaffold optimization.

9.6 A Convergent Approach to the Formation of Heterobimetallic Catalyst Systems

We have pursued an approach that is conceptually similar to that pursued by Reek, van Leeuwen, and coworkers (cited above) using metal complexation to generate novel metal-containing ligand structures. Of course, the use of substitution inert metal complexes, e.g., ferrocenyl and certain metal-arene complexes, as structural subunits within chelating ligands is well established. However, the use of more substitution-labile complexes, which offer greater potential for *in situ* and reversible formation of the complex, is less common. A convergent approach to ligand scaffold optimization using metal-directed multicomponent self-assembly is illustrated schematically in Figure 9.4.

Multicomponent reactions have garnered much attention as synthetic strategies due to their potential to increase diversity while simplifying the synthesis of structurally complex compounds. Multicomponent self-assembly can offer the same potential benefits in ligand synthesis. There are three basic steps. The first two steps parallel that of the typical modular approach to ligand design, except that they apply to separate halves (themselves "bifunctional ligands") of the ultimate bidentate ligand; (a) selecting the series of tether subunits begins to define the scaffold and (b) selecting the series of ligating groups begins to define both the nature of the metal–ligand interaction and the topography around the catalytic metal. The third step, the convergent self-assembly, is a reaction that turns out to be marvelously trivial (*vide infra*). It connects the bifunctional ligands to fully establish both the scaffold and the precise combination of ligating groups in the self-assembled ligand (SAL) **5**. Self-assembly leaves a second set of ligating groups now suitably disposed to bind a second metal. Addition of that catalytic metal (M_c) completes the bimetallic catalyst **6**. Other modular ligand designs attach ligating groups as substituents to a preformed scaffold. One key is that the ligand scaffold is established in the last step of the synthesis and, as such, allows us to rapidly assemble a diverse set of scaffolds and more thoroughly explore optimizing its structure. Its formation is based upon selectively forming a *heteroleptic* metal complex that can be carried out in a combinatorial fashion.

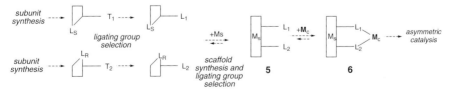

Figure 9.4 A modular approach to chiral ligand design using metal-directed heteroleptic self-assembly of chiral bidentate SALs **5** and bimetallic catalysts **6** incorporating both structural (M_s) and catalytic metals (M_c).

9.7
Chirality-Directed Self-Assembly: Selective Formation of Neutral, Heteroleptic Zinc(II) Complexes

Much of our work on the basic chirality-directed self-assembly reaction that we exploit has been published [13]; consequently, its discussion here is brief. Our approach came about via a rather circuitous route, prompted by observations made during a study into chiral Lewis acid catalyzed Diels–Alder cycloadditions [14] and the pioneering work of Kagan on nonlinear effects in asymmetric catalysis [15]. Selective heterochiral complex formation is often thought to be responsible for a positive nonlinear effect in asymmetric catalysis. It occurred to us that it also defines a simple strategy whereby ligand chirality can be used to predictably direct multicomponent self-assembly. Two types of molecular recognition are employed to direct multicomponent self-assembly [16], self-recognition [17] or self-discrimination [18]. These are typically reduced to practice by constructing ligands possessing complementary steric or electronic motifs to bias their metal-directed self-assembly [19]. We envisioned using the complementary chirality of box ligands to direct the selective formation of neutral, heteroleptic zinc(II) complexes.

The neutral, homochiral complex (*SS,SS*-**9**) is formed by stirring two equivalents of (4*S*,4′*S*)-**7** with Zn(OAc)$_2$ (Figure 9.5). Zn(OAc)$_2$ serves a dual role in the reaction, simultaneously delivering the metal center and the required base. When a racemic mixture of box ligands [i.e., one equivalent each of (4*S*,4′*S*)-**7** and (4*R*,4′*R*)-**7**] is combined with Zn(OAc)$_2$, three complexes could form, the homochiral complexes (*SS,SS*)- and (*RR,RR*)-**9** (i.e., chiral self-recognition) and the heterochiral complex (*SS,RR*)-**8** (i.e., chiral self-discrimination). The tetrahedral coordination geometry strongly favors self-discrimination; *only* the heterochiral complex (*SS,RR*)-**8** is

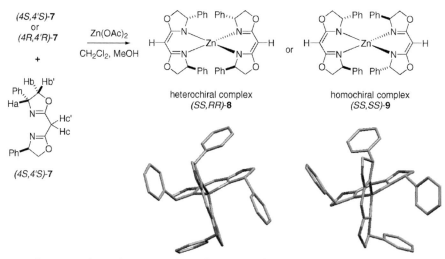

Figure 9.5 Selective formation and crystal structures of heterochiral and homochiral (box)$_2$Zn complexes.

9.7 Chirality-Directed Self-Assembly: Selective Formation of Neutral, Heteroleptic Zinc(II) Complexes

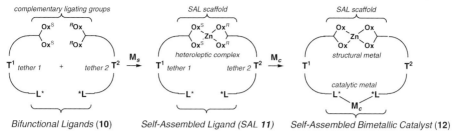

Figure 9.6 Chirality-directed self-assembly of chiral bidentate SALs and bimetallic catalyst systems.

observed in solution or upon crystallization. To better understand the factors favoring self-discrimination over self-recognition, the crystal structures of (SS,RR)-**8** and (SS, SS)-**9** were determined. The heterochiral complex (SS,RR)-**8** exhibits near perfect tetrahedral coordination while the homochiral complex (SS,SS)-**9**, for which there are two closely related conformers in the unit cell, distorts to minimize interactions between the phenyl substituents of the two box ligands. Otherwise, the bond lengths and angles are similar in the two complexes. A mixture of homochiral complexes (SS,SS)- and (RR,RR)-**9** rapidly equilibrates to the heterochiral complex (SS,RR)-**8** provided a proton source is present to initiate the exchange.

Our modular self-assembled ligand (SAL) design concept is illustrated in more detail below (Figure 9.6). We use box-tethered chiral monodentate ligands **10** as the bifunctional SAL subunits. Two independent bifunctional subunits bearing box subunits of opposite chirality self-assemble upon treatment with Zn(OAc)$_2$ to give a neutral complex, a chiral bidentate SAL **11**. Each independent half in this convergent route is in turn to be prepared from three modules: (a) an (R)- or (S)-box subunit, (b) a tether subunit, and (c) a ligating group; the latter is introduced in the penultimate step of the synthesis. Combinatorial self-assembly establishes both the final structure of the ligand scaffold and the exact combination of ligating groups in the SAL library. Metalloenzymes often use two types of metals to carry out their function. Structural metals help generate the three-dimensional structure, and a second metal binding site is often part of that structure. An independent, catalytic metal binds to this latter site to carry out the required chemistry. Our system attempts to mimic that strategy. We form a neutral, heteroleptic (box)$_2$Zn complex, wherein that zinc serves as a structural metal. At the same time, the process of self-assembly constructs a chiral binding site for a second, independent metal to bind and effect catalysis.

The neutral zinc complexes described in Figure 9.5 are proving to be quite remarkable compounds. Their preparation is simple and the synthetic yields are high (typically 70–90%). When one looks at the crude product, few if any side products are seen. Mechanical losses probably account for the less than perfect isolated yields obtained. The neutral metal complexes are typically freely soluble in various organic solvents (e.g., toluene, CH$_2$Cl$_2$, CHCl$_3$, THF, dioxane, and acetonitrile) and partially soluble in methanol. They are generally insoluble in hexanes, diethyl ether or cold methanol, and these latter organic solvents can be used to precipitate the

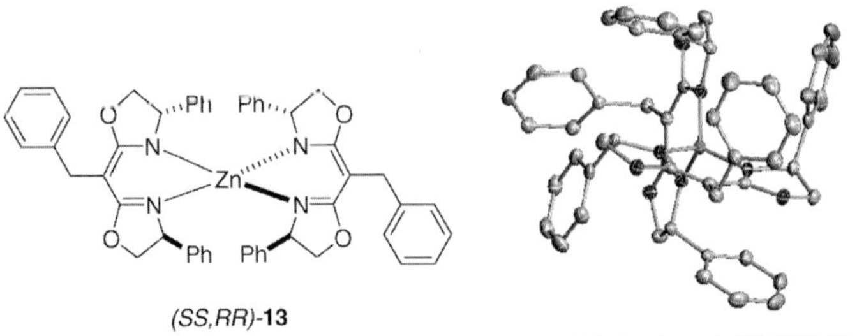

Figure 9.7 Crystal structure of (SS,RR)-**13** illustrates the size constraints for potential chelation.

complex from other solvents and/or wash the precipitated complex. In some cases the complex has been subjected to chromatography on silica (1% MeOH in CHCl$_3$ eluent) with little or no loss of zinc. The complexes are insoluble, but stable toward water, and can be subjected to extractive workup. Drying the organic phase with anhydrous Na$_2$SO$_4$ does not effect metal exchange.

Once formed, the heterochiral (box)$_2$Zn complex is stable towards exchange with several common ligands, e.g., 2,2′-isopropylidenebis(4-phenyl-2-oxazline), Ph$_3$P, BINAP, BINOL and diethyl tartrate. Furthermore, several potential ligands, i.e., 2,2′-isopropylidenebis(4-phenyl-2-oxazline), Ph$_3$P, Cy$_3$P, dppe, were tested in direct competition and do not to interfere with formation of the heterochiral complex. To get a better idea of the tether structure that would be needed to realize the approach in Figure 9.6, we prepared and crystallized the heterochiral (benzylBox)$_2$Zn complex (SS,RR)-**13** (Figure 9.7). Analyzing its crystal structure and manipulating that structure in Chem3D makes it is apparent that the appended benzyl groups are of sufficient length to extend beyond the bulk of the (box)$_2$Zn core, i.e., the benzyl groups can freely rotate and extend beyond the metal complex core. In fact, the core is quite compact; its end-to-end length is a little shorter than that of biphenyl. We reasoned that substituted benzyl or biphenyl tethers should extend a tethered ligating group beyond the core and permit chelation as depicted schematically in Figure 9.6. Note that the complex is inherently a rather rigid structural element, contrary to what one might suppose. Furthermore, since the two independent rotatable bonds connecting the two benzyl substituents (i.e., the intended tethers) lie along the same axis, there is only one rotational degree of freedom relating the relative positions of the ligating groups.

In light of the crystal structure of (SS,RR)-**13** and with an eye toward keeping the tether subunits relatively rigid, we prepared a series of substituted benzyl and biphenyl tethers for preparing SAL subunits (Figure 9.8). The synthetic route is illustrated for the preparation of **15F**. The key step is mono-alkylation [20] of the box subunit with a substituted benzyl bromide bearing a pendant silyl-protected hydroxyl substituent. In the initial studies, we prepared the derivative bearing a pendant TADDOL-derived monophosphite [21]. Deprotection of the silyl ether followed by coupling with [(R,R)-TADDOL]PCl [22] affords the desired bifunctional box-(TADDOL) phosphite conjugate **15F**. The six-step synthesis is quite efficient, and we typically prepare 5–10 mmol of the penultimate intermediate, alcohol **14**.

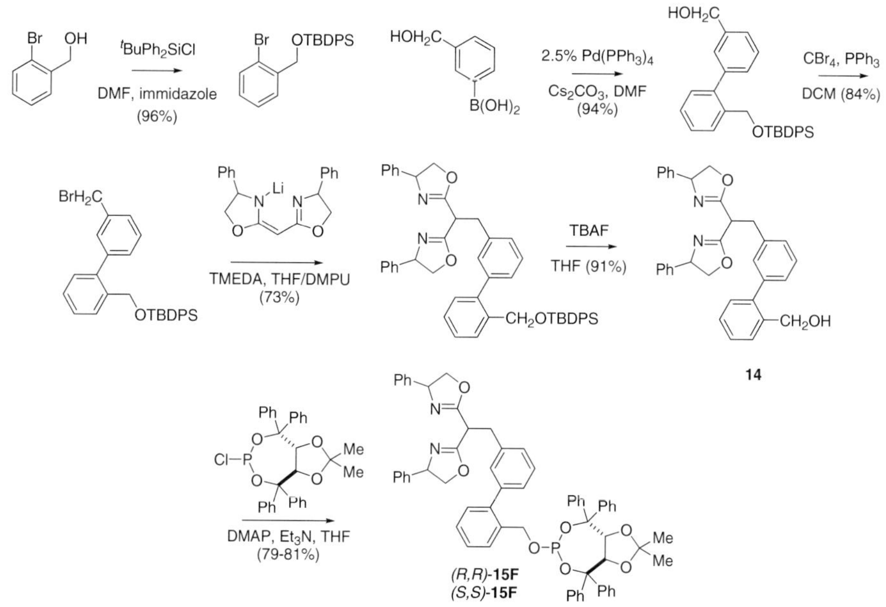

9.7 Chirality-Directed Self-Assembly: Selective Formation of Neutral, Heteroleptic Zinc(II) Complexes

Figure 9.8 Synthesis of a bifunctional box-(TADDOL)phosphite conjugate, **15F**.

Treating a mixture of (S,S)-**15F** and (R,R)-**15F** with Zn(OAc)$_2$ affords the pseudo-C_2-symmetric complex **SAL 16FF** (Figure 9.9). The reaction is trivial: mix the two diastereomeric bifunctional ligands (**15F**) with Zn(OAc)$_2$, remove the solvent (CH$_2$Cl$_2$/MeOH), triturate with methanol, and wash the resulting solid with methanol.

Having formed **16FF**, the crucial question is whether the pendant phosphites can coordinate a second metal. The ^{31}P NMR spectrum of **16FF** shows a single resonance at 130.2 ppm, similar to that of its precursor (S,S)-**15F** (130.7 ppm). Upon the addition of an equivalent of [(cod)$_2$Rh]BF$_4$, the ^{31}P NMR signal at 130.2 ppm is lost and a new doublet appears at 112.7 ppm ($J_{Rh,P}$ = 248.6 Hz). The results are consistent with the formation of the chiral heterobimetallic rhodium complex [(cod)Rh(**SAL 16FF**)]BF$_4$ (**17**) [23]. Demonstrating that **SAL 16FF** forms a rhodium(I) complex is perhaps surprising. These form large macrocyclic chelates that at first viewing may seem

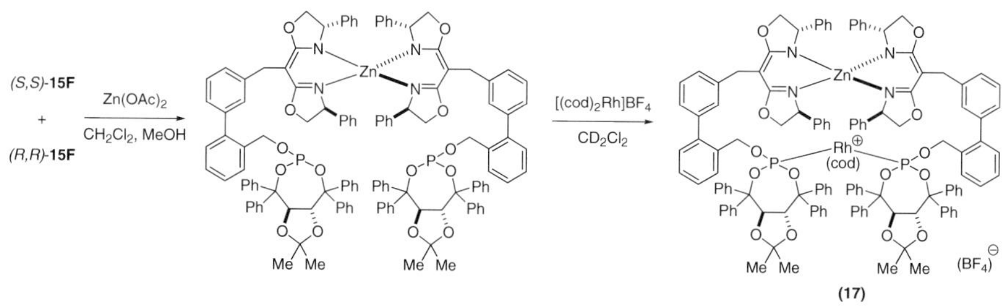

Figure 9.9 Preparation of the chiral diphosphite **SAL 16FF** and its rhodium complex **17**.

uncomfortably odd. That may simply be the price of innovation. Nonetheless, having prepared some, characterized their metal complexation as described above, and used them in catalysis, we find they handle and behave like other more typical ligand and catalyst systems. The successful use of macrocyclic chelates in catalysis now appears with increasing frequency in the literature.

9.8
In situ SAL Preparation

In our early studies, the SALs were individually prepared and isolated. In more recent studies two *in situ* procedures have been utilized. SALs are prepared *in situ* by adding Et$_2$Zn and a mixture of the (*R*)-box and (*S*)-box subunits of interest; the by-product is simply ethane. Alternatively, as illustrated in Figure 9.10 for the preparation of **19AA**, the heteroleptic complex is easily prepared via the *in situ* equilibration of a mixture of homochiral complexes. We have followed this exchange reaction via NMR; the equilibration is rapid (<5 min) in the presence of a proton source. These *in situ* procedures greatly improve throughput since the chiral ligands can be reliably generated in very small quantity (typically, a few micromoles).

9.9
Ligand Scaffold Optimization in Palladium-Catalyzed Asymmetric Allylic Amination

We recently published the results of a study in which we prepared a series of bifunctional ligands **20A–P** and used them in an asymmetric allylic amination (Figure 9.11, TADDOL = TDL) [24]. For the purposes of this initial study the ligating group was kept constant; the choice of the TADDOL-derived monophosphite ligating group was more or less arbitrary. Treating complementary combinations of bifunctional ligands **20A–P** with Zn(OAc)$_2$ is a simple way to prepare a library of **SALs**

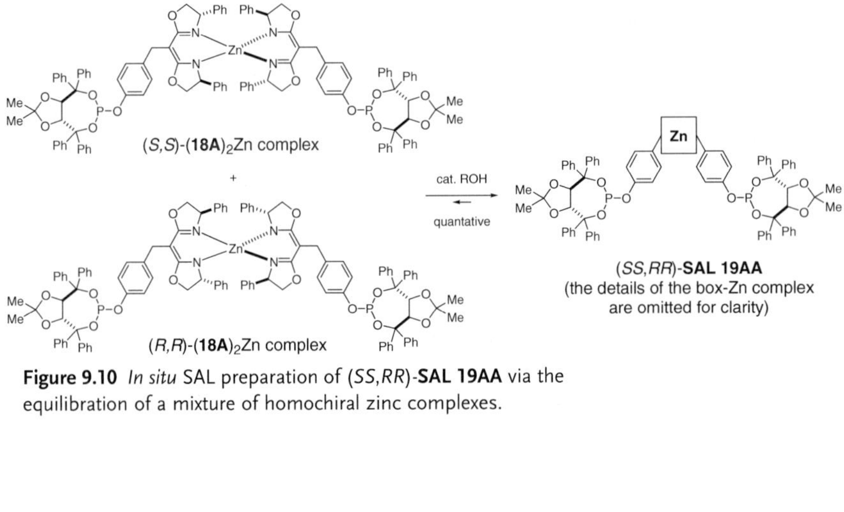

Figure 9.10 *In situ* SAL preparation of (*SS,RR*)-**SAL 19AA** via the equilibration of a mixture of homochiral zinc complexes.

9.9 Ligand Scaffold Optimization in Palladium-Catalyzed Asymmetric Allylic Amination

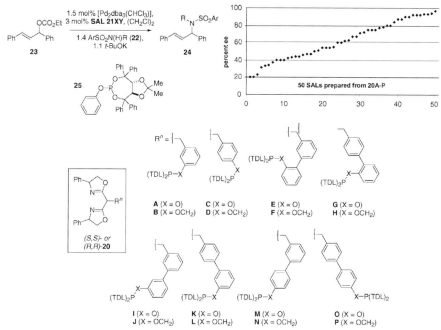

Figure 9.11 Remarkable variation in enantioselectivity from a library of 50 closely related **SALs 21XY**.

21XY. We prepared 50 of the 256 possible combinations of subunits **20A–P**. As each SAL **21XY** bears two pendant TADDOL-derived monophosphites the shape and electronic characteristics of the ligating groups are constant in this ligand library; only the ligand scaffold is varied. The idea was to determine whether the structure of the scaffold alone is sufficient to control the level of asymmetric induction exhibited by the derived catalyst systems. The SALs were screened in the palladium-catalyzed asymmetric allylic amination [25] of racemic carbonate **23** by N-methyl-p-toluenesulfonamide. Chiral diphosphites have been employed in a wide range of asymmetric reactions, including palladium-catalyzed allylations [26]; the latter are frequently used as a testing ground for new chiral ligand motifs. For reference, the corresponding allylic amination using two equivalents of [(R,R)-TADDOL]POPh (**25**, the same chiral ligating group without the SAL) gives the allylation product (R)-**24** (R = Me) in 48% ee.

The chiral diphosphites screened differ only in the structure of the self-assembled ligand scaffold, and while each gives predominantly (R)-**24** (R = Me) the variation in enantioselectivity is striking. Plotting the results obtained for the 50 combinations of subunits **A–P** in ascending order of percent enantiomeric excess we find that the ee varies almost linearly over a wide range (i.e., 20–97% ee). Two-thirds of the SALs screened give a higher level of asymmetric induction than the monomer [(R,R)-TADDOL]POPh (**25**). While the goal, of course, is to obtain enantiomerically pure material directly from the reaction, products obtained at the level of 90% ee (95 : 5 er) are often practical; they can often be enriched to enantiomeric purity via one recrystallization with minimal losses. Nine combinations of **SALs 21XY** effect the asymmetric allylation in 90% ee or higher.

The variation in enantiomeric excess demonstrates the ability to use very subtle changes in the scaffold to manipulate the ligand topography around palladium. As tabulated in Table 9.1 each of the top catalysts contains one of two closely related subunits, **F** or **H**. These two tethers differ only with respect to the substitution pattern (1,3- vs. 1,4-) on the phenyl ring closest to the box subunit. The most successful ligand, **SAL 21FH**, contains both subunits and affords 24 (R = Me) in 97% ee (entry 1). Surprisingly, the pseudo-C_2-symmetric derivatives **SAL 21FF** and **SAL 21HH** are less effective, giving 84% and 87% ee respectively (entries 12 and 11). Compared to the latter two, the combinations **SAL 21GH** and **SAL 21FG** (entries 8 and 10; 90% and 87% ee, respectively) are similar-to-perhaps slightly superior, and yet **SAL 21GG** is one of the worst combinations (entry 50, 20% ee). The presence of **F** or **H** in and of itself does not guarantee success; **SAL 21AF** is another of the worst combinations (entry 49, 20% ee). It is also interesting to note that two poor subunits can form a relatively favorable combination. Neither **SAL 21AA** (60% ee) nor **SAL 21KK** (31% ee) are very good, but the combination **SAL 21AK** at 75% ee (entry 17) is the best SAL that does not include an **F** or **H** subunit.

When substrate 22 is changed, even modestly, **SAL 21FH** is not always the best choice. This highlights another feature of the SAL approach to ligand scaffold optimization, easy ligand tunability. Chiral diphosphites **SAL 21FF**, **SAL 21FH**, and **SAL 21HH** were screened with four other N-substituted sulfonamides. In each case, SALs were identified that give the allylation product in high enantiomeric excess; for example (results for best SAL), N-(*n*-butyl) (**SAL 21FF**, 90% ee), -benzyl (**SAL 21FH**, 91% ee), -isopropyl (**SAL 21HH**, 95% ee), and –phenyl (**SAL 21FF**, 88% ee).

9.10
What has been Learned?

There are several theories to explain the origin of stereoselection in additions to chiral η^3-allylpalladium complexes and, in particular, several rely on electronic differentiation of the ligating groups at the two potential sites of attack [27]. Since all SALs tested have TADDOL-phosphites, we can ignore significant electronic differences as the source of their wide ranging relative effectiveness. The emerging picture is that **F** or **H** is very good at orienting the TADDOL-phosphite so as to efficiently enable (or dissuade, we can not distinguish) attack at its juxtaposed allylic site, while on the opposite ligation site any of several different tethers (**F, H, P, J, E, L,** or **G**) orient the TADDOL-phosphite so as to efficiently dissuade (or enable) attack. We think it is probably dissuade, since no combinations using only **P, J, E, L,** and **G** (from ten total such combinations tested) gave higher than 54% ee; recall that the monodentate model ligand, [(R,R)-TADDOL]POPh (25), gives 48% ee.

Table 9.1 Data summary: Screening **SALs 21XY** in the palladium-catalyzed allylic amination reaction with TsN(H)Me.

Entry	SAL	% ee
1	FH	97
2	FP	94
3	FJ	93
4	HJ	93
5	EH	92
6	FL	92
7	HL	92
8	GH	90
9	HP	90
10	FG	87
11	HH	87
12	FF	84
13	EF	80
14	FI	80
15	HK	78
16	FK	77
17	AK	75
18	AH	75
19	HI	75
20	IP	72
21	II	71
22	CF	67
23	IK	67
24	CH	63
25	EI	62
26	AA	60
27	CC	60
28	CK	55
29	GI	54
30	JP	54
31	PP	54
32	IJ	51
33	AC	50
34	CI	47
35	JK	47
36	JL	46
37	AI	44
38	GJ	44
39	CJ	43
40	EJ	42
41	BB	40
42	IL	40
43	JJ	40
44	EE	37
45	AJ	34
46	LL	33
47	KK	31
48	DD	23
49	AF	20
50	GG	20

9.11
Why such Wide Variation in Enantiomeric Excess given the Relatively Small Changes in Scaffold Structure?

At this stage, perhaps the best explanation seems to be a conceptual analogy to biological catalysts. Nature did not invent a new structure for every new reaction or substrate. Instead, a rather limited set of structures (i.e., amino acid side chains and enzyme cofactors) are positioned in different ways to define the shape and characteristics required for efficient catalysis. In our case, the shape of the TADDOL-phosphite subunit is invariant, and, yet, small changes in the scaffold reposition or reorient that shape to a more, or less, effective position with respect to asymmetric catalysis. While we cannot see, or yet predict, exactly how ligating groups move in response to changes in the scaffold, we can compensate by preparing focused ligand libraries and then work to understand why particular structures within the library are more or less effective.

9.12
Ligand Scaffold Optimization in Rhodium-Catalyzed Asymmetric Hydrogenation

Having used the palladium-catalyzed allylic amination reaction to demonstrate proof of principle, we have also begun to examine rhodium-catalyzed asymmetric hydrogenations. The need to carry out enantioselective hydrogenation is frequently encountered in chemical discovery and process research and, therefore, is relevant to a wide range of synthetic intermediates and of substantial industrial importance. It constitutes a second common testing ground for new ligand systems for asymmetric catalysis, and, as nicely summarized in a recent review by Blaser [28] (Solvias AG), there are significant drawbacks with the currently available catalyst systems. In addition, rhodium-catalyzed asymmetric hydrogenation constitutes another common testing ground for new ligand systems and offers us the opportunity to evaluate the SAL approach with a different metal catalyst; a preliminary account has been published [29]. We prepared and screened a library of 111 SAL combinations with enamides **26** and **28**. The SALs screened were largely drawn most from the set of 16 tether subunits discussed above (i.e., **20A–P**), except with the BIPHEP-phosphite ligating group (Figure 9.12). A few P, N-SAL combinations were also included; these combine a BIPHEP-phosphite with a chiral oxazoline. Once again, we see wide variation in enantioselectivity – from racemic to nearly 90% ee; based upon more recent results, further improvements are likely possible.

The enamides chosen are from two different classes of prototypical asymmetric hydrogenation substrates and they generally respond quite differently to the same series of SAL scaffolds. The graph in Figure 9.12 compares the percent ee obtained for the asymmetric hydrogenation of **26** (squares) with that obtained for **28** (diamonds) plotted in ascending order of % ee for the latter substrate. The

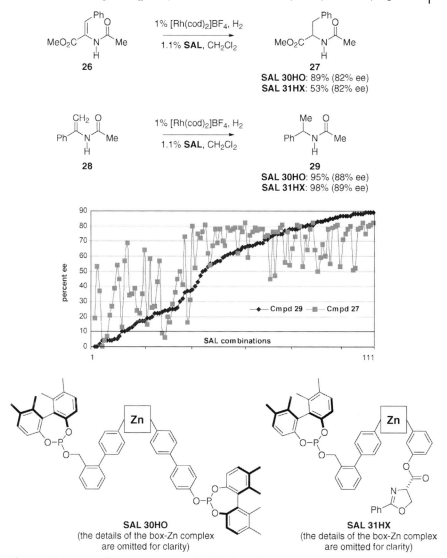

Figure 9.12 Ligand scaffold optimization using SALs in rhodium-catalyzed asymmetric hydrogenation.

scatter in the data indicates the two substrates often respond quite differently to the same SAL scaffold. Nonetheless, several SALs combinations, e.g., **SAL 30HO** and **SAL 31HX**, are essentially best case scaffolds for each of these substrates within the particular ligand library screened. In other words, among the many scaffolds that work well for one or the other substrate, SAL ligand scaffold optimization uncovered a few scaffolds that exhibit some substrate generality.

We found this to be the case in each of the reactions we have examined thus far, and it presents an important, rather unique opportunity to learn something of fundamental importance with respect to catalyst substrate generality. Why, within a series of closely related scaffolds, do some exhibit good, others poor, substrate generality? Admittedly, two substrates is only a modest step toward substrate generality, but hopefully it illustrates another unique feature of the approach. It is also noteworthy that, while the data are not shown here, we see similar trends for substrate conversion and product yield as a function of SAL scaffold. This must relate to turnover frequency and/or catalyst stability and, hopefully, it will prove possible to gain new insights into these issues of fundamental importance in catalysis in the course of future studies.

9.13
Concluding Remarks

Efficient construction of the ligand scaffold via self-assembly permits the rapid, often *in situ*, preparation of dozens, even hundreds, of unique ligand scaffolds. These self-assembled ligands (SALs) can be thought of as a focused ligand library with the unique feature that ligand scaffold optimization allows one to subtly manipulate the chiral catalytic pocket in ways not available with classical ligand designs. The catalyst systems described above are, presumably, macrocyclic chelates, and there is perhaps a conceptual analogy to biological catalysis. Nature uses a rather limited set of structures (i.e., amino acid side chains and enzyme cofactors) positioned in different ways by a self-assembled macromolecular scaffold to define the shape and characteristics required for each reaction catalyzed. In addition to the potential utility of such catalysts, exploring the reasons for the success/failure of the SALs identified during their study and potentially developing a mechanistic framework for understanding SAL structural diversity may lead not only to new fundamental insights into the reaction under investigation but, more generally, to other catalytic asymmetric reactions, to other types of catalysis, and to the challenge of self-assembled catalyst design.

Acknowledgments

Financial support for this research under grants from the Nebraska Research Initiative, ONR (05PR07809-00), NIH (STTR R41 GM074337), NSF (CHE-0316825) and ACS-PRF (47257-AC1) is gratefully acknowledged. We thank T.A. George (UNL Chemistry) for the loan of equipment and the NSF (CHE-0091975, MRI-0079750) and NIH (SIG-1-510-RR-06307) for the NMR spectrometers used in these studies carried out in facilities renovated under NIH RR016544. Special thanks are given to Jeffery M. Atkins and Paul Hrvatin whose initial explorations into chirality-directed self-assembly led to the design of the SAL approach.

References

1 Trost, B.M. (2004) Asymmetric catalysis: An enabling science. *Proc. Natl. Acad. Sci. U.S.A.*, **101**, 5348–5355; Noyori, R. (2002) Asymmetric catalysis: Science and opportunities (Nobel Lecture). *Angew. Chem., Int. Ed.*, **41**, 2008–2022.

2 Gennari, C. and Piarulli, U. (2003) Combinatorial libraries of chiral ligands for enantioselective catalysis. *Chem. Rev.*, **103**, 3071–3100.

3 Yoon, T.P. and Jacobsen, E.N. (2003) Privileged chiral catalysts. *Science*, **299**, 1691–1694.

4 Gilbertson, S.R. and Wang, X. (1999) The parallel synthesis of peptide based phosphine ligands. *Tetrahedron*, **55**, 11609–11618; Gilbertson, S.R. and Wang, X. (1996) The combinatorial synthesis of chiral phosphine ligands. *Tetrahedron Lett.*, **37**, 6475–6478.

5 Jiang, X.-B., Lefort, L., Goudriaan, P.E., de Vries, A.H.M., van Leeuwen, P.W.N.M., de Vries, J.G. and Reek, J.N.H. (2006) Screening of a supramolecular catalyst library in the search for selective catalysts for the asymmetric hydrogenation of a difficult enamide substrate. *Angew. Chem., Int. Ed.*, **45**, 1223–1227; Slagt, V.F., van Leeuwen, P.W.N.M. and Reek, J.N.H. (2003) Multicomponent porphyrin assemblies as functional bidentate phosphite ligands for regioselective rhodium-catalyzed hydroformylation. *Angew. Chem., Int. Ed.*, **42**, 5619–5623; Slagt, V.F., Van Leeuwen, P.W.N.M. and Reek, J.N.H. (2003) Bidentate ligands formed by self-assembly. *Chem. Commun.*, 2474–2475; Slagt, V.F., Roeder, M., Kamer, P.C.J., Van Leeuwen, P.W.N.M. and Reek, J.N.H. (2004) Supraphos: A supramolecular strategy to prepare bidentate ligands. *J. Am. Chem. Soc.*, **126**, 4056–4057.

6 Wu, C.-D., Hu, A., Zhang, L. and Lin, W. (2005) A homochiral porous metal-organic framework for highly enantioselective heterogeneous asymmetric catalysis. *J. Am. Chem. Soc.*, **127**, 8940–8941; Hua, J. and Lin, W. (2004) Chiral metallacyclophanes: Self-assembly, characterization, and application in asymmetric catalysis. *Org. Lett.*, **6**, 861–864; Jiang, H., Hu, A. and Lin, W. (2003) A chiral metallacyclophane for asymmetric catalysis. *Chem. Commun.*, 96–97; Lee, S.J., Hu, A. and Lin, W. (2002) The first chiral organometallic triangle for asymmetric catalysis. *J. Am. Chem. Soc.*, **124**, 12948–12949.

7 Weis, M., Waloch, C., Seiche, W. and Breit, B. (2006) Self-assembly of bidentate ligands for combinatorial homogeneous catalysis: Asymmetric rhodium-catalyzed hydrogenation. *J. Am. Chem. Soc.*, **128**, 4188–4189; Breit, B. and Seiche, W. (2006) Self-assembly of bidentate ligands for combinatorial homogeneous catalysis based on an a-t base pair model. *Pure Appl. Chem.*, **78**, 249–256; Seiche, W., Schuschkowski, A. and Breit, B. (2005) Bidentate ligands by self-assembly through hydrogen bonding: A general room temperature/ambient pressure regioselective hydroformylation of terminal alkenes. *Adv. Synth. Catal.*, **347**, 1488–1494; Breit, B. and Seiche, W. (2005) Self-assembly of bidentate ligands for combinatorial homogeneous catalysis based on an A-T base-pair model. *Angew. Chem., Int. Ed.*, **44**, 1640–1643; Breit, B. and Seiche, W. (2003) Hydrogen bonding as a construction element for bidentate donor ligands in homogeneous catalysis: Regioselective hydroformylation of terminal alkenes. *J. Am. Chem. Soc.*, **125**, 6608–6609.

8 Jonsson, S., Odille, F.G.J., Norrby, P.-O. and Waernmark, K. (2005) A dynamic supramolecular system exhibiting substrate selectivity in the catalytic epoxidation of olefins. *Chem. Commun.*, 549–551; Kovbasyuk, L., Pritzkow, H., Kraemer, R. and Fritsky, I.O. (2004) On/off

regulation of catalysis by allosteric control of metal complex nuclearity. *Chem. Commun.* 880–881; Carlier, P.R. (2004) Supramolecular chemistry: Threading the needle: Mimicking natural toroidal catalysts. *Angew. Chem., Int. Ed.*, **43**, 2602–2605; Kovbasyuk, L. and Kraemer, R. (2004) Allosteric supramolecular receptors and catalysts. *Chem. Rev.*, **104**, 3161–3187; Harada, T., Kanda, K., Hiraoka, Y., Marutani, Y. and Nakatsugawa, M. (2004) Asymmetric alkylation of aldehydes catalyzed by novel dinuclear bis-binolate titanium(iv) complexes. *Tetrahedron: Asymmetry*, **15**, 3879–3883.

9 Larsen, J., Rasmussen, B.S., Hazell, R.G. and Skrydstrup, T. (2004) Preparation of a novel diphosphine-palladium macrocyclic complex possessing a molecular recognition site. Oxidative addition studies. *Chem. Commun.*, 202–203; Braunstein, P., Clerc, G. and Morise, X. (2003) Cyclopropanation and Diels–Alder reactions catalyzed by the first heterobimetallic complexes with bridging phosphinooxazoline ligands. *New J. Chem.*, **27**, 68–72; Braunstein, P., Clerc, G., Morise, X., Welter, R. and Mantovani, G. (2003) Phosphinooxazolines as assembling ligands in heterometallic complexes. *Dalton Trans.*, 1601–1605.

10 Gianneschi, N.C., Nguyen, S.T. and Mirkin, C.A. (2005) Signal amplification and detection via a supramolecular allosteric catalyst. *J. Am. Chem. Soc.*, **127**, 1644–1645; Gianneschi, N.C., Cho, S.-H., Nguyen, S.B.T. and Mirkin, C.A. (2004) Reversibly addressing an allosteric catalyst in situ: Catalytic molecular tweezers. *Angew. Chem., Int. Ed.*, **43**, 5503–5507; Gianneschi, N.C., Bertin, P.A., Nguyen, S.T., Mirkin, C.A., Zakharov, L.N. and Rheingold, A.L. (2003) A supramolecular approach to an allosteric catalyst. *J. Am. Chem. Soc.*, **125**, 10508–10509.

11 Mikami, K., Tereda, M., Korenaga, T., Matsumoto, Y., Ueki, M. and Angeluad, R. (2000) Asymmetric activation. *Angew. Chem., Int. Ed.*, **39**, 3532–3556; Mikami, K., Matsukawa, S., Volk, T. and Terada, M. (1998) Self-assembly of several components into a highly enantioselective Ti catalyst for carbonyl-ene reactions. *Angew. Chem., Int. Ed.*, **36**, 2768–2771.

12 Reyes, S.J. and Burgess, K. (2006) Heterovalent selectivity and the combinatorial advantage. *Chem. Soc. Rev.*, **35**, 416–423; Sandee, A.J. and Reek, J.N.H. (2006) Bidentate ligands by supramolecular chemistry – the future for catalysis? *Dalton Trans.*, 3385–3391; Wilkinson, M.J., van Leeuwen, P.W.N.M. and Reek, J.N.H. (2005) New directions in supramolecular transition metal catalysis. *Org. Biomol. Chem.*, **3**, 2371–2383; Breit, B. (2005) Supramolecular approaches to generate libraries of chelating bidentate ligands for homogeneous catalysis. *Angew. Chem., Int. Ed.*, **44**, 6816–6825.

13 Takacs, J.M., Hrvatin, P.M., Atkins, J.M., Reddy, D.S. and Clark, J.L. (2005) The selective formation of neutral, heteroleptic zinc(II) complexes via self-discrimination of chiral bisoxazoline racemates and pseudoracemates. *New J. Chem.*, **29**, 263–265; Atkins, J.M., Moteki, S.A., DiMagno, S.G. and Takacs, J.M. (2006) Single enantiomer, chiral donor–acceptor metal complexes from bisoxazoline pseudoracemates. *Org. Lett.*, **8** (13), 2759–2762.

14 Takacs, J.M., Lawson, E.C., Reno, M.J., Youngman, M.A. and Quincy, D.A. (1997) Enantioselective Diels–Alder reactions: Room temperature bis(oxazoline)-zinc, magnesium, and copper triflate catalysts. *Tetrahedron: Asymmetry*, **8**, 3073–3078.

15 Kagan, H.B. (2001) Practical consequences of non-linear effects in asymmetric synthesis. *Adv. Synth. Catal.*, **343**, 227–233.

16 Kramer, R., Lehn, J.M. and Marquis-Rigault, A. (1993) Self-recognition in helicate self-assembly: Spontaneous formation of helical metal complexes from mixtures of ligands and metal ions. *Proc. Natl. Acad. Sci. U.S.A.*, **90**, 5394–5398.

17 Enemark, E.J. and Stack, T.D.P. (1998) Stereospecificity and self-selectivity in the

generation of a chiral molecular tetrahedron by metal-assisted self-assembly. *Angew. Chem., Int. Ed.*, **37**, 932–935; Rowland, J.M., Olmstead, M.M. and Mascharak, P.K. (2002) Chiral monomeric and homochiral dimeric copper(II) complexes of a new chiral ligand, n-(1,2-bis(2-pyridyl)ethyl)pyridine-2-carboxamide: An example of molecular self-recognition. *Inorg. Chem.*, **41**, 1545–1549.

18 Kim, H.-J., Moon, D., Lah, M.S. and Hong, J.-I. (2002) An enantiomerically pure propeller-shaped supramolecular capsule based on the stereospecific self-assembly of two chiral tris(oxazoline) ligands around three Ag(I) ions. *Angew. Chem., Int. Ed.*, **41**, 3174–3177.

19 Swiegers, G.F. and Malefetse, T.J. (2000) New self-assembled structural motifs in coordination chemistry. *Chem. Rev.*, **100**, 3483–3537.

20 Annunziata, R., Benaglia, M., Cinquini, M. and Cozzi, F. (2001) Synthesis of a bifunctional ligand for the sequential enantioselective catalysis of various reactions. *Eur. J. Org. Chem.*, 1045–1048.

21 Seebach, D., Hayakawa, M., Sakaki, J. and Schweizer, W.B. (1993) Derivatives of tetraaryl-2,2-dimethyl-1,3-dioxolane-4,5-dimethanol (TADDOL) containing nitrogen, sulfur, and phosphorus atoms. New ligands and auxiliaries for enantioselective reactions. *Tetrahedron*, **49**, 1711–1724.

22 Kranich, R., Eis, K., Geis, O., Muhle, S., Bats, J.W. and Schmalz, H.-G. (2000) A modular approach to structurally diverse bidentate chelate ligands for transition metal catalysis. *Chem.–Eur. J.*, **6**, 2874–2894.

23 Pamies, O., Net, G., Ruiz, A. and Claver, C. (2000) Chiral diphosphites as ligands for the rhodium- and iridium-catalysed asymmetric hydrogenation: Precatalyst complexes, intermediates and kinetics of the reaction. *Eur. J. Inorg. Chem.*, 1287–1294.

24 Takacs, J.M., Reddy, D.S., Moteki, S.A., Wu, D. and Palencia, H. (2004) Asymmetric catalysis using self-assembled chiral bidentate P,P-ligands. *J. Am. Chem. Soc.*, **126**, 4494–4495.

25 Hamada, Y., Seto, N., Takayanagi, Y., Nakano, T. and Hara, O. (1999) Asymmetric allylic substitution reaction with nitrogen and oxygen nucleophiles using the monodentate chiral phosphine 9-PBN. *Tetrahedron Lett.*, **40**, 7791–7794; Evans, D.A., Campos, K.R., Tedrow, J.S., Michael, F.E. and Gagne, M.R. (2000) Application of chiral mixed phosphorus/sulfur ligands to palladium-catalyzed allylic substitutions. *J. Am. Chem. Soc.*, **122**, 7905–7920; Pamies, O., van Strijdonck, G.P.F., Dieguez, M., Deerenberg, S., Net, G., Ruiz, A., Claver, C., Kamer, P.C.J. and van Leeuwen, P.W.N.M. (2001) Modular furanoside phosphite ligands for asymmetric Pd-catalyzed allylic substitution. *J. Org. Chem.*, **66**, 8867–8871; Agarkov, A., Uffman, E.W. and Gilbertson, S.R. (2003) Parallel approach to selective catalysts for palladium-catalyzed desymmetrization of 2,4-cyclopentenediol. *Org. Lett.*, **5**, 2091–2094.

26 Ansell, J. and Wills, M. (2002) Enantioselective catalysis using phosphorus-donor ligands containing two or three P–N or P–O bonds. *Chem. Soc. Rev.*, **31**, 259–268.

27 Vasse, J.-L., Stranne, R., Zalubovskis, R., Gayet, C. and Moberg, C. (2003) Influence of steric symmetry and electronic dissymmetry on the enantioselectivity in palladium-catalyzed allylic substitutions. *J. Org. Chem.*, **68**, 3258–3270.

28 Blaser, H.-U., Malan, C., Pugin, B., Spindler, F., Steiner, H. and Studer, M. (2003) Selective hydrogenation for fine chemicals: Recent trends and new developments. *Adv. Synth. Catal.*, **345**, 103–151.

29 Takacs, J.M., Chaiseeda, K., Moteki, S.A., Reddy, D.S., Wu, D. and Chandra, K. (2006) Rhodium-catalyzed asymmetric hydrogenation using self-assembled chiral bidentate ligands. *Pure Appl. Chem.*, **78** (2), 501–509.

10
Supramolecular Catalysis: Refocusing Catalysis
Piet W. N. M. Van Leeuwen and Zoraida Freixa

10.1
Introduction: A Brief Personal History

Supramolecular catalysis implies the use of noncovalent interactions in catalytic systems to achieve higher rates, more selective catalysts, or larger numbers of ligands than achieved so far by "covalent" systems. These interactions involve the following phenomena, in which some overlap may be noted:

- hydrogen bonding [1]
- $\pi - \pi$ stacking [2]
- hydrophobic interactions [3]
- anion–π interactions
- cation–π interactions [4]
- Lewis acid–metal ligand interactions
- ion–dipole interactions [5]
- cation–anion attractions
- Dewar–Chatt–Duncanson metal–ligand interactions
- van der Waals interactions
- charge-transfer interactions.

Usually, noncovalent interactions are regarded as being weaker than covalent ones, but if we have a closer look at ionic interactions it is realized that the Coulombic attraction between two counter ions can be as large as 500 kJ mol^{-1}. This is only true though in the absence of solvation, the energy of which may amount to the same order of magnitude, or even exceed this number by far when water is considered as the solvent. The reason to include such strong interactions in supramolecular interactions is both for practical reasons – as they are being used extensively to this end – and for a theoretical justification, as upon breaking of the bond two ions are formed rather than two radicals as happens to a covalent bond. The use is not restricted to the construction of ligands or small cavities, as metal organic frameworks (MOFs) cover a very large field today. The general phrasing that upon breaking a supramolecular bond the two electrons involved end up in one of the two fragments

Supramolecular Catalysis. Edited by Piet W. N. M. van Leeuwen
Copyright © 2008 WILEY-VCH Verlag GmbH & Co. KGaA, Weinheim
ISBN: 978-3-527-32191-9

cannot be used in a generalized sense because several other interactions (Lewis acid–base, metal alkene complexation) follow this principle. Another restriction should be made for metal-catalyzed reactions in general, especially transition metal catalyzed reactions. Key characteristic of metal catalysis is that "the metal brings the substrates together" before they enter in the process of forming or breaking covalent bonds. Thus, the many phenomena in the steps before the catalytic transformation starts (and the desorption ones after the catalytic transformation) is fully included in the definition of supramolecular catalysis. We can refine our definition by adding that the noncovalent interactions meant by "supramolecular catalysis" refer to those that are not included in the basic catalytic reaction, which, admittedly, leaves the matter ill-defined in several cases. Supramolecular chemistry started (Chapter 1) with host–guest chemistry mimicking enzyme catalysis, with the simplified characteristic that the host containing the catalyst will convert the guest into the product, the guest being complexed via a lock-and-key principle in the host. The reactions studied were often taken from known enzymatic reactions, such as ester hydrolysis, aldehyde condensations, but Diels–Alder reactions have also led to nice examples of accelerations and selectivity changes. Interestingly, most of these reactions also occur without catalyst and the effectiveness of the supramolecular catalyst is usually measured in this field by comparing its rate with that of the uncatalyzed reaction, often present as a background reaction. Organometallic catalysts cannot be compared with uncatalyzed reactions, as alkene polymerization, hydrogenation, and hydroformylation have no uncatalyzed counterparts apart from radical mechanisms in several instances. Therefore, in organometallic, supramolecular catalysis the new catalyst is compared with "classic" ones. Changes in selectivity can pose some problems in their evaluation, but comparison of rates is more tricky and apt to errors as, often, the kinetic equation changes from one catalyst to the other.

Over the years, our laboratories have been involved in a large number of projects on supramolecular catalysis, and as we will see, one of them "avant la lettre". The first interest in host–guest catalysis, for one of us, was aroused by Breslow's [6] and Tabushi's [7] publications on cyclodextrins containing metal complexes on one of the rims that were used as catalysts for ester hydrolysis. Nickel compounds were particularly interesting as certain nickel complexes were, and still are, of interest as ethylene oligomerization catalysts (Shell Higher Olefins Process, SHOP). SHOP, operated in a biphasic way, gives a wide Schulz–Flory distribution of alkenes. It was thought that perhaps the cyclodextrin cavity could lead to a change in selectivity through an effect of the limited space in the apolar cavity dissolved in a polar solvent; the slightest change could already be of great importance industrially. Cyclodextrins do show a size selective preference for alkene complexation, which has been exploited, for example, in two-phase hydrogenation [8] and Wacker oxidation [9] of mixtures. The molecular weight of a Schulz–Flory dominated reaction is governed by kinetics and, if any preference should occur for lower molecular weights, kinetics should provide the answer. Kinetic analysis shows that an equilibrium between growing chain and free product is one possible solution, which undoubtedly would lead to highly undesirable isomerization, but this potential drawback was taken for granted. In the Shell group R. L. Wife, former post-doc of R. Breslow and experienced in cyclodextrin modification, started to work on this topic (1978). The incompatibility of the phosphinophenoxy nickel complex and the hydroxyl groups of cyclodextrin prohibited catalysis, while protection of the hydroxyl groups with organosilyl groups limited the solubility in polar solvents, while in water they were hydrolyzed anyway. We then turned to the synthesis of new hosts, viz. 1,4-aryl-X cyclic hexamers, resembling cyclodextrins, connected in a stepwise manner, leaving a high flexibility for the introduction of functional groups for catalyst complexation, polar groups delivering solubility, and bulky groups for steric hindrance. In hindsight this was a mistake; the search for both new host molecules and new catalysts to achieve shape selective effects was too demanding (today's cross coupling techniques were not yet available!). We quoted this conclusion many times: "don't try to invent two things at the time; your boss will be happy with just one!". Many years later the Shell group embarked on a collaboration with the Nolte group who by then had developed their "clip" host molecules, thus reducing the problems for catalysis "simply" to connecting and positioning the catalyst and searching for an effect on catalysis of the host–guest interaction (see Chapter 6 and below). The work of Hein Coolen is reviewed in Chapter 6, while in Section 10.3 we comment on the work by Georg Dol, also part of the collaboration Nijmegen-Amsterdam on host–guest catalysis.

Shell research always had a great interest in shape selective catalysis; not only in heterogeneous catalysis (zeolites), but also homogeneous processes that might take advantage of this. As the SHOP process produces only 50% of 1-alkene products that can be used directly – the remainder enters a long recycle and leaves the as C_{12-13} internal olefins to be converted with the Shell hydroformylation process into detergent alcohols, via metathesis and isomerization steps – the incentive for a more efficient process is enormous and several other attempts were directed towards better selectivities. One other direction taken was the use of dendrimers, an approach very akin to supramolecular catalysis. In 1978 the aminodendrimers were introduced by Vögtle [10], and the Shell group started the development of silane-based dendrimers with the idea to use these as nano-microreactors, aiming again at a molecular weight controlling effect of an apolar core where the catalyst is situated and a polar outer sphere (as the name dendrimer had not been coined yet, the target molecules were called polymeric micelles!). Although the synthesis looked straightforward on paper, it took many years until eventually Van der Made solved the synthetic problems. Because of its potential importance to industry, the first publication appeared only in 1992! In the meantime the focus of research had changed and dendrimeric efforts were directed towards the palladium catalysis developed by Drent; in polyketone polymer synthesis dendrimeric catalysts were of interest to avoid precipitation of the polymer on the reactor walls (reactor fouling). The SHOP work was resumed at the University of Amsterdam and while a size selective catalyst remains elusive a beneficial site isolation effect has been reported by Müller and Ackerman [11], a project funded by the NRSC–C (National Research School Combination Catalysis).

Supramolecular catalysis received a strong boost in the Netherlands when the NRSC–C started in 1998 as it had, and still has, supramolecular catalysis as one of its main themes. Several groups in the Netherlands increased their effort in this field, including the Van Leeuwen, Kamer and Reek group in Amsterdam. The introduction by Reek of zinc porphyrin as a building block for making or modifying ligands was a successful choice (Chapter 8). The main theme today of the Reek group is the construction of catalytic systems (e.g., bidentate ligands) through the use of supramolecular interactions such as the interaction of pyridine donors with porphyrin or salen templates, and hydrogen bonding. In Section 10.2 we discuss the forerunners of hydrogen bonding in this field of bidentate ligand formation from monodentates used in metal catalysis, namely secondary phosphine oxides, which form, very selectively, dimeric ligands from monomeric ligands via hydrogen bond interactions. Presently, chiral derivatives of these are being studied by the Van Leeuwen and Freixa group in Tarragona; the two remaining topics in this chapter are from the Tarragona group. In Section 10.4 we review the selective formation of heterolytic pairs of monodentate ligands, steered by cation–anion interaction. Section 10.5 presents selective hydroformylation obtained by the assembly of ditopic ligands containing hard coordination sites for the assembly metal and a soft phosphorus center for the catalytic rhodium metal.

By way of introduction, in each part we mention a few literature references, which are a rather personal choice, as the number we could choose from is enormous. Note for instance that the review by Feiters in 1996 [12], named supramolecular catalysis, contains already over 500 references, and today, when we refocus history within the light of supramolecular events and when we add the publications published since then, we arrive at a very large number to choose from. In our concluding remarks (Section 10.6) we will come back to this.

10.2
Secondary Phosphines or Phosphites as Supramolecular Ligands

Secondary phosphine oxide (SPO) complexes with transition metals have been known for 40 years. The particular one that was first used in a catalytic reaction by the Shell group in the early 1980s, reported in 1975 by Roundhill [13], was complex **1**, a platinum hydride containing two SPOs connected by a strong hydrogen bond and a triphenylphosphine to complete the coordination sphere. SPOs have a very strong tendency to occur in pairs such as these in many metal complexes, acting as a bidentate monoanion. It occurred to us that this complex might be of interest in catalysis, as platinum hydrides of related structure, containing a neutral diphosphine as the ligand, are active hydrogenation and hydroformylation catalysts. Many years later, after the Breit group (Chapter 2) published their highly active and selective hydroformylation catalysts, based on rhodium and bidentate ligands assembled via hydrogen bonding, it was Boerner [14] in a mini-review who pointed out that actually the first use of bidentates obtained via hydrogen bonding in catalysis is represented by the work on SPO platinum complexes. The "avant la lettre" supramolecular results had not been presented as such, but we were aware already that the hydrogen bonds in such systems belong to the strongest ones known [15]. Complex **1** indeed is an active hydroformylation catalyst for 1-heptene, showing comparable activity to, and a higher

10.2 Secondary Phosphines or Phosphites as Supramolecular Ligands

Table 10.1 Hydroformylation of 1-heptene with Pt(cod)$_2$ as precursor and SPOs.[a]

Ligands	Aldehyde	Alcohol	Alkane	2-Heptene	Rate
	% Conversion (% linear)				
1 PPh$_2$OH 1 PPh$_3$	9.8 (>90)	8.6 (>90)	0.9	30	18
2 PPh$_2$OH	3.6 (>90)	1.8 (>90)	2.1	80	3
4 PPh$_2$OH	1.7 (>90)	5.4 (>90)	3.6	60	7
2 PPh$_3$	<0.1				
1 PPh$_2$OH 1 dppe	24.0 (>90)	3.1 (>90)	0.9	33	27
1 PPh$_2$OH 1 PCy$_3$	3.6	2.0	1.0	50	5
Complex 1	7.9 (90)	9.6 (80)	1.2	27	11
Complex 1[b]	8.1 (55)	5.85 (73)	4.8	–	5

[a] Solvent benzene, 50 bar of H$_2$/CO = 2:1, 100 °C, 1 hour, 0.1 mmol of Pt(cod)$_2$, 10 mmol of 1-heptene, rates in moles of oxo products per mol catalyst per hour.
[b] 94 bar, 2.5 h, 2-heptene as substrate.

selectivity for linear product than, the commercial cobalt system of those days [16]. It shares with the cobalt system the hydrogenation activity to give in part alcohols in addition to aldehydes and also the undesired alkane formation that amounts to a loss of feedstock. Table 10.1 gives some typical results. The catalysts are also active as isomerization catalysts, again similar to cobalt systems, and are active for internal heptene hydroformylation leading to linearities around 60% (last entry Table 10.1).

The reaction could be carried out stepwise and reaction of hydride **1** with ethylene under pressure at 90 °C gives the ethyl platinum complex **2**. Treatment of **2** with CO at ambient temperature and pressure rapidly gives the propionyl platinum complex **3**. The reaction with ethene requires higher pressures and temperatures, presumably because dissociation of triphenylphosphine is required to create a vacant site, while surely for CO and platinum a five-coordinated species may be involved. Reaction of **3** with dihydrogen under catalytic conditions gives the aldehyde product and platinum hydride **1** is obtained again. Under stoichiometric conditions this is more difficult because the back reaction occurs, giving **1** but only small amounts of aldehyde. Isolation of the catalyst after reaction gives a mixture of the acyl complex and hydride **1**. This may indicate that indeed the aldehyde formation is rate determining, but the result strongly depends on the way the work-up is done, as pressure release at high temperatures gives back the hydride. The use of chiral SPOs in the hydroformylation

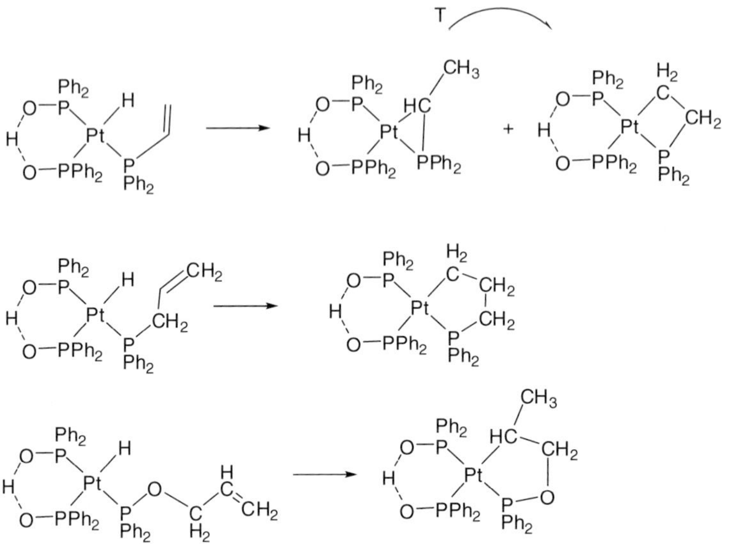

Scheme 10.1 Intramolecular reactions with alkenylphosphines proceed smoothly [18].

failed to give enantiomeric excess in the hydroformylation of styrene [17], which was surprising since in every aspect the catalysts behaves as classic diphosphine catalysts, amongst which are highly enantioselective catalysts. Most likely, racemization of the chiral SPO had occurred (*vide infra*). The catalytic hydroformylation of ethylene leads to mixtures of propanal, the expected product, and pentan-3-one. The last product can be obtained via ethylene insertion in propionyl complex **3**, which, interestingly, subsequently reacts with hydrogen rather than with a second molecule of CO and higher oligomers are not observed.

The first example in Scheme 10.1 shows that vinylphosphine kinetically first gives the three-membered heterocycle (75%), which upon heating transforms into the thermodynamic product, the four-membered heterocyclic complex. From the other examples we see that, if the possibility exists, the preferred route is formation of a five-membered ring [18].

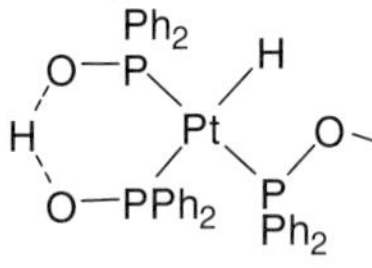

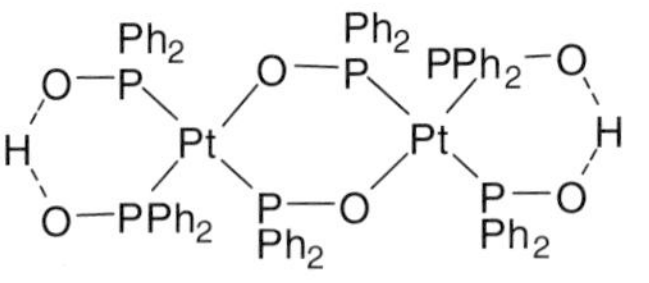

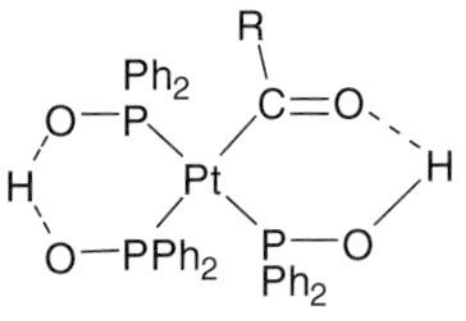

In the absence of triphenylphosphine a tris-SPO platinum complex forms when SPO is reacted with $Pt(cod)_2$, which is insoluble and for which the structure is not known; it could be either a hydride (**4**) or an oligomeric or a dimeric SPO complex (**5**) lacking one molecule of hydrogen. The reaction with alkene, CO and H_2 gives complex **6**, which was identified as the acyl tris-SPO complex by NMR. It shows an

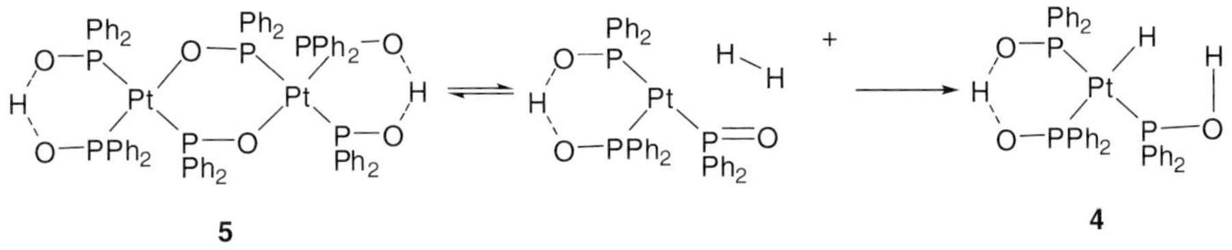

Scheme 10.2 Heterolytic cleavage of dihydrogen on platinum SPO complex.

interesting behavior in that the hydrogen bond to the acyl group switches from one SPO to the other one cis of the acyl group.

It has been speculated that the platinum phosphine oxide moiety is ideally set up for heterolytic splitting of dihydrogen (Scheme 10.2) [19]. In such a reaction the valence state of platinum does not change, remaining as Pt(II). The process shown in Scheme 10.2 has hardly any energy barrier according to DFT calculations and thus is a highly interesting phenomenon [20]. In practice there is a barrier, though, because the catalyst resting state is a complex that either contains one more ligand that must dissociate first, or it is a dimer **5** that must dissociate to the monomeric unsaturated species.

Previously reported platinum catalysts contain monodentate phosphines or bidentate phosphines and weakly coordinating anions (we have argued on the basis of ^{31}P spectroscopy characteristics that the hitherto popular SnCl$_3^-$ anion used in platinum catalysis should also be regarded as a weakly coordinating anion; indeed triflate complexes of diphosphines also give active catalysts).

In addition to hydroformylation catalysts the platinum SPO complexes are even better hydrogenation catalysts for aldehydes (as already happened during hydroformylation) – even more so when CO is omitted from the system. Apparently, carbon monoxide blocks coordination sites for the aldehyde and hydrogen substrates. Again, in this process the facile heterolytic splitting of hydrogen may play a role. This reaction requires the presence of carboxylic acids and perhaps the release of alkoxides from platinum requires a more reactive proton than the one located on the nearby SPO. For example, **4** hydrogenates 2-methylpropanal at 95 °C and 40 bar of hydrogen to the alcohol with a turnover frequency of 4500 mol mol^{-1} h^{-1}. The formation of a platinum–oxygen bond is substantiated by the intramolecular reaction of diphenylphosphinobenzaldehyde with **4**, which gives the six-membered ring compound **7** containing a Pt–O rather than a Pt–C bond.

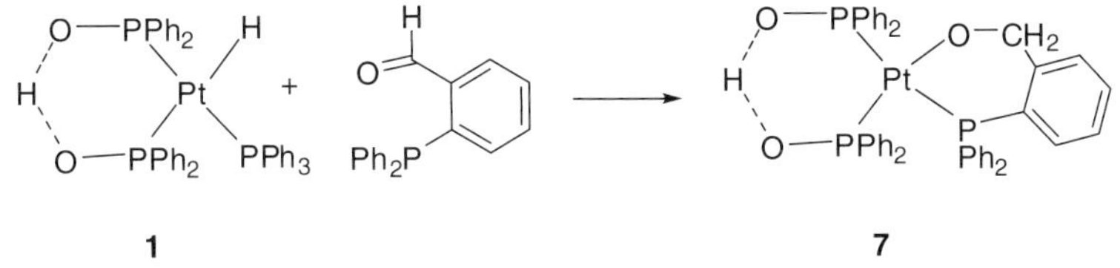

The SPOs had initially been chosen as ligands for platinum hydroformylation because their ligand properties are similar to those of phosphines despite their

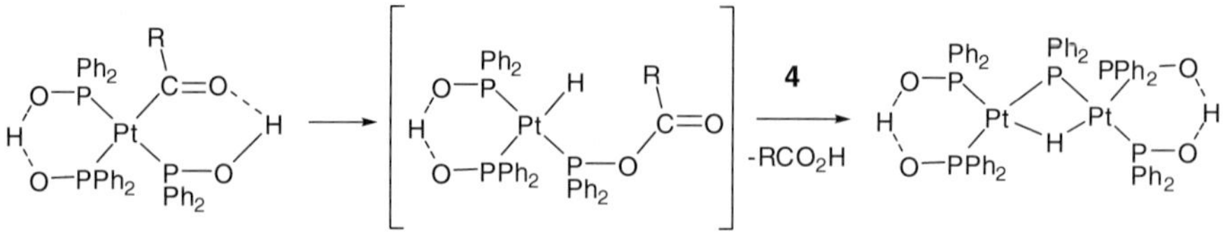

Scheme 10.3 Decomposition of platinum SPO complexes during hydroformylation.

negative charge and because they were thought to be less prone to decomposition and oxidation than phosphines; the latter suffer from carbon–phosphorus cleavage that leads to inactive phosphido complexes. Strangely, prolonged hydroformylation reactions with SPOs also led to phosphido bridged platinum dimers of structure 8; only one of them is shown in Scheme 10.3. It turned out that the highly stable phosphorus oxygen double bond was cleaved via acylation of the SPO at oxygen by a transfer of the acyl group from platinum, an intermediate in the hydroformylation reaction shown above. Acid was indeed observed as a byproduct and the reaction of 4 with acetic anhydride led to 8.

More recently, many uses of SPOs in catalysis have been reported and they have been reviewed by Ackerman [21]. They are widely used in cross coupling reactions in which the anionic character may be more important than the hydrogen bonding or protonation phenomena. We single out a few reactions in which the proton transfer capacity of SPO may be important. A platinum complex of dimethylphosphine oxide is an efficient hydrolysis catalyst for the hydration of nitriles, as reported by Parkins [22]. Most impressive are the high turnover numbers (up to 77 000) and the rate of the reaction (1500 mol mol^{-1} h^{-1}) achieved for the hydrolysis of acrylonitrile. Dimethyl SPO based catalysts are always much faster than diphenyl SPO based ones for this reaction. As a mechanism, Parkins reported the transfer of the SPO oxygen to carbon, and the transfer of the hydrogen atom of the SPO to the nitrogen atom of the substrate (Scheme 10.4). The nucleophilic attack of oxygen is reminiscent of the acylation of an SPO in the hydroformylation reaction above (Scheme 10.3). Interestingly, hydride platinum complexes containing phosphine ligands (PMe$_3$ and water) are also efficient catalysts for this conversion [23]. Thus, evidence for the involvement of the POH function in catalysis, both in hydroformylation and hydration is lacking, although the SPO systems are much more active.

Hindered tertiary nitriles can be hydrolyzed under neutral and mild conditions to the corresponding amides using the above platinum(II) catalysts with dimethylphosphine oxide or other SPOs as ligands. The procedure also works well for nitriles containing acid- or base-sensitive groups with unprecedented selectivity. The catalyst loading can be as low as 0.5 mol.%. Reactions are carried out at 80 °C but take place even at room temperature. Jiang and de Vries also attempted an asymmetric version of the hydrolysis reaction, but the attempt failed, presumably due to racemization of the chiral SPO under the stated reaction conditions [24]. The same group reported that *tert*-butylphosphinoylbenzene turned out to be a versatile ligand in the iridium-catalyzed hydrogenation of β-branched dehydroamino esters and in the rhodium-catalyzed hydrogenation of an enol carbamate.

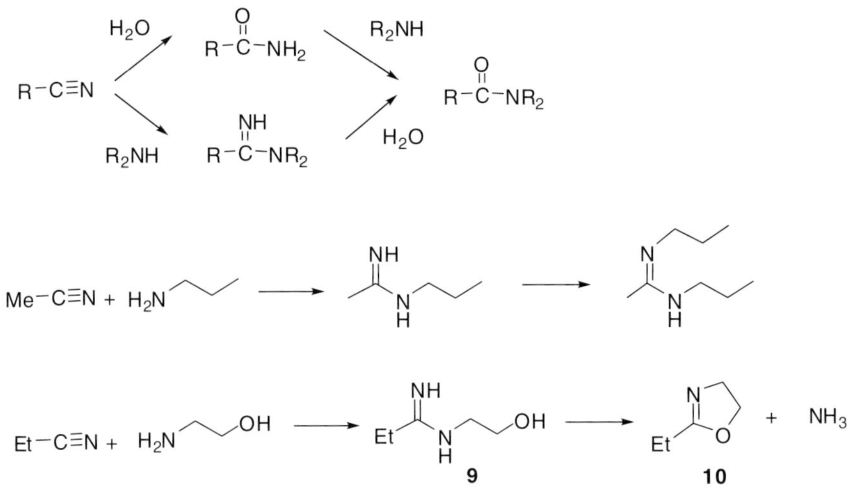

Scheme 10.4 Nitrile hydrolysis catalyzed by platinum SPO complexes.

Platinum-catalyzed hydrolytic amidation of unactivated nitriles was reported by Cobley and coauthors. The platinum(II) complex, [(Me$_2$PO···H···PMe$_2$)PtH (PMe$_2$OH)], efficiently catalyzes the direct conversion of unactivated nitriles into N-substituted amides with both primary and secondary amines. Possible mechanisms for this reaction are discussed and evidence for initial amidine formation is reported. Isolated yields vary from 51 to 89% [25].

An interesting extension to this finding resulted when 2-aminoethanol was reacted with propionitrile. Initial formation of amidine **9** occurs, with subsequent ring-closure giving the final product, 2-ethyl-2-oxazoline **10**.

10.3
Host–Guest Catalysis

Biological systems have provided much of the inspiration for the development of supramolecular chemistry (see Chapters 1 and 6). One of the main challenges in this area is to mimic the enzymatic systems and to understand the assembly processes involved. These natural systems have an extremely high selectivity and catalytic efficiency. Although there are many successful results in this area [26], a complete understanding of these systems is still lacking.

The first examples of the so-called supramolecular catalysis are based on bio-inspired molecular recognition, which is an essential attribute of biochemical systems. Structures such as receptors, antibodies, and enzymes can all recognize a feature that is important for their specific functions, often in the presence of species of quite similar structure. The ability to discriminate depends exclusively on the structural properties of these biological macromolecules. Recent progress in bioorganic chemistry has shown that many of these functions can be incorporated into smaller, synthetically more accessible structures as model systems [27].

In supramolecular chemistry, molecular recognition has evolved over the last 35 years and now much effort is directed towards the complexation of anionic [28], zwitterionic [29], ion-pairs [30] and neutral guests for various purposes, including catalysis [31]. Host molecules can be constructed covalently, or they can themselves also be assembled in a supramolecular fashion. This strategy, called receptor site self-assembly, has been exploited in recent years. Especially, dynamic host formation in the presence of a substrate is highly interesting [32].

In itself, selective guest recognition does not seem very useful for man-made catalysis yet, as the substrate supplied will usually be pure and while recognition between 1-hexene and 1-dodecene in a hydroformylation reaction is interesting the issue can be easily avoided by separating the substrates by distillation. Recognition may be important if by such an event the concentration of one of the intermediates in the catalytic cycle can be raised; this is useful if the concentration of this species occurs in the overall rate equation. Secondly, host–guest complexation can serve to enhance the selectivity towards a certain product, or induce chirality. In addition, an entatic state effect might occur when a substrate is bound in the cavity; complexation of the substrate makes the catalyst more active by deforming its surroundings. A potential drawback is strong binding of the product ("product inhibition" in enzyme catalysis), and certainly if our efforts are directed towards selectivity for a certain product the chances are that this product will have the highest binding constant towards our host.

Some early successful examples of application of such a selective structures as catalysts to accelerate some given reaction are presented below.

Breslow's β-cyclodextrine ribonuclease model system represents one of the best examples concerning the construction of small enzyme-like molecules [33]. Breslow functionalized the β-cyclodextrine with two imidazole moieties (Figure 10.1). Selectively, catechol cyclic phosphate carrying a 4-*tert*-butyl group (Figure 10.1a) binds into the cavity of the catalyst (Figure 10.1b) in water solution, and is then hydrolyzed by the

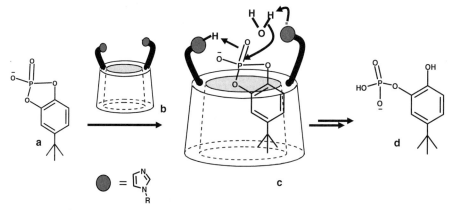

Figure 10.1 Host–guest complexation for enhanced hydrolysis (Breslow).

combined action of one imidazole ring acting as a base and the other one, protonated, acting as an acid. The reaction proceeds in a controlled specific direction, forming the ring-opening ester product (Figure 10.1d) 100 times faster than the non-catalyzed reaction [33c] in a selective fashion.

As mentioned in the introduction (Section 10.1) we returned to supramolecular catalysis in a collaboration with the Nolte group using the host molecule developed by them, often referred to as a "molecular clip", consists of a diphenylglycoluril framework to which two aromatic side-walls are connected via four methylene linkers (Chapter 6, Figure 6.1a, compound **1**) [34]. These hosts are excellent for neutral aromatic guest molecules, such as dihydroxybenzenes. Guest molecules like resorcinol derivatives are clammed between the aromatic walls of the host. The binding is established via hydrogen bonds, between the hydroxyl groups of the guest and the carbonyl groups of the host, and π–π stacking interactions between the aromatic walls of the host and the aromatic ring of the guest. The binding strength towards these types of guests can be up to $K_a > 10^5$ M^{-1} [35]. Initially our goal was to attach two phosphines on top of the clip molecules and employ hydroxyl-allyl-benzenes as substrates in hydroformylation, striving for regional selectivity caused by substrate–receptor interactions. Synthetic difficulties in the phosphine synthesis led us to the synthesis of phosphites, as described in Chapter 6, and the use of hydrogenation and isomerization reactions as our test reaction. While hydroformylation shows almost always a dependence on the substrate concentration, and hence most likely on substrate to catalytic host-binding, this is not always the case in hydrogenation, and the effect on the rate found was very moderate indeed, but binding guests showed an acceleration and non-binding guest a slight retardation of the catalysis [36].

In Amsterdam, eventually, Georg Dol succeeded in making phosphine substituted "basket" molecules **11** based on diphenylglucoluril. Both tetradentate and bidentate phosphines were synthesized. In hindsight the syntheses look straightforward, but much effort went into this as for many steps the conditions had to be discovered, which is not reflected in the publications. This rather tedious job is a plea for supramolecular strategies to synthesize host molecules more rapidly and in larger variation!

Several metal complexes, both cis for $PtCl_2$ and trans for $PdCl_2$, RhCl(CO) and $PdClCH_3$, have been synthesized and identified for diphosphine **11** [37]. Catalyst **12** has been used for the hydroformylation of various allylbenzene substituted substrates (**S1–9**) [38], binding more or less strongly to the diphenylglycoluril host in the presence also of guests **G1–4** [39]. The orientation of the substrates in the cavity, the two hydroxyls pointing downwards at the bottom and the alkene towards the rhodium center, seems to fulfill the minimum requirements and, according to CPK models and simple MM2 studies, it does not seem that large rearrangements are needed to arrive at alkene-rhodium complex formation. Whether the transition state is easily accessible is hard to predict with these models. With few exceptions the hydroformylation reaction is first order in rhodium and in alkene and thus the raised effective molarity should accelerate the reaction (Chapter 1).

The association constants were determined using NMR titrations, monitoring the protons of the aromatic walls of **11** and **13**. These signals shift to a lower ppm value when an aromatic guest is bound. Since the association constants of **11** and **13** were found to be nearly the same, **13** was used in further studies with other guests (Table 10.2, entries 1 and 2). Table 10.2 shows that the strongest host–guest complex is formed between **13** and a resorcinol derivative containing an electron withdrawing group, viz. MeO$_2$C (**G2**). Electron-withdrawing groups increase the acidity of the hydroxyl groups and the π–π stacking interaction [40] between the guest and the aromatic walls of the host, resulting in a higher binding affinity between host and guest. Electron-donating groups on resorcinol [C$_5$H$_{11}$ (**G1**) and C$_3$H$_5$ (**S9**)] result in a negative effect on the association constant.

Table 10.2 Association constants of hosts **13** and **11** with various guests in CDCl$_3$.

Entry	Host	Guest	K (M^{-1})	CIS (ppm)
1	13	G1	2240 (125)	−0.51
2	11	G1	2380 (150)	−0.51
3	13 + Na^{+a}	G1	10500 (2700)	−0.48
4	13	G2	9870 (800)	−0.56
5	13	G3	2550 (100)	−0.53
6	13	G4	N.d.^{b}	–
7	13	S9	1490 (25)	−0.52
8	13	S8	90 (10)	−0.33
9	13	S1–S7	N.d.^{c}	–

^{a} 10 Equivalents of the dodecylbenzene sulfonate salt were added.
^{b} Estimated to be (1 × 10^5) based on a literature value of a similar compound.
^{c} N.d. = not determined, and K estimated to be <100 M^{-1}.

For the hydroformylation, (PPh$_3$)$_3$Rh(H)(CO) with host **11** was used as the catalyst. An excess of PPh$_3$ (stemming from the catalyst precursor) was needed to avoid isomerization, as was found when phosphine-free precursors were used (at the concentrations used even bidentates should be added in excess to prevent substantial exchange with carbon monoxide). Linear to branched ratios of 2 : 1 were obtained and no isomerized alkene could be detected. These results are similar to those obtained by Kalck and coworkers [41]. As expected, catalysis for **11** is slower than that for (PPh$_3$)$_3$Rh(H)(CO) as the host is a bidentate phosphine; catalysis with (PPh$_3$)$_3$Rh(H)(CO) strongly depends on the concentrations of rhodium and PPh$_3$ and comparison of the rates of the two systems does not make sense.

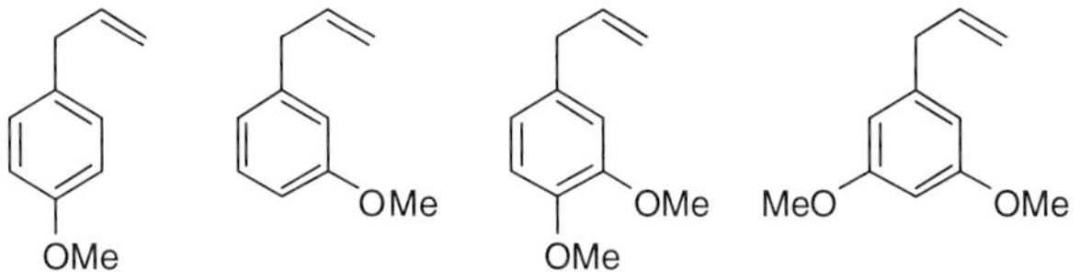

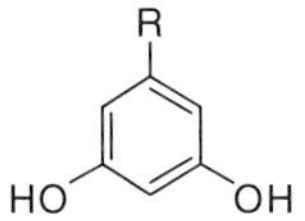

G1: R= C$_5$H$_{11}$
G2: R= CO$_2$Me
G3: R= p-tolyl
G4: R= CN

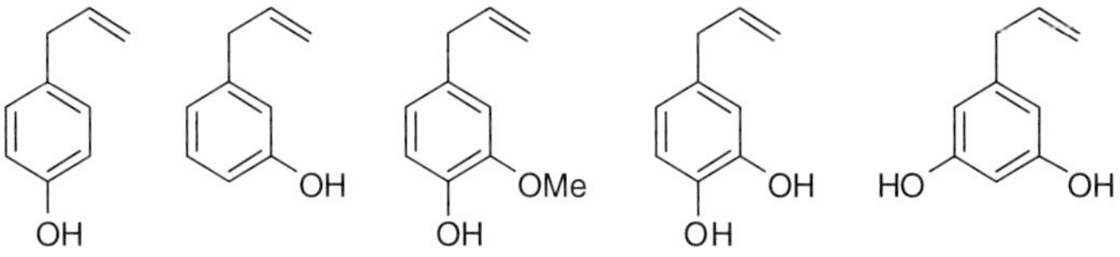

Table 10.3 shows that the more strongly bound substrates **S8** and **S9** give the highest rate, although an increase by a factor of 4 is only obtained, while for the PPh$_3$ catalyst their rate is lower. The latter was ascribed to the formation of inactive rhodium phenolates. **S8** and **S9** also gave higher linearities in host–guest catalysis. The moderate increase in rates for **S8** and **S9** is disappointing since the reaction is usually well behaved and a perfect first order in alkene concentration is found in phosphine catalysts [42]. It was shown by monitoring the reaction that the initial rate for **S9** was high, but at 30% conversion product inhibition took place.

An intriguing side reaction was observed that led to a benzocyclohexene derivative; reactions conducted with presumed intermediates proved that the mechanism leading to their formation was the rhodium acyl intermediate (Scheme 10.5). Most strikingly, though not understood, is that this side-reaction is found for the strongest binding substrates **S8** and **S9**. Thus, we may regard this as a selectivity effect brought about by the host, perhaps by bringing the reacting centers closer together.

Table 10.3 Comparison of the effect of ligand (L) in the hydroformylation of **S1–S9** using PPh$_3$ or **11** in combination with catalyst precursor (PPh$_3$)$_3$Rh(H)(CO).

Substrate	Conversion[a] (%) L = PPh$_3$	l/b	Conversion (%) L = 11	l/b
S1	27.3	1.8	6.6	1.8
S2	22.4	1.5	5.6	1.6
S3	27.6	2	7.4	1.9
S4	26.6	2	9.4	1.8
S5	21.5	1.8	12	2.2
S6	21.5	0.7	11	0.8
S7	26	1.9	10.4	2
S8	16.9	_[b]	21	_[b]
S9	12.5	_[b]	26	_[b]

[a] Conversion after 30 min reaction time, toluene 20 mL, $T = 60\,°C$, $p(CO/H_2) = 20$ bar; CO/H$_2$ = 1, [PPh$_3$]/[Rh] = 3, [11]/[Rh] = 1.5, [substrate]/[Rh] = 350–400, [Rh] = 2×10^{-4} M.
[b] Due to side product formation the l:b ratio could not be calculated, but <5% of the product mixture consisted of branched aldehyde.

In addition, in competition experiments **S9** is converted more quickly than any of the other substrates. Inhibition was also observed when guests **G1–4** were added, with only a slight effect for **G1** but strong inhibition by the strongly binding guest **G4**.

Table 10.4 shows the conversions ($t = 0.5$ h) of **S9** in the presence of various additives. The effect of the addition of resorcinol derivatives on the conversion of a "bound" substrate was also investigated (Table 10.4). First, olivetol (**G1**) was added to the experiments with **S4** and **S9**. Addition of one equivalent of **G1** per **S4** had no effect on the conversion of the latter substrate. When five equivalents were added the conversion increased from 9.4% (without any additives present) to 14.4%. A rather surprising result since we know that hydroxyl groups cause a decrease in conversion when triphenylphosphine is used as ligand in combination with precursor (PPh$_3$)$_3$Rh (H)(CO). Even more surprising was the observation that one equivalent of **G1** added to **S9** resulted in an increase in conversion from 26% to 37%. Addition of five

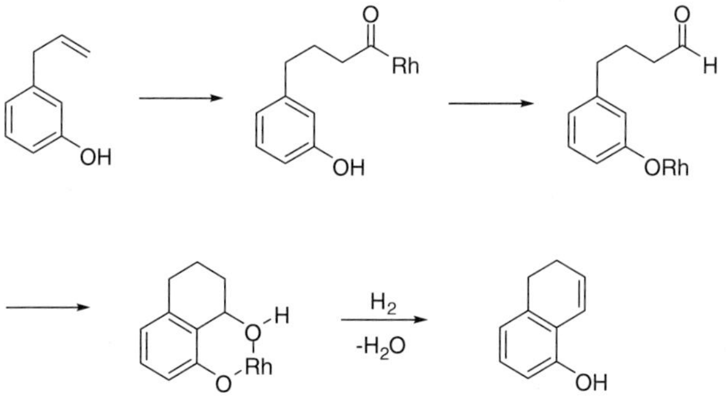

Scheme 10.5 Formation of side product in host **11**.

Table 10.4 Effect of additives on the hydroformylation of **S9** using host **11** in combination with catalyst precursor $(PPh_3)_3Rh(H)(CO)$.

Additive (equiv. per S9)	Conversion S9 (%)[a,b]
–	26
G1 (1)	37
G1 (5)	29.3
G2 (0.2)	30.2
G2 (1)	27.7
G2 (5)	4.5
G3 (1)	39.7
G4 (0.2)	5.9
Product of **9** (1)	11.5

[a] Conversion after 0.5 h reaction time, toluene 20 mL, $T = 60\,°C$, $p(CO/H_2) = 20$ bar; $CO/H_2 = 1$, [**11**]/[Rh] = 1.5, [substrate]/[Rh] = 350–400, [Rh] = 2×10^{-4} M.
[b] Due to side product formation the l:b ratio could not be calculated, but <5% of the product mixture consisted of branched aldehyde (according to ^{1}H NMR).

equivalents did not result in a further increase; instead a decrease in conversion to 29.3% was found. Perhaps this behavior can be ascribed to cooperative binding (Chapter 6, Figure 6.3).

A general conclusion that can be drawn for this approach is that indeed the synthetic effort that goes into a covalently attached catalyst to a host molecule is enormous, and at the end it gives us only one catalyst. The risk therefore in this hit or miss approach is substantial and it is important to build into the plans sufficient modularity (see Chapter 9, the work of the Takacs group) or choose a supramolecular approach also for making the host molecules. The latter is still somewhat limited, since the geometry of a host and changes in the building blocks lead to discontinuous rather than to continuous changes.

10.4
Ionic Interactions as a Means to Form Heterobidentate Assembly Ligands

The use of heterocombinations of monodentate ligands in bis-ligated complexes has become an extremely powerful tool recently. Combinatorial approaches and high-throughput experimentation can be applied far more expediently when covalent synthesis of bidentates becomes redundant. In the absence of steric or electronic preferences for either combination, mixing of two monodentate ligands will give a statistical mixture of the three combinations. The approach has been proved to be useful in asymmetric hydrogenation [43], asymmetric C–C bond formation [44], and hydroformylation [45].

While the formation of heterocombinations can be left at mere chance, almost simultaneously the idea emerged that the population of the heterocombinations could be increased by noncovalent interactions between the monophosphorus compounds. This approach is based on the utilization of monodentate ligands functionalized by complementary binding sites. Such monophosphorus compounds will form assemblies with one another, and the equilibrium of the complexes described above can be shifted to the desired hetero-complex. The first examples present the formation and application of heterobidentate ligands assembled by Lewis-type acid–base interactions (Chapter 8) [46]. The building blocks are monophosphites having Zn(II)-porphyrin binding motifs and different monophosphorus ligands with nitrogen donor functionalities. The assembly ligands have been tested in asymmetric allylic alkylation [46a,b], hydroformylation [46b], and asymmetric hydrogenation [46c]. Most recently complementary H-bond motifs have been exploited to assemble monophosphorus ligands (Chapter 2) [47]. It has been demonstrated that the applied aminopyridine and isoquinolone functionalities form a pair of strong hydrogen bonds, and the attached phosphine and phosphonite units coordinate to the active metal center in a heterobidentate fashion. These ligand systems have been utilized in rhodium-catalyzed hydroformylation of 1-octene [47a], as well as in the asymmetric hydrogenation of several substrates [47b].

It occurred to us that ionic interactions might be a highly suitable binding motif to enforce the formation of heterobidentate ligand combinations [48]. The assembly ligand $14^-/15^+$ has been formed from the well-known TPPMS (**14**, monosulfonated triphenylphosphine sodium salt) and 3-(diphenylphosphinyl)aniline hydrochloride (**15**) by a simple ion-exchange reaction (Scheme 10.6). The coordination behavior of the ion-pair $14^-/15^+$ has been tested with various transition metal complexes. Other

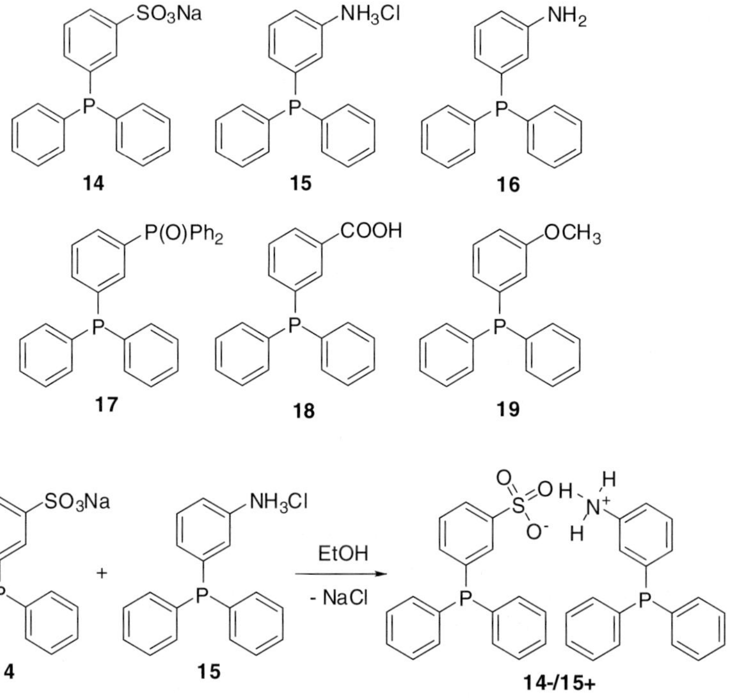

Scheme 10.6 Phosphorus ligands involved in the study of ionic interactions.

Table 10.5 Reaction of triphenylphosphine derivatives and Pt(1,5-cod)Cl$_2$.[a]

Phosphine	Solvents	Products	^{31}P{^{1}H} (ppm)	Product distribution (%)
14	CD$_3$OD	cis-PtL$_2$Cl$_2$	17.07 (s, J_{PtP} = 3702 Hz);	78
		PtL$_3$Cl	15.40 (t, J_{PP} = 19.8 Hz,	16
			J_{PtP} = 3664 Hz, 1P),	
		cis-PtL$_2$(CD$_3$OD)$_2$	25.91 (d, J_{PP} = 19.8 Hz,	6
			J_{PtP} = 2511 Hz, 2P);	
			22.92 (s, J_{PtP} = 3123 Hz)	
15	CDCl$_3$	cis-PtL$_2$Cl$_2$	17.18 (s, J_{PtP} = 3696 Hz)	100
16	CDCl$_3$	cis-PtL$_2$Cl$_2$	17.56 (s, J_{PtP} = 3694 Hz)	100
17	d_7-DMF	cis-PtL$_2$Cl$_2$	18.00 [s, J_{PtP} = 3662 Hz,	100
			P(III)], 27.71 [s, P(V)]	
18	CDCl$_3$:CD$_3$OD = 5:1	cis-PtL$_2$Cl$_2$	17.14 (s, J_{PtP} = 3684 Hz)	95
		trans-PtL$_2$Cl$_2$	23.80 (s, J_{PtP} = 2640 Hz)	5
19	CDCl$_3$	cis-PtL$_2$Cl$_2$	17.72 (s, J_{PtP} = 3674 Hz)	100

[a] 0.04 mmol of Pt(1,5-cod)Cl$_2$ and 0.084 mmol of TPP derivative were transferred in a Schlenk tube, then the solvent was added, and the mixture was stirred for 10–30 min. L = phosphine.

monophosphines (16–19) holding functional groups capable of hydrogen bonds have also been tested, to compare the efficiency of different types of noncovalent interactions.

All monophosphorus ligands used are meta-substituted triphenylphosphine derivatives. Their structural similarity ensures that they will also have similar binding properties to transition metal central atoms, and the distribution of the complexes will be influenced by the interactions of the functional groups only.

Firstly, reactions of phosphines 14–19 with Pt(1,5-cod)Cl$_2$ were studied. Notably, using this precursor, under the applied conditions, each of the monophosphines form cis-Pt(phosphine)$_2$Cl$_2$ as the major product (Table 10.5; Figure 10.3A below). These results suggest that the geometry of the platinum complexes does not refer necessarily to attractive ligand–ligand interactions [49].

The statistical distribution of cis-PtL_aL_bCl$_2$, cis-Pt(L$_a$)$_2$Cl$_2$, cis-Pt(L$_b$)$_2$Cl$_2$ has been modeled using a mixture of anisyldiphenylphosphine (19) and triphenylphosphine (TPP). Significant attractive interaction could not be expected, and, accordingly, cis-Pt(19)(TPP)Cl$_2$, cis-Pt(19)$_2$Cl$_2$, cis-Pt(TPP)$_2$Cl$_2$ were observed by ^{31}P{^{1}H} NMR in a ratio of 2:1:1 (Table 10.6). Although in the case of 16/18, 17/18 and 15/17 ligand mixtures formation of COOH···NH$_2$ and COOH···O=P, NH$_3^+$···O=P hydrogen bonds might have been anticipated [50], under the applied circumstances no increase of the heterocombinations was achieved. An interesting phenomenon was observed, however, during the reaction of Pt(1,5-cod)Cl$_2$ and a 1:1 mixture of m-(diphenylphosphinyl)aniline (16) and its hydrochloride (15). The ^{31}P{^{1}H} NMR spectrum displays only one singlet, at 16.7 ppm with a coupling constant of $^1J_{Pt-P}$ = 3684 Hz. These NMR characteristics suggest that a fast inter-/intramolecular proton-exchange among the nitrogen atoms render the two cis-positioned phosphines equivalent on the NMR time-scale. An X-ray study of the crystals obtained from the reaction mixture confirms the formation of cis-Pt(15)(16)Cl$_2$, although, in the solid state, NH$_3^+$···Cl$^-$ interactions are observed instead of NH$_3^+$···NH$_2$ hydrogen bonds (Figure 10.2).

Table 10.6 Attempts to form assemblies of TPP derivatives via hydrogen bonds.[a]

Phosphines	Functional groups	Solvents	Products	Distribution (%)
TPP, 19[b]	–, OCH$_3$	[CDCl$_3$] : {DMF} = [1 mL] : {1 mL}	cis-PtL$_a$L$_b$Cl$_2$;	50
			cis-Pt(L$_a$)$_2$Cl$_2$;	25
			cis-Pt(L$_b$)$_2$Cl$_2$	25
16, 18[b]	NH$_2$, COOH	[CDCl$_3$] : {DMF} = [1 mL] : {1 mL}	cis-PtL$_a$L$_b$Cl$_2$;	50
			cis-Pt(L$_a$)$_2$Cl$_2$;	25
			cis-Pt(L$_b$)$_2$Cl$_2$	25
16, 18[c]	NH$_2$, COOH	CDCl$_3$: THF = 1 mL : 0.1 mL	cis-PtL$_a$L$_b$Cl$_2$;	50
			cis-Pt(L$_a$)$_2$Cl$_2$;	25
			cis-Pt(L$_b$)$_2$Cl$_2$	25
17, 18[b]	P=O, COOH	[CDCl$_3$] : {DMF} = [1 mL] : {1 mL}	cis-PtL$_a$L$_b$Cl$_2$;	50
			cis-Pt(L$_a$)$_2$Cl$_2$;	25
			cis-Pt(L$_b$)$_2$Cl$_2$	25
17, 18[c]	P=O, COOH	[CDCl$_3$] = 1 mL	cis-PtL$_a$L$_b$Cl$_2$;	50
			cis-Pt(L$_a$)$_2$Cl$_2$;	25
			cis-Pt(L$_b$)$_2$Cl$_2$	25
15, 17[b]	P=O, NH$_3^+$	[CDCl$_3$] : {DMF} = [1 mL] : {1 mL}	cis-PtL$_a$L$_b$Cl$_2$;	50
			cis-Pt(L$_a$)$_2$Cl$_2$;	25
			cis-Pt(L$_b$)$_2$Cl$_2$	25
15, 16[b]	NH$_2$, NH$_3^+$	[CDCl$_3$] : {DMF} = [1 mL] : {1 mL}	cis-PtL$_a$L$_b$Cl$_2$[d]	100

[a] 0.04 mmol of Pt(1,5-cod)Cl$_2$, and 0.042 mmol of both phosphines.
[b] The solution of the ligand mixture (solvent in square brackets, [ ]) was added to the solution of the precursor (solvent in braces, { }), and the reaction mixture was stirred for 10–30 min.
[c] The solution of the ligand mixture was added to the solid precursor, and the reaction mixture was stirred for 10–30 min.
[d] The two coordinated phosphines are equivalent on the NMR time-scale.

In contrast to the mixtures of TPP derivatives capable of hydrogen bonds only, the phosphine ion-pair 14$^-$/15$^+$ did increase the formation of the hetero-complex, cis-Pt(14$^-$/15$^+$)Cl$_2$, up to as high as 97% (Table 10.7).* The byproduct is the ion-pair formed from the homocombinations: [cis-Pt(14$^-$)$_2$Cl$_2$][cis-Pt(15$^+$)$_2$Cl$_2$] (14$^-$ = 14–Na$^+$, 15$^+$ = 15–Cl$^-$).† Its relative proportion depends on the solvents used, the

* ^{31}P{^{1}H} NMR characteristics of cis-Pt(14$^-$/15$^+$)Cl$_2$ varies slightly with the solvent. Representative ^{31}P{^{1}H} NMR data in CDCl$_3$: CD$_3$OD (v/v = 5): 15.50 (d, J_{PP} = 15.9 Hz, $^1J_{PtP}$ = 3607 Hz); 20.87 (d, J_{PP} = 15.9 Hz, $^1J_{PtP}$ = 3771 Hz).

† ^{31}P{^{1}H} NMR characteristics of [cis-Pt(14$^-$)$_2$Cl$_2$][cis-Pt(15$^+$)$_2$Cl$_2$] varies slightly with the solvent. The resonances of the anionic and cationic complex overlap in several solvent systems. Representative ^{31}P{^{1}H} NMR data in CD$_3$OD : DMF (v/v = 1): 17.18 (s, J_{PtP} 3686 Hz); 17.36 (s, J_{PtP} = 3690 Hz).

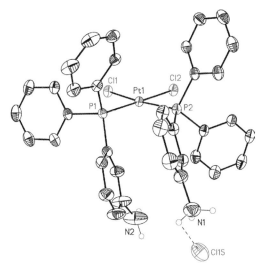

Figure 10.2 X-ray structure of *cis*-Pt(**15**)(**16**)Cl$_2$.

concentration of the reaction mixtures, and the way how the complexes are made (Table 10.7). It is particularly remarkable that the ion-pair **14**$^-$/**15**$^+$ formed 77% of the heterocombination even in strongly polar, protic solvents, such as a 1 : 1 mixture of MeOH–DMF (Table 10.7). The X-ray structure analysis confirms the cis-coordination and the expected NH$_3^+$···$^-$O$_3$S ionic bond (Figure 10.3B). This ionic bond is not necessarily the cause of the *cis*-coordination. Both the NMR data and the X-ray structure of the *cis*-Pt(**19**)$_2$Cl$_2$ (Figure 10.3A) reveal that triarylphosphines lacking complementary functional groups can form *cis*-Pt(phosphine)$_2$Cl$_2$ complexes with high selectivity [51].

To survey the coordination properties of the ionic assembly ligand **14**$^-$/**15**$^+$, Pd (1,5-cod)(CH$_3$)Cl was chosen as precursor. The reaction of phosphines with this transition metal complex is greatly influenced by their structure. Diphosphines of small bite-angle give *cis*-complexes, while trans-spanning diphosphines and monophosphines favor the trans-coordination mode. The different coordination modes result in rather distinctive ^{31}P{^{1}H} NMR spectra. The reaction of **14**$^-$/**15**$^+$ and Pd (1,5-cod)(CH$_3$)Cl could be studied in relatively apolar halogenated hydrocarbons (CD$_2$Cl$_2$ and CDCl$_3$) without addition of polar co-solvents. In the ^{31}P{^{1}H} NMR spectrum the chemical shifts of the coordinated phosphines are very similar, resulting in a single broad resonance at 34.4 ppm. This pattern excludes *cis*-coordination, and the trans-position of the phosphines has also been proved by X-ray studies (Figure 10.4B). Most interestingly, the structure reveals that in this coordination mode the ionic bond can be as efficient as in the case of cis-coordination. The conformation of the coordinated ionically assembled ligand is quite different to that of monophosphines lacking complementary binding motifs (Figure 10.4A), and rather resembles the coordination of trans-spanning bisphosphines.

Table 10.7 Reactions of the ionic assembly $14^-/15^+$ and Pt(1,5-cod)Cl$_2$.[a]

Method[b]	Solvents	Products	Distribution (%)
A	CDCl$_3$: CD$_3$OD 1 mL : 0.2 mL	cis-Pt($14^-/15^+$)Cl$_2$;	83
		[cis-Pt(14^-)$_2$Cl$_2$]	
		[cis-Pt(15^+)Cl$_2$]	17[c]
A	Deuterated DMF (1 mL)	cis-Pt($14^-/15^+$)Cl$_2$;	80
		[cis-Pt(14^-)$_2$Cl$_2$]	10
		[cis-Pt(15^+)Cl$_2$]	10
A	THF : d7-DMF = 9 mL : 1 mL	cis-Pt($14^-/15^+$)Cl$_2$;	97
		[cis-Pt(14^-)$_2$Cl$_2$]	1.5
		[cis-Pt(15^+)Cl$_2$]	1.5
B	[CDCl$_3$] : {CDCl$_3$: CD$_3$OD} = [0.5 mL] : {0.5 mL : 0.2 mL}	cis-Pt($14^-/15^+$)Cl$_2$;	85
		[cis-Pt(14^-)$_2$Cl$_2$]	
		[cis-Pt(15^+)Cl$_2$]	15[c]
B	[DMF] : {DMF} = [0.5 mL] : {0.5 mL}	cis-Pt($14^-/15^+$)Cl$_2$;	80
		[cis-Pt(14^-)$_2$Cl$_2$]	10
		[cis-Pt(15^+)Cl$_2$]	10
B	[DMF] : {DMF} = [1 mL] : {1 mL}	cis-Pt($1^-/2^+$)Cl$_2$;	83
		[cis-Pt(1^-)$_2$Cl$_2$]	8.5
		[cis-Pt(2^+)Cl$_2$]	8.5
B	[CDCl$_3$] : {DMF} = [1 mL] : {1 mL}	cis-Pt($14^-/15^+$)Cl$_2$;	85
		[cis-Pt(14^-)$_2$Cl$_2$]	7.5
		[cis-Pt(15^+)Cl$_2$]	7.5
B	[CD$_3$OD] : {DMF} = [1 mL] : {1 mL}	cis-Pt($14^-/15^+$)Cl$_2$;	77
		[cis-Pt(14^-)$_2$Cl$_2$]	11.5
		[cis-Pt(15^+)Cl$_2$]	11.5
B	[CHCl$_3$: CDCl$_3$] : {CHCl$_3$: DMF} = [4 mL : 1 mL] : {4 mL : 1 mL}	cis-Pt($14^-/15^+$)Cl$_2$;	90
		[cis-Pt(14^-)$_2$Cl$_2$]	5
		[cis-Pt(15^+)Cl$_2$]	5

[a] 0.04 mmol of Pt(1,5-cod)Cl$_2$ and 0.042 mmol of $14^-/15^+$.
[b] Method A: The precursor and the assembly ligand were transferred in a Schlenk tube, then the solvent was added, and the mixture was stirred for 10–30 min. Method B: The solution of the assembly ligand $14^-/15^+$ (solvent in square brackets, []) was added to the solution of the precursor (solvent in braces, {}), and the reaction mixture was stirred for 10–30 min.
[c] The resonances of the anionic and cationic complex overlap.

Finally, the reaction of $14^-/15^+$ and [Rh(CO)$_2$Cl]$_2$ was studied. To the best of our knowledge, the coordination behavior of bisphoshines assembled by noncovalent interactions has not been tested with dimeric precursors up to now. Despite the supposedly more complex reaction mechanism, a single product was observed in the ^{31}P{^{1}H} NMR spectrum, with $14^-/15^+$ coordinated as a heterobidentate ligand. The characteristics of the strongly distorted ABX spin-system (36.1 ppm, $^1J_{Rh-P}$ = 125 Hz, $^2J_{P-P}$ = 357 Hz; 38.7 ppm, $^1J_{Rh-P}$ = 125 Hz, $^2J_{P-P}$ = 357 Hz) indicate a trans-position of the two phosphorus atoms. An X-ray structure analysis of Rh($14^-/15^+$)(CO)Cl (Figure 10.5) is very similar to that of the palladium complex Pd($14^-/15^+$)(CH$_3$)

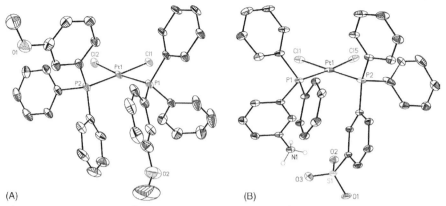

Figure 10.3 X-Ray structure of cis-Pt(**19**)$_2$Cl$_2$ (A) and cis-Pt(**14**$^-$/**15**$^+$)Cl$_2$ (B).

Cl, confirming that the trans coordination mode allows the formation of a strong NH$_3^+$···$^-$O$_3$S electrostatic interaction.

In conclusion, the *ionic bond* as driving force for the formation of a heterobidentate bisphosphine seems a useful tool. Sulfonation of one ligand, and nitration followed by reduction to amine, may afford a relatively easy entry into this chemistry. Several

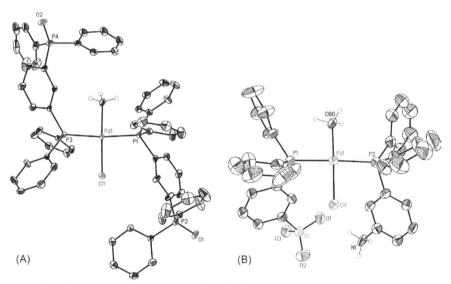

Figure 10.4 X-Ray structure of *trans*-Pd(**17**)$_2$(CH$_3$)Cl (A): no significant interaction between the functionalized TPP derivatives; X-ray structure of *trans*-Pd(**14**$^-$/**15**$^+$)(CH$_3$)Cl (B): **14**$^-$/**15**$^+$ acts as a trans-spanning assembly ligand, maintaining the ionic interaction in the trans-position.

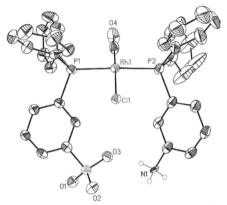

Figure 10.5 X-Ray structure of trans-Rh(**14**$^-$/**15**$^+$)(CO)Cl.

test reactions have shown that the cationic and anionic monodentate phosphorus ligands form an ion-pair even in strongly polar, protic medium. The ligand design, position of the ionic functional groups, allows both cis- and trans-coordination of **14**$^-$/**15**$^+$ in square planar transition metal complexes.

10.5
Ditopic Ligands for the Construction of Bidentate Phosphine Ligands

Several noncovalent interactions used for the construction of bidentate ligands suffer from relatively small interaction energies, and, for instance, in the case of hydrogen bonds as the binding motive the complexation energies involved in the binding of the catalytic metal to phosphorus donor atoms are much larger. Nevertheless, hydrogen bonding can be highly effective. For example, Breit (see Chapter 2) reported on "*in situ*" generated bidentate ligands using self-assembly through hydrogen bonding of monodentate ligands [52]. As a base for hydrogen bonding, the tautomeric equilibrium of 2-pyridone and 2-hydroxypyridine was used. The formation of "chelating" complexes is mediated by the conjugated dual hydrogen-bonding but the phosphine metal coordination is the main driving force for complex formation. Highly active and selective catalysts were obtained.

Recently, we [53] and others [54] simultaneously reported an example of a complex in which the transition metal dictates the coordination mode, viz. urea-functionalized phosphine **20**, which forms a trans palladium complex, complemented by hydrogen bonding of the urea fragments (Figure 10.6). Bear in mind that any monophosphine, and even wide bite angle diphosphines, give trans complexes with a hydrocarbyl palladium halide. More interestingly, the urea moieties can function as a host for another halide ion.

Palladium dichloride forms more easily both cis and trans complexes with monophosphines and it provides a better example. Yam et al. described cis to trans isomerization of palladium dichloride and a similar ligand to **20** containing an

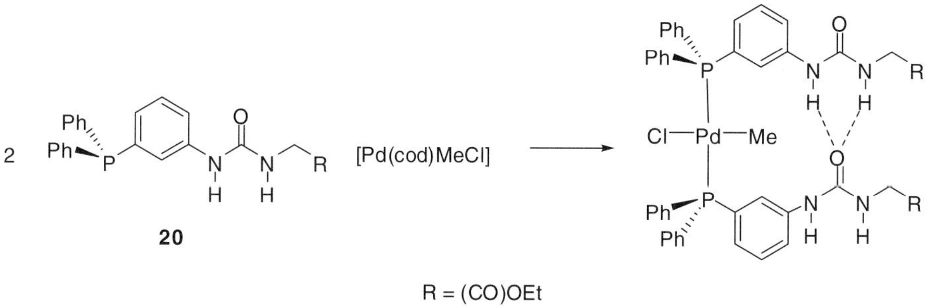

10.5 Ditopic Ligands for the Construction of Bidentate Phosphine Ligands

Figure 10.6 Urea motif for making bidentate ligands.

acetanilide at one of the phenyl groups in the para position, in which chloride addition to a solution of the trans complex transformed it into the cis one, with the acetanilide fragments serving as the chloride host [55].

Still, the hydrogen bond energies are small and as a result the phosphine–metal interactions may often determine which complex will form preferentially. It may be useful therefore to use the strong interactions known in coordination chemistry not only for the catalytically active transition metal part but also for the construction of the supramolecular architectures [56–58]. In recent years this has been extensively explored. These structures are spontaneously generated by simply mixing the components (anionic or neutral ligands and metal salts) in solution. Knowledge of coordination chemistry together with symmetry considerations applied to the linker fragments allows the rational design of various nanoscale systems with desirable shapes, sizes, and, ultimately, functions.

The ability of a metallic ion to organize a flexible ligand around its coordination sphere has led to the design of several intramolecularly organized recognition sites [59]. For example, the flexible bis(aminomethyl)pyridine derivative (**21**) (Figure 10.7), developed by Scrimin, is organized by Cu^{2+} binding to the tridentate binding site. This leads to enhanced complexation of a second Cu^{2+} ion at the distal amine site. The resulting bis Cu^{2+} complex shows selective hydrolysis of β-amino acid esters due to the cooperative effect of the two metal ions [60,61].

Figure 10.8 shows the example of a ruthenium(II) bis-terpyridine complex that acts as the scaffold for a biscarboxylate recognition site **22** [62].

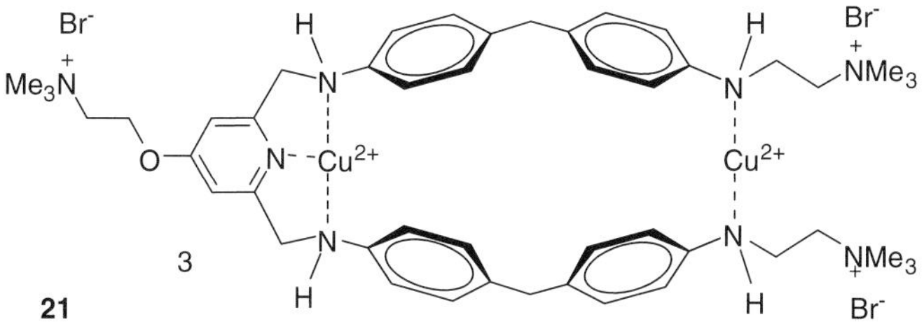

Figure 10.7 Cooperative binding of Cu in two different sites.

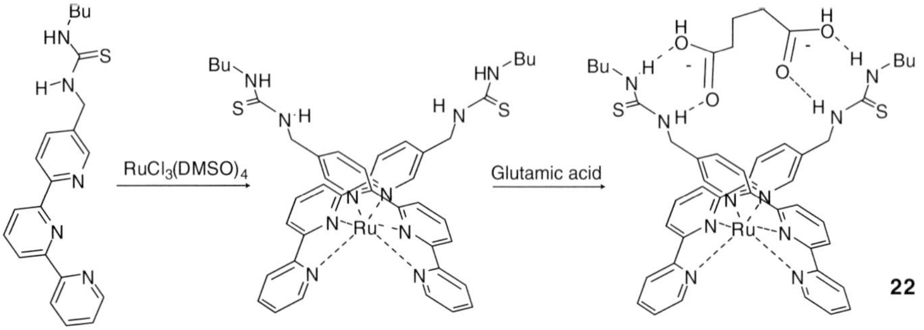

Figure 10.8 Metal template self-assembly of artificial recognition sites.

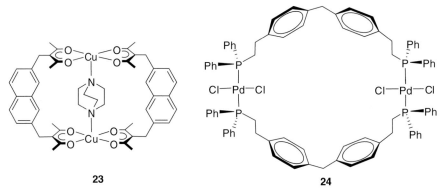

Figure 10.9 Bimetallic supramolecular structures.

The use of two or more metal centers in the self-assembly receptors also offers the possibility of closing a macrocyclic structure, conferring new properties on these potential molecules, which were pioneered by Fujita and, meanwhile, developed into an intriguing area on its own. The self-assembled macrocyclic host **23** for DABCO is an early simple example (Figure 10.9) [63]. Another example is the macrocyclic dinuclear Pd(II) complex **24** reported by Fujita [64], which was prepared by simply mixing $PdCl_2$ and a diphosphine bridging ligand (Figure 10.9) [65].

The first example of a self-assembled macrocyclic complex possessing cis-protected Pd(II) blocks is the tetranuclear square compound **25** (Figure 10.10) [66].

Yam and coworkers have reported a new trans–cis isomerization of dichloropalladiun(II) phosphanylcrown complexes that is induced by the binding of alkali-metal ions (**26–28**, Figure 10.11) [67]. The addition of K^+ ions to **26** produced an increase in the population of the cis isomer.

Crown-ether containing triarylphosphines have been studied in the Shell laboratories in an attempt to find a favorable effect of metal ion complexation in rhodium-catalyzed hydroformylation [68]; no effect was found, although precedents exist in which the metal ion acts as a Lewis acid that activates coordinated carbon monoxide towards methyl migration [69].

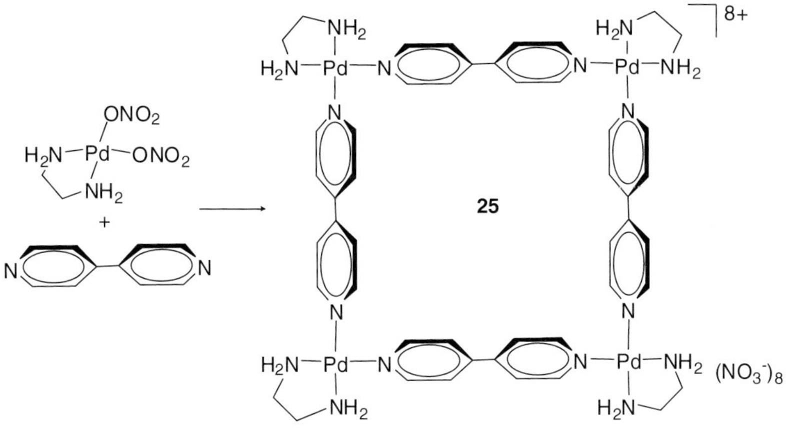

10.5 Ditopic Ligands for the Construction of Bidentate Phosphine Ligands

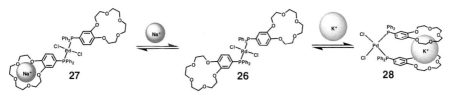

Figure 10.10 Tetrametallic supramolecular structure.

Figure 10.11 Effect of cations on cis–trans equilibrium.

While today the construction of such bidentate phosphine ligands with the use of an assembly metal is referred to as supramolecular chemistry, this is not to say that there are no examples in older literature utilizing this principle. Actually, there are many and using as a search term "hetero bimetallic complexes" (a named coined in the early 1980s) leads us to a plethora of complexes. For instance in complex **29**, reported by Rauchfuss in 1982, one would consider copper as the assembly metal and platinum as the potential catalytic metal [70].

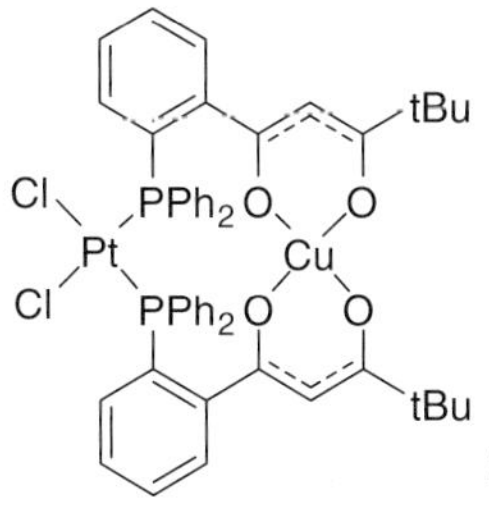

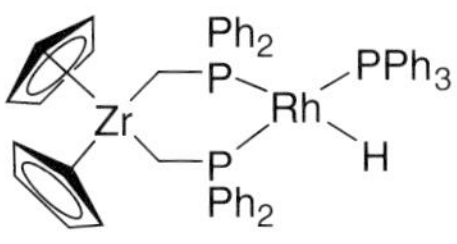

Early and late transition metal combinations were explored by Casey [71] and many others. Reaction of the bis(phosphinomethyl) complex $Cp_2Zr(CH_2PPh_2)_2$ (Cp = cyclopentadienyl) on $Rh(H)(PPh_3)_4$ gave the bimetallic hydride $Cp_2Zr(CH_2PPh_2)_2Rh(H)(PPh_3)$ (**30**), the crystal structure of which was determined [72].

The most remarkable structural features of **30** are the trans chelating arrangement of the bis[(diphenylphosphino)methyl]zirconium on rhodium together with pentacoordination around zirconium due to a strong interaction of Zr on the Rh–H bond with a Zr···H distance of 2.28(5) Å. The bimetallic system is an efficient precursor for selective catalytic hydroformylation of 1-hexene [73]. One might argue that the zirconium template has actually a covalent nature, as our choices become less clear now (e.g., what to do with dppf one might ask, which is a diphosphine containing iron as the assembly metal!).

Recently, several reports have appeared in which self-assembly is used to generate families of novel bidentate ligands [74], and catalytic systems [75]. Takacs et al. reported the synthesis of bidentate phosphines ligands (Chapter 9) based on self-assembly of two ditopic units, binding a structural metal to form a heterodimeric chelating ligand system [76]. Large libraries of ligands can be conveniently obtained this way and asymmetric allylic alkylation was studied as a test reaction. The ees obtained cover a surprisingly wide range and we conclude that, despite the similarity of the ligands, the backbone variation gave rise to a diverse library.

Mirkin and coworkers reported on catalytic molecular tweezers used in the asymmetric ring opening of cyclohexene oxide. In this case the early transition metal is the catalyst and rhodium functions as the structural inductor metal. The catalyst consists of two chromium salen complexes, the reaction is known to be bimetallic, and a switchable rhodium complex, using carbon monoxide as the switch. Indeed, when the salens are forced in close proximity in the absence of CO the rate is twice as high and the effect is reversible [77].

The early–late transition metal complex of Raymond et al. is interesting in that it requires both metal atoms to form the basic C_3 structure (ideally D_{3h} may form). Titanium complexation to three catechol units leads to a tridentate ligand, and, only when palladium bromide is added, trans coordination to palladium gives the agglomerate **31** [78].

We were particularly interested in complex **32** reported by Boerner [79], an early–late transition metal complex that was used in hydroformylation, aiming at effects that the second metal (titanium) might bring about. In this instance the chiral titanium fragment may induce chirality, but only the racemic product was obtained. The activity of the catalyst was rather moderate and, perhaps, the rhodium center was too close to the ionic titanium center, thereby hindering the formation of rhodium hydrides.

In the project described below [80] we wanted to expand on the approaches outlined above, but aimed at a catalytic center more remote from the assembly metal yet still possessing a relatively rigid structure, enabling the use of molecular mechanics to design the assemblies. For bidentate phosphines, one of the parameters most widely used is the natural bite angle. This natural bite angle (β_n), introduced by Casey and Whiteker [81], can be defined as the "ligand-preferred" P–M–P angle, and can be easily obtained by using simple molecular mechanics calculations [82]. The relationship between the bite angle of chelating diphosphines and the activity and/or selectivity of the generated systems has been studied and reviewed for many catalytic reactions, and the effect is often impressive [83].

Despite the examples already known of ligands having wide bite angles, their synthesis with covalent bonds remains tedious and supramolecular chemistry is a powerful tool for the expedient generation of libraries of ligands (Chapters 8 and 9) with different bite angles and other properties.

Salen type ligands **32** as used by Boerner were selected as bidentate monoanions for the assembly part. This should allow us to control the bite angle of the chelating diphosphine ligand (Figure 10.12), an area of particular interest to us, and which is especially useful in obtaining selective and active catalysts. The present approach gives the opportunity to access libraries of ligands closely related to our designed entities. The assembly metal hybridization (tetrahedral, square-planar, octahedral) gives further means of variation at low synthetic effort.

These assembly ligands will be tested in suitable catalytic reactions that leave the assemblies intact. Salt-forming reactions are not attractive as the salts might interact with the assembly, nor is the use of catalytic metals that compete with the assembly metal for the salen type positions in the ditopic ligand; ideally, all potential problems can be avoided if the same metal could be used. Rhodium-catalyzed hydroformylation of 1-octene is a suitable reaction, with the only disadvantage that high pressures are needed, but hydrogen or CO do not interfere with our assemblies. Metal salts do not interfere with the rhodium hydrides involved in the hydroformylation catalysis, as for instance the most effective industrial process today for propene hydroformylation

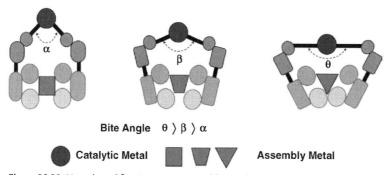

Figure 10.12 Ligand modification via an assembly metal.

(Ruhrchemie-Rhone Poulenc) functions in water in the presence of large amounts of sodium sulfonates.

Scheme 10.7 depicts the ditopic ligands synthesized and published so far. The phosphine moiety is sufficiently out of the way of the assembly part, but its position on the aryl group restricts the bite angle upon complex formation. In the X-ray structures (see below) the phosphorus–phosphorus distance is far too long for the ligands to act in bidentate fashion, but rotation of the 3-$(Ph_2P)C_6H_4$ group brings the phosphorus atoms to the desired position. Its position meta with respect to nitrogen is favored over the para position, which limits complex formation drastically. Molecular mechanics indicated the preferred bite angles of such ligands were close to 110°. Many catalytic reactions carried out with catalysts containing bidentate phosphorus ligands are highly sensitive to the bite angle of the diphosphine. Examples are rhodium-catalyzed hydroformylation [84], nickel- and palladium-catalyzed hydrocyanation [85], and palladium-catalyzed cross-coupling reactions [86], in which either the rate or the selectivity was considerably improved by the use of, for instance, Xantphos, which has a natural bite angle [81] of ~110°.

Ligands **33–35** are based on 3-diphenylphosphino-2-hydroxy-5-methylbenzaldehyde (**45**), already reported (without the 5-methyl group) by Boerner who used it, assembled with an asymmetric diamine and $Ti(OiPr)_4$, to give Salenophos, the ligand in **32**, for asymmetric rhodium-catalyzed hydroformylation. For the synthesis of **33–35** the respective amines were reacted with **45**. Ditopic ligands **36–38** are made via a condensation of the respective aldehyde and 3-diphenylphosphinoaniline. In general the Schiff base condensations were carried out in refluxing toluene in the presence of molecular sieves. The assembled bidentate phosphine ligands **39–44** were made by reaction of **33–38** with $Zn[N\{Si(Me)_3\}_2]_2$, $Zn(OAc)_2$, $ZnCl_2$ and n-butyllithium, or $Ti(OiPr)_4$. Molecular modeling calculations showed that, in particular, assemblies **42–44** might be interesting as wide bite angle ligands. On the one hand, this approach allows screening of large numbers of catalysts but, on the other hand, one might choose sophisticated guesses; indeed, **39–41** are less promising as we will see below, both for structural and for reactivity reasons. Note that **40** and **41** are isostructural with SPANphos (**46**), which gave several wide-bite angle complexes [87], in which the spiro carbon atom fragment $(CH_2)_2C(O)_2$ has been replaced by the zinc fragment $(=NR)_2Zn(O)_2$.

For several of these bidentate phosphine assemblies crystal structures were obtained. Figure 10.13 (Table 10.8) shows the crystal structure of octahedrally coordinated titanium complex **39**.

The molecular structure of **40** (Figure 10.14) shows two independent ditopic ligands. The zinc metal is four-coordinate to the Schiff base ligands (each of them

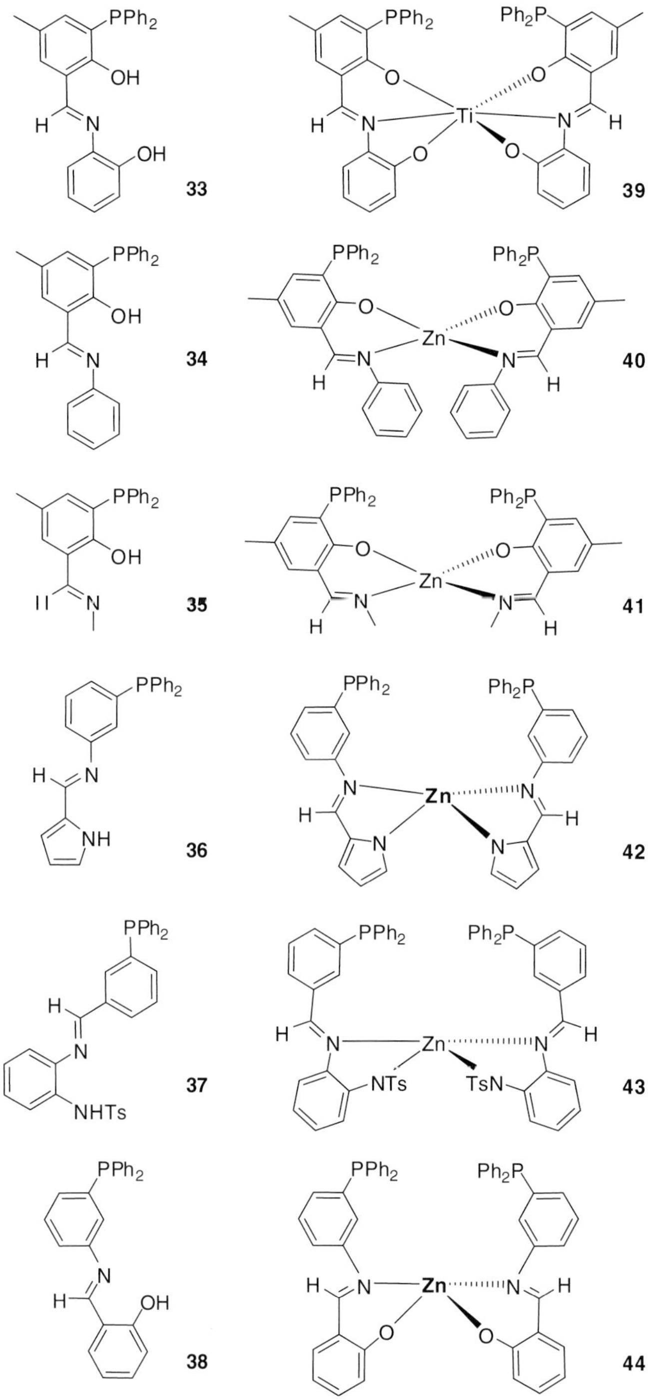

Scheme 10.7 Ditopic ligands synthesized.

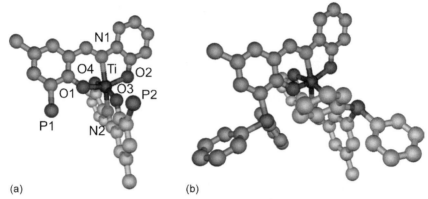

Figure 10.13 Two different views of the molecular structure of **39**. Hydrogen atoms (a, b) and phenyl groups of the phosphines (a) are omitted for clarity.

forming six-membered chelate rings), forming a distorted tetrahedral geometry around the zinc. Consequently, this zinc configuration led the phosphines to an adequate disposition to coordinate in a chelating manner to the catalytic metal. Table 10.9 lists selected bond distances and angles for **40**.

The crystal structure of **42** reveals two independent ditopic ligands coordinated to a zinc metal through both nitrogen atoms. The structure of **42** (Figure 10.15) shows a tetrahedral environment around the zinc center, with a C_2 symmetry, each ligand forming a five-membered chelate ring. Also in this case, both phosphines moieties have the appropriate position to coordinate a catalytic metal. Table 10.10 presents selected bond distances and angles.

The molecular structure of **43** shows that the complex adopts a distorted tetrahedral structure around the zinc center with a C_2 symmetry (Figure 10.16). The ligand arrangement shows a similar disposition of the nitrogen atoms and phosphines to that in **42**, offering the same possibility to coordinate to the active metal as a chelating ligand. Table 10.11 lists selected bond distances and angles.

The molecular structure of **44** (Figure 10.17) revealed that two ditopic ligands are coordinated to Zn. As expected, coordination takes place through the nitrogen and

Table 10.8 Selected bond distances (Å) and angles (°) for **39**.

Distance (Å)	
P1–P2	6.08
Angle (°)	
O1–Ti–O3	88.11
O1–Ti–N1	81.95
N1–Ti–O2	76.31
O3–Ti–N2	81.58
N2–Ti–O4	76.48
Ni–Ti–N2	158.49

10.5 Ditopic Ligands for the Construction of Bidentate Phosphine Ligands | 285

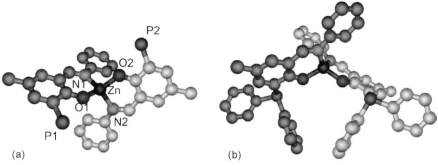

Figure 10.14 Two different views of the molecular structure of **40**. Hydrogen atoms (a, b) and phenyl groups of the phosphines (a) are omitted for clarity.

Table 10.9 Selected bond distance (Å) and angles (°) for **40**.

Bond	Distances (Å)
P1–P2	8.10
	Angle (°)
O1–Zn–O2	116.97
N1–Zn–N2	121.67
N1–Zn–O1	98.66
N2–Zn–O2	96.61

the oxygen atoms; each ligand forms a six-membered chelate ring to the metal. The structure shows a distorted tetrahedral geometry around the Zn center with a C_2 symmetry. Table 10.12 presents selected bond distances and angles. Additionally, the tetrahedral zinc configuration gave the complex a special disposition of the free phosphines that is, in principle, adequate to coordinate as a chelate to a catalytic metal.

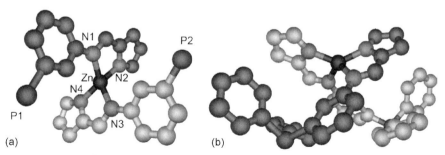

Figure 10.15 Two different views of the molecular structure of **42**. Hydrogen atoms (a, b) and phenyl groups of the phosphines (a) are omitted for clarity.

Table 10.10 Selected bond distances (Å) and angles (°) for **42**.

Bond	Distance (Å)
P1–P2	7.97
	Angle (°)
N1–Zn–N3	116.9
N2–Zn–N4	132.66
N1–Zn–N2	84.23
N3–Zn–N4	84.93

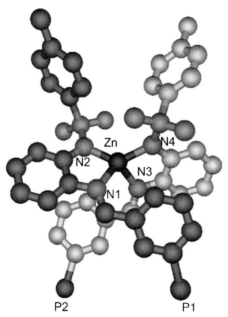

Figure 10.16 Molecular structure of **43**.

Table 10.11 Selected bond distances (Å) and angles (°) for **43**.

Bond	Distance (Å)
P1–P2	6.46
	Angle (°)
N1–Zn–N2	100.76
N3–Zn–N4	130.71
N1–Zn–N3	81.89
N2–Zn–N4	82.06

10.5 Ditopic Ligands for the Construction of Bidentate Phosphine Ligands

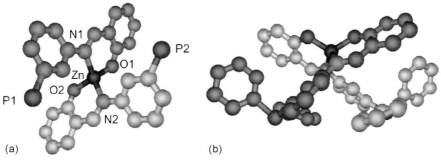

Figure 10.17 Two different views of molecular structure of **44**. Hydrogen atoms (a, b) and phenyl groups of the phosphines (a) are omitted for clarity.

Table 10.12 Selected bond distances (Å) and angles (°) for **44**.

Bond	Distance (Å)
P1–P2	7.7
	Angle (°)
N1–Zn–N2	118.8
O1–Zn–O2	117.2
N1–Zn–O1	98.0
N2–Zn–O2	97.4

The new diphosphines **39/42–44** react with various common rhodium and palladium precursors [Rh(nor)$^+$, Rh(acac), PdMeCl] to give both cis and trans complexes [88].

The catalytic reaction studied was the hydroformylation of 1-octene (Scheme 10.8).

When the rhodium complexes are coordinated by monodentate ligands they frequently show high activity and moderate selectivity for the linear aldehyde [89], but rhodium catalyst that contain bidentate ligands with wide bite angles frequently show a lower activity but an increased selectivity for the linear product [42].

To study the effect of the assembly of monophosphines **33–38** into diphosphines **39–44** with zinc or titanium on the catalytic properties the rhodium-catalyzed hydroformylation of 1-octene was studied. Optimal phosphine to rhodium ratios (at 1 mM of Rh) are different for mono and diphosphines and average convenient

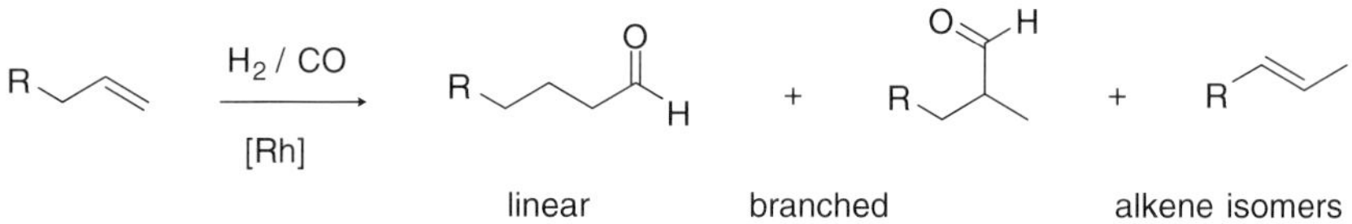

Scheme 10.8 Hydroformylation reaction.

Table 10.13 Rhodium-catalyzed hydroformylation of 1-octene.[a]

Entry	Ligand L	L/Rh	Conversion[b] (%)	Linear aldehyde[c] (%)	l:b	Isomerization (%)	TOF[c] × 10^{-3}
1	36	4	100	59	2.9	20	1.5
2	36	20	100	70	3.2	8	1.2
3	42	2	98	76	5.1	9	0.9
4	42	10	91	79	6.1	8	0.9
5	37	20	100	68	3.4	12	0.9
6	43	2	100	71	3.5	9	0.8
7	43	10	92	81	13[d]	13	0.8
8	38	4	100	62	2.8	16	1.4
9	38	20	100	68	3.6	13	1.2
10	44	2	100	74	4.9	11	1.5
11	44	10	95	84	21	12	0.9
12	PPh$_3$	20	98	65	2.5	9	1.8
13	Xantphos	2.2		97	54	0.5	0.8

[a] Conditions: $T = 80\,°C$, $CO/H_2 = 1$, $p(CO/H_2) = 10$ bar, [Rh] = 1 mM in toluene, substrate/Rh = 670.
[b] Percentage of 1-octene converted was determined at 2 h of reaction.
[c] Turnover frequency in mol mol^{-1} h^{-1} was determined at 40% alkene conversion.
[d] l:b Changes during reaction (25–11) (0.4–3 h).

concentrations were used [90]. The results are summarized in Table 10.13. An incubation time of 3 h was applied before 1-octene was added to the catalyst solution under pressure as the *in situ* IR measurements showed that in most cases the formation of HRh(CO)$_2$(phosphine)$_2$ from Rh(CO)$_2$(acac) and the ligands took about 3 h at 80 °C and 10 bar of syn gas. Ditopic ligands **1–3** and their assemblies **7–9** gave poor catalysts (not shown in Table 10.13), most likely due to formation of salen complexes or phosphinophenolate complexes of rhodium. *In situ* IR of **4** showed the presence of i.a. dicarbonyl rhodium(I) species at 2074 and 2023 cm^{-1}, in accordance with literature data of similar complexes [91].

Ligands **36–38** show a catalytic performance close to that of PPh$_3$ (entry 12), in terms of both selectivity and rate. Isomerization decreases when more ligand is used (entries 1 and 2) and the linear branched ratio goes up. Consistently, ligands **36–38** give higher l:b ratios than PPh$_3$, which may point to an intramolecular interaction between the two salen or salen-like fragments via hydrogen bond interactions (e.g., 3.4 for entry 5 versus. 2.5 for entry 12). The assembled bidentate diphosphines **42–44** with their calculated natural bite angles of 110–120° gave, indeed, higher l:b ratios than the monodentates. The initial l:b ratio for **43** was as high as 25 (entry 7); **44** gave a more stable system with an l:b ratio of 21. The rates are about half of those of the monodentates, as is also the case for Xantphos (entry 13, compared with PPh$_3$), which typically gives higher l:b ratios ranging from 25 to 60. Isomerization remains relatively high for **42–44**, which may be due to the low pressure applied and/or to the presence of other Rh(I) species. Indeed, the *in situ* high-pressure IR spectra of complexes of **42** show the same species as **36** in low concentrations. The *in situ* IR spectra of **38** and **44** are interesting in that they both show the characteristic bands for a HRh(CO)$_2$(arylphosphine)$_2$ species at 2043, 1987 (broad, two peaks), and 1951 cm^{-1} (**38** shows a few more absorptions), but in the spectrum of **44** the intensity of the bisequatorial diphosphine species (2043 and 1987 cm^{-1}) is considerably stronger (fulfilling our expectations), at the cost of the equatorial–apical species (1987 and 1951 cm^{-1}). This explains nicely the increased preference for the formation of linear aldehyde of **44** [90].

In summary, the authors extended the number of ditopic ligands that bind with their hard donor atoms N and O to an assembly metal, in the present contribution zinc and titanium, and to a soft metal such as rhodium(I) through their P donor atom. While the class of compounds is not new, it has been shown that by proper selection catalytically active and selective species can be generated in a facile manner. MM2 calculations had shown that assembly by tetrahedral zinc may lead to wide bite angle diphosphines. It was found that the assemblies indeed give high selectivities for linear product in the rhodium-catalyzed hydroformylation of 1-octene. The method presented is extremely versatile as both the building blocks of the ditopic ligands and the assembly metal can be varied extensively. The synthesis of the assembly usually involves three steps only. Furthermore, the assembly metal fragments can be modified by additional donor molecules or additional anionic fragments, also including chirality.

10.6
Conclusions and Outlook

Host–guest catalysis has led to several beautiful systems and its design and synthesis is an art, but many contributions in this book have shown that for catalysis "simple is beautiful". As noted in Chapter 1, the enormous effort exerted in host–guest catalysis has brought us only a few impressive accelerations of reactions, and perhaps the resulted selectivity changes have attracted more attention. As stated by Ballester and Vidal (Chapter 1) the lock-and-key idea at the heart of host–guest chemistry only works well in catalysis if the transition state geometry is extremely close to the host–guest structure obtained, and, if not, the complexation may even turn out to be counterproductive. While the few examples for which accelerations were found need no explanation, most data remain unexplained. The diphenylglucoluril cavity developed by Nolte [92] and equipped with rhodium catalysts for hydrogenation and isomerization (Chapter 6) and hydroformylation (Section 10.3) is a strongly binding host for resorcinol derivatives, and host–guest binding does not hold us back from rate accelerations. The catalyst and the reactive alkene group are both situated at the top of the assembly, but this is as far as the design goes. Synthetic possibilities and problems encountered strongly limited the outcome. Transition state design has not even been attempted. Thus, it is important that the synthesis of hosts (the same holds for ligand design) allows very subtle variations at the end of the synthetic route so as to obtain a series of catalytic systems within the range of our design. Changes in the ligand from phosphine to phosphite bring about changes in the kinetics of the reaction to be studied and this may destroy the effect we are after. Hydroformylation remains a good choice for a reaction in host–guest catalysis or supramolecular catalysis, as no ionic species are involved and solvation effects and hard to predict entropy [93] changes are reduced to a minimum (Chapters 2, 8, 10). The diphenylglucoluril outer fragments can be varied and substrate variation gives further flexibility, but the synthetic effort remains time consuming. In recent years, self-assembly of cavities has been extensively explored by Fujita [94], Raymond (Chapter 7) and others. Their generally charged nature may be either an advantage or a disadvantage, depending on the reaction to be studied. Neutral capsules formed by zinc porphyrins and salens (Chapter 8) are readily synthesized and are useful in hydroformylation. Self-assembly is not intrinsically simpler than covalent synthesis (!), but the literature cited shows that if it is well planned it has a lot of mileage. One step further brings us to dynamic combinatorial libraries, which may be held together by supramolecular forces or by reversibly formed covalent bonds [95].

For the synthesis of bidentate ligands, supramolecular approaches have led to a renaissance in homogeneous catalyst discovery (Chapters 2, 4, 8, 9, 10), and in a few cases even monodentate ligands have been modified in a supramolecular fashion (Chapter 8, Section 8.2). Combinations of monodentate ligands can be left to chance and in several instances this has led to successful, new catalysts [96]. Such heterocombinations can form spontaneously for steric or electronic reasons or the reactivity of the combinations can be different such that on certain occasions highly enantioselective catalysts are obtained. There are many ways to synthesize the desired heterocombinations selectively and the ionic modification outlined in Section 10.4 is only one of them; since nitration (followed by reduction to amines) and sulfonation are robust methods, the ionic route may prove useful. Hydrogen bonding between different donor–acceptors (Chapters 2 and 8), Lewis acid–base interactions (Chapter 8), and assembly via a metal salt (Chapters 4,9, and Section 10.5) are other ways to achieve this goal.

At the end of the day (actually a decade or more) one might ask if supramolecular catalysts may find application in industry or the laboratory. Many of the supramolecular bidentate ligand systems are not necessarily more expensive than some of the bidentates that have proven their utility in industry [97]. For high-value added products one can easily imagine the use of the new generation of catalysts. Further more, the added advantages of having libraries at ones disposal that can be used in rapid screening can accelerate the discovery process, which is important in the development of routes for the manufacture of pharmaceutical (and other) intermediates [98]. Low value added products are less likely to use catalysts made from more components, unless a specific feature of a new selectivity can be added.

The use of host–guest catalysis in industrial applications would seem further away, as even the construction of a host by supramolecular means still remains costly. In terms of space–time yields heterogeneous counterparts may be more attractive, for instance (periodic) mesoporous (organo) silicas (for PMOs [99]). Highly selective processes that do not occur otherwise may find their way to applications. The capacity of a host to distinguish between guests (substrates) is as yet not a useful property for industrial applications as mixtures are not commonly used as feedstock (apart from mixtures of lower alkenes in a few bulk processes, in which a classic catalyst is more reactive towards one of the substrates and the remainder of the stream is used in another process). The recognition of substrates is highly interesting in learning more about these phenomena and they find applications in sensors, but practical applications in catalysis seem rather remote. The use of renewable bio-feedstocks, however, may create new possibilities for substrate selective catalysts as isolation of one substrate from bio-mass sources may be difficult and selective conversion of one or more would be another way of mimicking nature.

References

1 Rebek, J., Jr (1990) Molecular recognition with model systems. *Angew. Chem., Int. Ed. Engl.*, **29**, 245–255.

2 (a) Harmata, M. and Barnes, C.L. (1990) Molecular clefts. 3. The crystal structure of a chiral molecular tweezer and its guest. *J. Am. Chem. Soc.*, **112**, 5655–5657. (b) Zimmerman, S.C. and VanZuyl, C.M. (1987) Rigid molecular tweezers: Synthesis, characterization, and complexation chemistry of a diacridine. *J. Am. Chem. Soc.*, **109**, 7894–7896.

3 Otto, S. and Engberts, J.B.F.N. (2003) Hydrophobic interactions and chemical reactivity. *Org. Biomol. Chem.*, **1**, 2809–2820.

4 Sheppod, T.J., Petti, M.A. and Dougherty, D.A. (1986) Tight, oriented binding of an aliphatic guest by a new class of water-soluble molecules with hydrophobic binding sites. *J. Am. Chem. Soc.*, **108**, 6085–6087.

5 Pederson, C.J. (1967) Cyclic polyethers and their complexes with metal salts. *J. Am. Chem. Soc.*, **89**, 7017–7036.

6 (a) Breslow, R. and Overman, L.E. (1970) "Artificial enzyme" combining a metal catalytic group and a hydrophobic binding cavity. *J. Am. Chem. Soc.*, **92**, 1075–1077. (b) Breslow, R. (1995) Biomimetic chemistry and artificial enzymes – Catalysis by design. *Acc. Chem. Res.*, **28**, 146–153.

7 Tabushi, I. and Shimizu, N. (1979) Cyclodextrin metal complexes. *Jpn. Kokai Tokkyo Koho, Chem. Abstr.*, **90**, 39197.

8 Reetz, M.T. and Waldvogel, S.R. (1997) β-Cyclodextrin-modified diphosphanes as ligands for supramolecular rhodium catalysts. *Angew. Chem., Int. Ed. Engl.*, **36**, 865–867.

9 Harada, A., Hu, Y. and Takahashi, S. (1986) Cyclodextrin-palladium chloride. New catalytic system for selective oxidation of olefins to ketones. *Chem. Lett.*, 2083–2084.

10 Buhleier, E., Wehner, W. and Voegtle, F. (1978)f. "Cascade"- and "nonskid-chain-like" syntheses of molecular cavity topologies. *Synthesis*, 155.

11 Müller, C., Ackerman, L.J., Reek, J.N.H., Kamer, P.C.J. and van Leeuwen, P.W.N.M. (2004) Site-isolation effects in a dendritic nickel catalyst for the oligomerization of ethylene. *J. Am. Chem. Soc.*, **126**, 14960–14963.

12 Feiters, M.C. (1996) in *Comprehensive Supramolecular Chemistry*, (eds J.L. Atwood, J.E.D. Davies, D.D. MacNicol, F. Vögtle, D.N. Reinhoudt and J-.M. Lehn), Elsevier Science Ltd, Pergamon, Elmsford, vol. 10.

13 Beaulieu, W.B., Rauchfuss, T. and Roundhill, D.M. (1975) Interconversion reactions between substituted phosphinous acid-phosphinito complexes of platinum(II) and their capping reactions with boron trifluoride-diethyl etherate. *Inorg. Chem.*, **14**, 1732–1734.

14 Dubrovina, N.V. and Boerner, A. (2004) Enantioselective catalysis with chiral phosphine oxide preligands. *Angew. Chem., Int. Ed.*, **43**, 5883–5886.

15 Hunter, C.A. (2004) Quantifying intermolecular interactions: Guidelines for the molecular recognition toolbox. *Angew. Chem., Int. Ed.*, **43**, 5310–5324.
16 van Leeuwen, P.W.N.M. and Roobeek, C.F. (1983) *Eur. Pat. Appl.*, **EP 82**, 576, (Chem. Abstr. 1983, 99 121813).
17 Nefkens, S.C.A. (1992) Thesis, ETH, Zürich.
18 van Leeuwen, P.W.N.M., Roobeek, C.F., Frijns, J.H.G. and Orpen, A.G. (1990) Characterization of the intermediates in the hydroformylation reaction catalyzed by platinum diphenylphosphinous acid complexes. *Organometallics*, **9**, 1211–1222.
19 van Leeuwen, P.W.N.M., Roobeek, C.F., Wife, R.L. and Frijns, J.H.G. (1986) *J. Chem. Soc., Chem. Commun.*, 31.
20 Santos, E., Freixa, Z., Bo, C., van Leeuwen, P.W.N.M.,to be published.
21 Ackerman, L. (2006) Air- and moisture-stable secondary phosphine oxides as preligands in catalysis. *Synthesis*, 1557.
22 Ghaffar, T. and Parkins, A.W. (1995) A new homogeneous platinum containing catalyst for the hydrolysis of nitriles. *Tetrahedron Lett.*, **36**, 8657–8660.
23 Jensen, C.M. and Trogler, W.C. (1986) Kinetics and mechanism of nitrile hydration catalyzed by unhindered hydridobis(phosphine)platinum(11) complexes. Regioselective hydration of acrylonitrile. *J. Am. Chem. Soc.*, **108**, 723–729.
24 Jiang, X., Minnaard, A.J., Feringa, B.L. and de Vries, J.G. (2004) Platinum-catalyzed selective hydration of hindered nitriles and nitriles with acid- or base-sensitive groups. *J. Org. Chem.*, **69**, 2327–2331.
25 Cobley, C.J., den Heuvel, M., Abbadi, A. and de Vries, J.G. (2000) Platinum catalysed hydrolytic amidation of unactivated nitriles. *Tetrahedron Lett.*, **41**, 2467–2470.
26 (a) Kirby, A.J. (1996) Enzyme mechanisms, models, and mimics. *Angew. Chem., Int. Ed.*, **35**, 707–724. (b) Breslow, R. (1995) Biomimetic chemistry and artificial enzymes: Catalysis by design. *Acc. Chem. Res.*, **28**, 146–153. (c) Hamilton, A.D., Luo, Z. and Liu, S.J. (1997) Rapid and highly selective cleavage of ribonucleoside 2',3'-cyclic monophosphates by dinuclear CuII complexes. *Angew. Chem., Int. Ed.*, **36**, 2678–2680.
27 (a) Lehn, J.-M. (1988) Supramolecular chemistry – scope and perspectives molecules, supermolecules, and molecular devices (Nobel Lecture). *Angew. Chem., Int. Ed.*, **27**, 89–112. (b) Cram, D.J. (1988) The design of molecular hosts, guests, and their complexes (Nobel Lecture). *Angew. Chem., Int. Ed.*, **27**, 1009–1020. (c) Fouquey, C., Lehn, J.-M. and Levelut, A.-M. (1990) Molecular recognition directed self-assembly of supramolecular liquid crystalline polymers from complementary chiral components. *Adv. Mater.*, **2**, 254–257.
28 Beer, P.D. and Gale, P.A. (2001) Anion recognition and sensing: The state of the art and future perspectives. *Angew. Chem., Int. Ed.*, **40**, 486–516.
29 Metzger, A., Gloe, K., Stephan, H. and Schmidtchen, F.P. (1996) Molecular recognition and phase transfer of underivatized amino acids by a foldable artificial host. *J. Org. Chem.*, **61**, 2051–2055.
30 (a) Beer, P.D., Hopkins, P.K. and McKinney, J.D. (1999) Cooperative halide, perrhenate anion-sodium cation binding and pertechnetate extraction and transport by a novel tripodal tris(amido benzo-15-crown-5) ligand. *Chem. Commun.*, 1253–1254. (b) Shukla, R., Kida, T. and Smith, B.D. (2000) Effect of competing alkali metal cations on neutral host's anion binding ability. *Org. Lett.*, **2**, 3099–3102.
31 Chang, S.K. and Hamilton, A.D. (1988) Molecular recognition of biologically interesting substrates: Synthesis of an artificial receptor for barbiturates employing six hydrogen bonds. *J. Am. Chem. Soc.*, **110**, 1318–1319.
32 Corbett, P.T., Leclaire, J., Vial, L., Wietor, J..-L., West, K.R., Sanders, J.K.M. and Otto, S. (2006) Dynamic combinatorial chemistry. *Chem. Rev.*, **106**, 3652–3711.

33 (a) Wenz, G. (1994) Cyclodextrins as building blocks for supramolecular structures and functional units. *Angew. Chem.*, **33**, 803–822. (b) Anslyn, E.V. and Breslow, R. (1989) Proton inventory of a bifunctional ribonuclease model. *J. Am. Chem. Soc.*, **111**, 8931–8932. (c) Breslow, R. and Schmuck, C. (1996) Goodness of fit in complexes between substrates and ribonuclease mimics: Effects on binding, catalytic rate constants, and regiochemistry. *J. Am. Chem. Soc.*, **118**, 6601–6605.

34 Rowan, A.E., Elemans, J.A.A.W. and Nolte, R.J.M. (1999) Molecular and supramolecular objects from glycoluril. *Acc. Chem. Res.*, **32**, 995–1006.

35 Sijbesma, R.P., Kentgens, A.P.M., Lutz, E.T.G., van der Maas, J.H. and Nolte, R.J.M. (1993) Binding features of molecular clips derived from diphenylglycoluril. *J. Am. Chem. Soc.*, **115**, 8999–9005.

36 (a) Coolen, H.K.A.C., van Leeuwen, P.W.N.M. and Nolte, R.J.M. (1992) A RhI-centered cage compound with selective catalytic properties. *Angew. Chem., Int. Ed. Engl.*, **31**, 905–907. (b) Coolen, H.K.A.C., Meeuwis, J.A.M., van Leeuwen, P.W.N.M. and Nolte, R.J.M. (1995) Substrate selective catalysis by rhodium metallohosts. *J. Am. Chem. Soc.*, **117**, 11906–11913.

37 Dol, G.C., Kamer, P.C.J., Nolte, R.J.M., van Leeuwen, P.W.N.M.,unpublished.

38 Dol, G.C., Kamer, P.C.J. and Van Leeuwen, P.W.N.M. (1998) Synthesis of 5-substituted resorcinol derivatives via cross-coupling reactions. *Eur. J. Org. Chem.*, 359–364.

39 Dol, G.C. (1998) Thesis, Amsterdam.

40 Reek, J.N.H., Priem, A.H., Engelkamp, H., Rowan, A.E., Elemans, J.A.A.W. and Nolte, R.J.M. (1997) Binding features of molecular clips. Separation of the effects of hydrogen bonding and pi-pi interactions. *J. Am. Chem. Soc.*, **119**, 9956–9964.

41 Frances, J.-M., Thorez, A. and Kalck, P. (1984) *N. J. Chim.*, **8**, 213.

42 van Leeuwen, P.W.N.M. and Claver (eds), C. (2000) *Rhodium-Catalyzed Hydroformylation*, Kluwer Academic Publishers, Dordrecht.

43 (a) Reetz, M.T., Sell, T., Meiswinkel, A. and Mehler, G. (2003) Ein neuartiges prinzip in der kombinatorischen asymmetrischen übergangsmetall-katalyse: Mischungen von chiralen einzähnigen P-liganden. *Angew. Chem.*, **115**, 814–817. A new principle in combinatorial asymmetric transition-metal catalysis: Mixtures of chiral monodentate P ligands. *Angew. Chem., Int. Ed.*, **42**, 790–793. (b) Reetz, M.T. and Mehler, G. (2003) Mixtures of chiral and achiral monodentate ligands in asymmetric Rh-catalyzed olefin hydrogenation: reversal of enantioselectivity. *Tetrahedron Lett.*, 4593–4596. (c) Peña, D., Minnaard, A.J., Boogers, J.A.F., de Vries, A.H.M., de Vries, J.G. and Feringa, B.L. (2003) Improving conversion and enantioselectivity in hydrogenation by combining different monodentate phosphoramidites; a new combinatorial approach in asymmetric catalysis. *Org. Biomol. Chem.*, **1**, 1087–1089. (d) Reetz, M.T. and Li, X. (2004) Combinatorial approach to the asymmetric hydrogenation of beta-acylamino acrylates: use of mixtures of chiral monodentate P-ligands. *Tetrahedron*, **60**, 9709–9714. (e) Reetz, M.T., Mehler, G., Meiswinkel, A. and Sell, T. (2004) Mixtures of chiral monodentate phosphites, phosphonites and phosphines as ligands in Rh-catalyzed hydrogenation of N-acyl enamines: Extension of the combinatorial approach. *Tetrahedron: Asymmetry*, **15**, 2165–2167. (f) Reetz, M.T., Fu, Y. and Meiswinkel, A. (2006) Nonlinear effects in Rh-catalyzed asymmetric olefin hydrogenation using mixtures of chiral monodentate P ligands. *Angew. Chem., Int. Ed.*, **45**, 1412–141. (g) Minnaard, A.J., Feringa, B.L., Lefort, L. and de Vries, J.G. (2007) Asymmetric hydrogenation wing monodentate phosphoramidite ligands. *Acc. Chem. Res.*, **40**, 1267–1277.

44 Duursma, A., Hoen, R., Schuppan, J., Hulst, R., Minnaard, A.J. and Feringa, B.L.

(2003) First examples of improved catalytic asymmetric C–C bond formation using the monodentate ligand combination approach. *Org. Lett.*, **5**, 3111–3113.

45 Reetz, M.T. and Li, X. (2005) The influence of mixtures of monodentate achiral ligands on the regioselectivity of transition-metal-catalyzed hydroformylation. *Angew. Chem., Int. Ed.*, **44**, 2962–2964.

46 (a) Slagt, V.F., Röder, M., Kamer, P.C.J., van Leeuwen, P.W.N.M. and Reek, J.N.H. (2004) Supraphos: A supramolecular strategy to prepare bidentate ligands. *J. Am. Chem. Soc.*, **126**, 4056–4057. (b) Reek, J.N.H., Röder, M., Goudriaan, P.E., Kamer, P.C.J., van Leeuwen, P.W.N.M. and Slagt, V.F. (2005) Supraphos: A supramolecular strategy to prepare bidentate ligands. *J. Organomet. Chem.*, **690**, 4505–4516. (c) Jiang, X.-J., Lefort, L., Goudriaan, P.E., de Vries, A.H.M., van Leeuwen, P.W.N.M., de Vries, J.G. and Reek, J.N.H. (2006) Screening of a supramolecular catalyst library in the search for selective catalysts for the asymmetric hydrogenation of a difficult enamide substrate. *Angew. Chem.*, **118**, 1245–1249. (d) (2006) Screening of a supramolecular catalyst library in the search for selective catalysts for the asymmetric hydrogenation of a difficult enamide substrate. *Angew. Chem., Int. Ed.*, **45**, 1223–1227.

47 (a) Breit, B. and Sache, W. (2005) Optically active conjugated polymers prepared from achiral monomers by polycondensation in a chiral nematic solvent. *Angew. Chem.*, **117**, 4401–4404; (2005) Optically active conjugated polymers prepared from achiral monomers by polycondensation in a chiral nematic solvent. *Angew. Chem., Int. Ed.*, **44**, 1640–1643. (b) Weis, M., Waloch, C., Seiche, W. and Breit, B. (2006) Self-assembly of bidentate ligands for combinatorial homogeneous catalysis: Asymmetric rhodium-catalyzed hydrogenation. *J. Am. Chem. Soc.*, **128**, 4188–4189.

48 Gulyas, H., Benet-Buchholz, J., Escudero-Adan, E.C., Freixa, Z. and van Leeuwen, P.W.N.M. (2007) Ionic interaction as a powerful driving force for the formation of heterobidentate assembly ligands. *Chem.–Eur. J.*, **13**, 3424–3430.

49 (a) For a recent computational study on the factors controlling the relative stabilities of cis- and trans-PtX_2L_2 isomers see: Harvey, J.N., Heslop, K.M., Orpen, A.G. and Pringle, P.G. (2003) Factors controlling the relative stabilities of cis- and trans-[PtX2L2] isomers: Chatt and Wilkins – 50 years on. *Chem. Commun.*, 278–279. (b) Eberhard, M.R., Heslop, K.M., Orpen, A.G. and Pringle, P.G. (2005) Nine-membered trans square-planar chelates formed by a bisbi analogue. *Organometallics*, **24**, 335–337.

50 (a) Hunter, C.A. (2004) Zwischenmolekulare wechselwirkungen in lösung: Eine vereinfachende quantifizierungsmethode. *Angew. Chem.*, **116**, 5424–543. (b) (2004) Quantifying intermolecular interactions: Guidelines for the molecular recognition toolbox. *Angew. Chem., Int. Ed.*, **43**, 5310–5324.

51 For a recent computational study on the factors controlling the relative stabilities of cis- and trans-PtX_2L_2 isomers, see: (a) Harvey, J.N., Heslop, K.M., Orpen, A.G. and Pringle, P.G. (2003) Factors controlling the relative stabilities of cis- and trans-[PtX2L2] isomers: Chatt and Wilkins – 50 years on. *Chem. Commun.*, 278–279. (b) Eberhard, M.R., Heslop, K.M., Orpen, A.G. and Pringle, P.G. (2005) *Nine-membered trans square-planar chelates formed by a bisbi analogue Organometallics*, **24**, 335–337.

52 Breit, B. and Seiche, W. (2003) Hydrogen bonding as a construction element for bidentate donor ligands in homogeneous catalysis: Regioselective hydroformylation of terminal alkenes. *J. Am. Chem. Soc.*, **125**, 6608–6609.

53 Knight, L.K., Freixa, Z., van Leeuwen, P.W.N.M. and Reek, J.N.H. (2006) Supramolecular trans-coordinating phosphine ligands. *Organometallics*, **25**, 954–960.

54 Duckmanton, P.A., Blake, A.J. and Love, J.B. (2005) Palladium and rhodium

ureaphosphine complexes: Exploring structural and catalytic consequences of anion binding. *Inorg. Chem.*, **44**, 7708–7710.

55 Lu, X.-X., Tang, H.-S., Ko, C.-C., Wong, J.K.-Y., Zhu, N. and Yam, V.W.-W. (2005) Anion-assisted trans-cis isomerization of palladium(II) phosphine complexes containing acetanilide functionalities through hydrogen bonding interactions. *Chem. Commun.*, 1572–1574.

56 (a) Leininger, S., Olenyuk, B. and Stang, P.J. (2000) Self-assembly of discrete cyclic nanostructures mediated by transition metals. *Chem. Rev.*, **100**, 853–907. (b) Fujita, M. (1998) Metal-directed self-assembly of two- and three-dimensional synthetic receptors. *Chem. Soc. Rev.*, **27**, 417–425. (c) Albrecht, M. (1998) Dicatechol ligands: Novel building-blocks for metallo-supramolecular chemistry. *Chem. Soc. Rev.*, **27**, 281–287.

57 (a) Schubert, U.S. and Eschbaumer, C. (2002) Macromolecules containing bipyridine and terpyridine metal complexes: Towards metallosupramolecular polymers. *Angew. Chem., Int. Ed.*, **41**, 2892–2926. (b) Swiegers, G.F. and Malefetse, T.J. (2000) New self-assembled structural motifs in coordination chemistry. *Chem. Rev.*, **100**, 3483–3537. (c) Ciferri, A. (2002) Supramolecular polymerizations. *Macromol. Rapid Commun.*, **23**, 511–529. (d) Ikkala, O. and ten Brinke, G. (2002) Functional materials based on self-assembly of polymeric supramolecules. *Science*, **295**, 2407–2409. (e) Liu, G.-X. and Puddephatt, R.J. (1996) Divergent route to organoplatinum or platinum-palladium dendrimers. *Organometallics*, **15**, 5257–5279.

58 (a) Hagrman, P.J., Hagrman, D. and Zubieta, J. (1999) Organic-inorganic hybrid materials: From "simple" coordination polymers to organodiamine-templated molybdenum oxides. *Angew. Chem., Int. Ed.*, **38**, 2638–2684. (b) Holliday, B.J. and Mirkin, C.A. (2001) Strategies for the construction of supramolecular compounds through coordination chemistry. *Angew. Chem., Int. Ed.*, **40**, 2022–2043. (c) Seidel, S.R. and Stang, P.J. (2002) High-symmetry coordination cages via self-assembly. *Acc. Chem. Res.*, **35**, 972–983. (d) Batten, S.R. and Robson, R. (1998) Interpenetrating nets: Ordered, periodic entanglement. *Angew. Chem., Int. Ed.*, **37**, 1460–1494.

59 Inouye, M., Konishi, T. and Isagawa, K. (1993) Artificial allosteric receptors for nucleotide bases and alkali-metal cations. *J. Am. Chem. Soc.*, **115**, 8091–8095.

60 Scrimin, P., Tecilla, P., Tonellato, U. and Vignana, M. (1991) A water-soluble tweezers-like metalloreceptor: Binding and selective catalytic properties. *J. Chem. Soc., Chem. Commun.*, 449.

61 Another allosteric two binding domain system: Schneider, H.-J. and Ruf, D. (1990) A synthetic allosteric system with high cooperativity between polar and hydrophobia binding sites. *Angew. Chem., Int. Ed. Engl.*, **29**, 1159–1160.

62 Goodman, M.S., Jubian, V. and Hamilton, A.D. (1995) Metal templated receptors for the effective complexation of dicarboxylates. *Tetrahedron Lett.*, **36**, 2551–2554.

63 Maverick, A.W. and Klavetter, F.E. (1984) Cofacial binuclear copper complexes of a bis(β-diketone) ligand. *Inorg. Chem.*, **23**, 4129–4130.

64 Fujita, M., Yazaki, J., Kuramochi, T. and Ogura, K. (1993) Self-assembly of a macrocyclic dinuclear Pd(II)-phosphine complex. *Bull. Chem. Soc. Jpn.*, **66**, 1837–1839.

65 Self-assembled complexes: (a) Pryde, A.J., Shaw, B.L. and Weeks, B. (1973) Large ring compounds involving trans-bonding bidentate ligands. *J. Chem. Soc., Chem. Commun.*, 947. (b) Schwabacher, A.W., Lee, J. and Lei, H. (1992) Self-assembly of a hydrophobic binding site. *J. Am. Chem. Soc.*, **114**, 7597–7598. (c) Hunter, C.A. and Sarson, L.D. (1994) Self-assembly of a dimeric porphyrin host. *Angew. Chem., Int. Ed. Engl.*, **33**, 2313–2316.

66 Fujita, M., Yazaki, J. and Ogura, K. (1990) Preparation of a macrocyclic polynuclear complex, [(en)Pd(4,4′-bpy)]4(NO3)8,1 which recognizes an organic molecule in aqueous media. *J. Am. Chem. Soc.*, **112**, 5645–5647.

67 Yam, V.W.-W. and Lu, X.-X.C.-C. (2003) First observation of alkali metal ion induced trans-cis isomerization of palladium(II) phosphane complexes containing crown ether moieties. *Angew. Chem.*, **115**, 3507–3510.

68 Van Zon, A., Torny, G.J. and Frijns, J.H.G. (1983) *Recueil: J. Royal Nethl. Chem. Soc.*, **102**, 326.

69 Butts, S.B., Strauss, S.H., Holt, E.M., Stimson, R.E., Alcock, N.W. and Shriver, D.F. (1980) Activation of coordinated carbon monoxide toward alkyl and aryl migration (CO insertion) by molecular Lewis acids and X-ray structure of the reactive intermediate Mn(C(OAlBrBr2)CH3)(CO)4. *J. Am. Chem. Soc.*, **102**, 5093–5100.

70 Wrobleski, D.A. and Rauchfuss, T.B. (1982) Synthetic approaches to coordinatively unsaturated heterobimetallic complexes. *J. Am. Chem. Soc.*, **104**, 2314–2316.

71 Casey, C.P., Jordan, R.F. and Rheingold, A.L. (1983) Metal-metal-bonded zirconium-ruthenium and zirconium-iron complexes. *J. Am. Chem. Soc.*, **105**, 665–667, and references therein. For more recent: reviews see: (a) Stephan, D.W. (1989) Early-late heterobimetallics. *Coord. Chem. Rev.*, **95**, 41–107. (b) Wheatley, N. and Kalck, P. (1999) Structure and reactivity of early-late heterobimetallic complexes. *Chem. Rev.*, **99**, 3379–3419.

72 (a) Choukroun, R. and Gervais, D. (1982) A new heterobimetallic Zr–Rh complex, [{(η^5-C5H5)2ZrCl(CH2PPh2)}2Rh (CO)Cl], leading to an unexpected terminal carbonyl-zirconium bond. *J. Chem. Soc., Chem. Commun.*, 1300. (b) Senocq, F., Randrianalimanana, C., Thorez, A., Kalck, P., Choukroun, R. and Gervais, D. (1984) A novel d^0–d^8 heterobimetallic complex obtained by co-ordination of a zirconium (IV) diphosphine to dinuclear rhodium(I) moiety leading to a hydroformylation-active species. *J. Chem. Soc., Chem. Commun.*, 1376.

73 Choukroun, R., Iraqi, A., Gervais, D., Daran, J.C. and Jeannin, Y. (1987) A semibridging hydrido zirconium-rhodium complex: A possible way to catalytic hydrogen transfer on do-de systems. *Organometallics*, **6**, 1197–1201.

74 (a) Slagt, V.F., van Leeuwen, P.W.N.M. and Reek, J.N.H. (2003) Bidentate ligands formed by self-assembly. *Chem. Commun.*, 2474–2475. (b) Braunstein, P., Clerc, G., Morise, X., Welter, R. and Mantovani, G. (2003) Phosphinooxazolines as assembling ligands in heterometallic complexes. *Dalton Trans.*, 1601–1605. (c) Mokuolu, Q.F., Duckmanton, P.A., Hitchcock, P.B., Wilson, C., Blake, A.J., Shukla, L. and Love, J.B. (2004) Early-late, mixed-metal compounds supported by amidophosphine ligands. *Dalton Trans.*, 1960–1970. (d) Breit, B. and Seiche, W. (2003) Hydrogen bonding as a construction element for bidentate donor ligands in homogeneous catalysis: Regioselective hydroformylation of terminal alkenes. *J. Am. Chem. Soc.*, **125**, 6608–6609. (e) Jiang, H., Hu, A. and Lin, W. (2003) Chiral metallacyclophane for asymmetric catalysis. *Chem. Commun.*, 96–97. (f) Takacs, J.M., Hrvatin, P.M., Atkins, J.M., Reddy, D.S. and Clark, J.L. (2005) The selective formation of neutral, heteroleptic zinc(II) complexes via self-discrimination of chiral bisoxazoline racemates and pseudoracemates. *New J. Chem.*, **29**, 263–265. (g) Nagashima, H., Sue, T., Oda, T., Kanemitsu, A., Matsumoto, T., Motoyama, Y. and Sunada, Y. (2006) Dynamic titanium phosphinoamides as unique bidentate phosphorus ligands for platinum. *Organometallics*, **25**, 1987–1994. (h) Yam, V.-W.-W., Lu, X.-X. and Chi-Chiu, K. (2003) First observation of alkali metal ion induced trans-cis isomerization of

palladium(II) phosphane complexes containing crown ether moieties. *Angew. Chem.*, **115**, 3507–3510.

75 (a) Gianneschi, N.C., Bertin, P.A., Nguyen, S.T., Mirkin, C.A., Zakharov, L.N. and Rheingold, A.L. (2003) A supramolecular approach to an allosteric catalyst. *J. Am. Chem. Soc.*, **125**, 10508–10509. (b) Lee, S.J., Hu, A. and Lin, W. (2002) First chiral organometallic triangle for asymmetric catalysis. *J. Am. Chem. Soc.*, **124**, 12948–12949. (c) Weis, M., Waloch, C., Seiche, W. and Breit, B. (2006) Self-assembly of bidentate ligands for combinatorial homogeneous catalysis: Asymmetric rhodium-catalyzed hydrogenation. *J. Am. Chem. Soc.*, **128**, 4188–4189. (d) Slagt, V.F., van Leeuwen, P.W.N.M. and Reek, J.N.H. (2003) Multicomponent porphyrin assemblies as functional bidentate phosphite ligands for regioselective rhodium-catalyzed hydroformylation. *Angew. Chem., Int. Ed.*, **42**, 5619–5623.

76 Takacs, J.M., Reddy, D.S., Moteki, S.A., Palencia, D. and Wu, H. (2004) Asymmetric catalysis using self-assembled chiral bidentate P,P-ligands. *J. Am. Chem. Soc.*, **126**, 4494–4495.

77 Gianneschi, N.C., Cho, S.-H., Nguyen, S.T. and Mirkin, C.A. (2004) Reversibly addressing an allosteric catalyst in situ: Catalytic molecular tweezers. *Angew. Chem., Int. Ed.*, **43**, 5503–5507.

78 Sun, X., Johnson, D.W., Caulder, D.L., Powers, R.E., Raymond, K.N. and Wong, E.H. (1999) Exploiting incommensurate symmetry numbers: Rational design and assembly of M2M3′L6 supramolecular clusters with C3h symmetry. *Angew. Chem., Int. Ed.*, **38**, 1303–1307.

79 (a) Kless, A., Kadyrov, R., Boerner, A., Holz, J. and Kagan, H.B. (1995) A new chiral multidentate ligand for asymmetric catalysis. *Tetrahedron Lett.*, **36**, 4601–4602. (b) Kless, A., Lefeber, C., Spannenberg, A., Kempe, R., Baumann, W., Holz, J. and Boerner, A. (1996) The first chiral early-late heterobimetallic complex – A titanium (IV)-palladium(II) complex based on salenophos. *Tetrahedron*, **52**, 14599–14606.

80 Rivillo, D., Gulyas, H., Benet-Buchholz, J., Escudero-Adan, E.C., Freixa, Z. and van Leeuwen, P.W.N.M. (2007) Catalysis by design: Wide-bite-angle diphosphines by assembly of ditopic ligands for selective rhodium-catalyzed hydroformylation. *Angew. Chem., Int. Ed.*, **46**, 7247–4750.

81 Casey, C.P. and Whiteker, G.T. (1990) *Isr. J. Chem.*, **30**, 299.

82 Dierkes, P. and van Leeuwen, P.W.N.M. (1999) The bite angle makes the difference: A practical ligand parameter for diphosphine ligands. *J. Chem. Soc., Dalton Trans.*, 1519–1529.

83 (a) Freixa, Z. and van Leeuwen, P.W.N.M. (2003) Bite angle effects in diphosphine metal catalysts: Steric or electronic? *Dalton Trans.*, 1890–1901. (b) van Leeuwen, P.W.N.M., Kamer, P.C.J., Reek, J.N.H. and Dierkes, P. (2000) Ligand bite angle effects in metal-catalyzed C–C bond formation. *Chem. Rev.*, **100**, 2741–2769. (c) Kamer, P.C.J., van Leeuwen, P.W.N.M. and Reek, J.N.H. (2001) Wide bite angle diphosphines: Xantphos ligands in transition metal complexes and catalysis. *Acc. Chem Res.*, **34**, 895–904. (d) Kamer, P.C.J., Reek, J.N.H. and van Leeuwen, P.W.N.M. (1998) Designing ligands with the right bite. *CHEMTECH*, **28**, 27–33.

84 Kranenburg, M., van der Burgt, Y.E.M., Kamer, P.C.J., van Leeuwen, P.W.N.M., Goubitz, K. and Fraanje, J. (1995) New diphosphine ligands based on heterocyclic aromatics inducing very high regioselectivity in rhodium-catalyzed hydroformylation – Effect of the bite angle. *Organometallics*, **14**, 3081–3089.

85 Kranenburg, M., Kamer, P.C.J., van Leeuwen, P.W.N.M., Vogt, D. and Keim, W. (1995) Effect of the bite angle of diphosphine ligands on activity and selectivity in the nickel-catalysed hydrocyanation of styrene. *J. Chem. Soc. Chem. Commun.*, 2177–1778.

86 (a) Harris, M.C., Geis, O. and Buchwald, S.L. (1999) Sequential N-arylation of

primary amines as a route to alkyldiarylamines. *J. Org. Chem.*, **64**, 6019–6022. (b) Guari, Y., van Es, D.S., Reek, J.N.H., Kamer, P.C.J. and van Leeuwen, P.W.N.M. (1999) An efficient, palladium-catalysed, amination of aryl bromides. *Tetrahedron Lett.*, **40**, 3789–3790.

87 Jiménez-Rodríguez, C., Roca, F.X., Bo, C., Benet-Buchholz, J., Escudero-Adán, E.C., Freixa, Z. and van Leeuwen, P.W.N.M. (2006) SPANphos: trans-spanning diphosphines as cis chelating ligands!. *Dalton Trans.*, 268–278.

88 Rivillo, D., Freixa, Z. and van Leeuwen, P.W.N.M., to be published.

89 (a) Buhling, A., Kamer, P.C.J. and van Leeuwen, P.W.N.M. (1995) Rhodium-catalyzed hydroformylation of higher alkenes using amphiphilic ligands. *J. Mol. Catal. A*, **98**, 69–80. (b) Trzreciak, A.M. and Ziólkowski, J.J. (1999) *Coord. Chem. Rev.*, **190**, 883.

90 (a) van Leeuwen, P.W.N.M., Casey, C.P. and Whiteker, G.T. (2000) *Catalysis by Metal Complexes*, Kluwer, Dordrecht, vol. 22, ch 4 (Rhodium catalyzed hydroformylation), pp. 63–105. (b) van der Veen, L.A., Keeven, P.H., Schoemaker, G.C., Reek, J.N.H., Kamer, P.C.J., van Leeuwen, P.W.N.M., Lutz, M. and Spek, A.L. (2000) Origin of the bite angle effect on rhodium diphosphine catalyzed hydroformylation. *Organometallics*, **19**, 872–883.

91 Ho, J.H.H., St Clair Black, D., Messerle, B.A., Clegg, J.K. and Turner, P. (2006) Reactive indolyl complexes of group 9 metals. *Organometallics*, **25**, 5800–5810.

92 Rowan, A.E., Elemans, J.A.A.W. and Nolte, R.J.M. (1999) Molecular and supramolecular objects from glycoluril. *Acc. Chem. Res.*, **32**, 995–1006.

93 Schmidtchen, F.P. (2006) Reflections on the construction of anion receptors – Is there a sign to resign from design? *Coord. Chem. Rev.*, **250**, 2918–2928.

94 (a) For a recent example of catalysis see: Nishioka, Y., Yamaguchi, T., Yoshizawa, M. and Fujita, M. (2007) Unusual [2 + 4] and [2 + 2] cycloadditions of arenes in the confined cavity of self-assembled cages. *J. Am. Chem. Soc.*, **129**, 7000. (b) For reviews, see: Maurizot, V., Yoshizawa, M., Kawano, M., Fujita, M., Control of molecular interactions by the hollow of coordination cages. *Dalton Trans.*, **2006**, 2750–2756. (c) Yoshizawa, M. and Fujita, M. (2005) Self-assembled coordination cage as a molecular flask. *Pure Appl. Chem.*, **77**, 1107.

95 (a) Brady, P.A. and Sanders, J.K.M. (1997) Thermodynamically-controlled cyclisation and interconversion of oligocholates: Metal ion templated 'living' macrolactonisation. *J. Chem. Soc., Perkin Trans.*, **1**, 3237. (b) Corbett, P.T., Leclaire, J., Vial, L., West, K.R., Wietor, J.-L., Sanders, J.K.M. and Otto, S. (2006) Dynamic combinatorial chemistry. *Chem. Rev.*, **106**, 3652–3711. (c) Ludlow, R.F., Liu, J., Li, H., Roberts, S.L., Sanders, J.K.M. and Otto, S. (2007) Host–guest binding constants can be estimated directly from the product distributions of dynamic combinatorial libraries. *Angew. Chem., Int. Ed.*, **46**, 5762–5764.

96 (a) Reetz, M.T., Sell, T., Meiswinkel, A. and Mehler, G. (2003) A new principle in combinatorial asymmetric transition-metal catalysis: Mixtures of chiral monodentate P ligands. *Angew. Chem., Int. Ed.*, **42**, 790–793. (b) Reetz, M.T. and Bondarev, O. (2007) Mixtures of chiral phosphorous acid diesters and achiral P ligands in the enantio- and diastereoselective hydrogenation of ketimines. *Angew. Chem., Int. Ed.*, **46**, 4523–4526.

97 (a) Blaser, H.-U., Indolese, A. and Schnyder, A. (2000) Applied homogeneous catalysis by organometallic complexes. *Curr. Sci.*, **78**, 1336–1344. (b) Blaser, H.-U., Spindler, F. and Studer, M. (2001) Enantioselective catalysis in fine chemicals production. *Appl. Catal. A: Gen.*, **221**, 119–143.

98 de Vries, J.G. and de Vries, A.H.M. (2003) The power of high-throughput experimentation in homogeneous catalysis research for fine chemicals. *Eur. J. Org. Chem.*, 799–811.

99 Hunks, W.J. and Ozin, G.A. (2005) Challenges and advances in the chemistry of periodic mesoporous organosilicas (PMOs). *J. Mater. Chem.*, **15**, 3716–3724.

Index

a

A–T base pairing 39f.
acetylation 121ff.
acyl transfer reactions 8, 113
alcoholysis of aryl acetates 117
alkaline-earth metal salts 114
allylic alkylation 220, 270, 280
– asymmetric 220, 270, 280
allylic amination 23, 226f., 244f.
amidase/esterase 116, 131
anchoring of catalysts 225f.
asymmetric hydrogenation 37, 46ff., 53, 248, 269
– N-(3,4-dihydro-2-naphthalenyl)-acetamide 222, 224f.
– enamides 248
3-aza-Cope rearrangement 15, 175
azacrown 129
azacrown decorated calixarenes 133

b

bimetallic catalysts 128
bimolecular reactions 7
bioamphiphiles 157
biomacromolecules 22
bite angle 21, 60, 281ff.
building blocks 93, 97, 200f., 204f., 225f.
– (box)$_2$Zn 21, 241f.
– Porphyrins 21, 97, 153ff., 200f.
– zinc(II) salphen 205

c

C-H bond activation 170
calixcrown-5 barium complex 135ff.
Candida Antarctica (CALB) 158
catalyst libraries 29, 210
catalytic efficiency 121f., 132, 263

cation-anion attractions 113, 255
chelation through self-assembly 30
chiral diphosphites cyclodextrin 245f.
CO/styrene polymerisation 208
coordination polymers 57f., 72ff.
cooperative binding 146f., 268
crown ethers 113, 123
– thiol-pendant 124
cucurbit [6] uril 11ff.
cyclization 95f., 106
– thermodynamics 166, 185
cytochrome P450 mimics 153

d

deacetylation 121f.
dendrimers 227, 257
Diels-Alder reaction 7f., 14, 175, 256
diphenylglycoluril 144, 265f.
1,3-dipolar cycloaddition 12ff.

e

effective molarity 136, 199, 256
electrophilic catalysis 118
– by barium 118
– by strontium 118
encapsulation 9f., 65, 166ff., 199ff.
– CpRu(p-cymene)$^+$ 168
– CpRu(η^4-diene)(H$_2$O)$^+$ 169
– Cp*(PMe3)Ir(Me)Otf 168
– Cp*(PMe$_3$)Ir(Me)(η^2-olefin)$^+$ 170
– *bis*(dimethylphosphinomethane) 183
– (PMe$_3$)$_2$Rh(H$_2$O)$_2$$^+$ 162
– [(P-P)Rh(diene)][BF$_4$] 173
– template-ligand assisted 199ff.
– tris-*meta* pyridylphosphine 201ff.
encapsulation-driven protonation 186
entropy effects 2f., 95, 176

enzyme model systems 1
epoxidation 6, 149f., 154f.,
– manganese porphyrin 150, 153ff.
– olefins 6, 18, 256f.
– α-pinene 155
– single molecule studies 155
ester methanolysis 123
ethanolysis 114ff.
– anilide 129f., 277
– esters 121ff., 263, 277
– trifluoroacetanilide 115

f
ferri-protoporphyrin 158

g
glucoluril 289f.
grid architectures 63
guest encapsulation 166, 185

h
Heck reaction 203
heterobidentate 210, 214, 269ff.
– ligated 217
heterocapsule 210
heterocomplexes 214
heterodimeric ligand 31f.
– bidentate 32, 39
– ionic interactions 210, 269ff.
heterolytic splitting dihydrogen 261f.
high-throughput screening 225
homobidentate 22, 211
hydration 39ff., 262
– alkynes 39ff.
– ruthenium complexes 39
– nitriles 45ff.
hydroformylation 17ff., 37ff., 49ff., 203ff., 227, 258ff., 280ff.
– allylbenzene 265
– encapsulation effects 203f.
– heptene-1 258f.
– heptene-2 259
– 1-octene 21, 51
– 2-octene 206f.
– 3-octene 206f.
– SPO platinum complexes 258
– styrene 21f., 154f., 214
hydrogen bonding 9ff., 20, 29ff., 58, 258, 276
– A-T base pair 40
– Aminopyridine/isoquinolone 38f.
– complementary 18, 33, 38f.
– secondary phosphine oxides (SPOs) 258ff.
– strength 31
– urea 223ff.

hydrogenation 18, 22, 37f., 46ff., 145ff., 222ff., 248ff., 256, 262ff.
– *N*-acetamidoacrylate 22, 37f., 48
– acetamidoacrylate 22, 47f., 225
– aldehydes 18, 39f., 106ff., 170ff., 207
– dimethyl itaconate 37f., 48f.
– methyl α-(Z)-*N*-acetamido cinnamate 37f., 47, 49
– styrene derivatives 6, 145
hydrolysis 3f., 159ff., 179ff., 256, 262
– acetals 188
– acrylonitrile 162
– 2,2-dimethoxypropane 188
– iminium cation 181
– *p*-nitrophenylcholine carbonate 4
– orthoformates 185ff.

i
Ibuprofen 5f.
– oxidation 5f.
ion-dipole interactions 113, 255
isomerization 145, 173f., 207
– allyl alcohols 173

l
ligand library 30f.
ligand scaffold optimisation 235ff.

m
membrane-based catalysts 153ff.
metal-grids 64, 85
metallocycles 53ff.
methanolysis 120ff.
– *pNPOAc* 120ff.
– thiol-mediated 127
Michael addition 18
Michaelis complex 14, 136, 139
Michaelis-Menten 146, 160, 186f.
molecular clip 2f., 5, 7
molecular polygons 93f., 106f.
molecular triangles 95, 101f.

n
nanovessels 10
nucleophilic catalysts 6f., 9, 11, 13, 15
– calixcrown 119, 123f.
– transacylase 6f., 9, 11, 13, 15
nucleophilic-electrophilic-general acid catalysis 123

o
organocatalysts 18
oxidation catalysts 146, 155ff.
– copper 146ff.

p

phenyl acetate 3f., 114f.
– ethanolysis of 114f.
– methanolysis of 114
phototunable dinuclear catalyst 135f.
– azobenzene 135f.
preorganization 2, 147, 167, 176, 179, 214
product inhibition 129f., 175

r

rhodium complexes 18, 21, 145, 173, 203, 213, 225

s

self-assembled ligand (SAL) 239ff.
– modular design 239ff.
self-assembly 17ff., 30ff., 49ff., 93f., 102ff., 159, 210
– assembly metal 279ff., 289
– hydrogen bonding 20, 33, 37
– metal-directed 20f., 238ff.
– pyridine phosphorus ligands 213
– 2-pyridone/2-hydroxypyridine 17, 33
– supramolecular bidentate ligands 18, 210, 212, 217, 228
self-replication 9
Shell Higher Olefins Process (SHOP process) 256f.

sigmatropic rearrangement 167, 175f. 179
single enzyme catalysis 159ff.
SPANphos 282
π–π-stacking 9, 68
streptavidin-biotin 159
structure-activity relations 31
– single CALB (lipase B from *Candida antarctica*) enzymes 160
SUPRAphos ligand library 221

t

thiolysis 125
– *pNPOAc* 125
TPPMS/3(diphenylphosphinyl)aniline 270
transacylation 116, 118
transition state 2ff., 8ff., 114, 117f., 150, 165, 167, 179
– desolvation 3, 170
– solvation 3, 173, 176
trifunctional catalysis 123

w

Watson-Crick base pairing 39f.

x

xantphos 30, 228, 282, 288